The Chemistry of Soils

The Chemistry of Soils

Garrison Sposito

University of California

at Berkeley

New York · Oxford
OXFORD UNIVERSITY PRESS
1989

Oxford University Press

Oxford   New York   Toronto
Delhi   Bombay   Calcutta   Madras   Karachi
Petaling Jaya   Singapore   Hong Kong   Tokyo
Nairobi   Dar es Salaam   Cape Town
Melbourne   Auckland

and associated companies in
Berlin   Ibadan

Library of Congress Cataloging-in-Publication Data

Sposito, Garrison, 1939–
  The chemistry of soils/Garrison Sposito.
    p. cm.
  Bibliography: p.
  Includes index.
  ISBN 0-19-504615-3
  1. Soil chemistry.   I.   Title.
S592.5.S656 1989
631.4′1—dc19   88-11768 CIP

The epigraph on page 1 was translated by Archibald Cologuhoun (1960).
Reprinted with the permission of Pantheon Books, a Division of Random
House, Inc., New York.

4 6 8 9 7 5 3
Printed in the United States of America
on acid-free paper

For Mary

ὅ τι καλὸν φίλον ἀεί

Preface

More than most of the branches of soil science, soil chemistry has reflected with an ever-decreasing lag time the conceptual breakthroughs and technological growth of its parent discipline, pure chemistry. This physical science, especially in the areas of precise spectroscopic instrumentation and computational technology, has experienced explosive change over the past decade. Spectroscopic techniques, for example, have increased the sensitivity of routine analytical methods by three orders of magnitude, and small laboratory computers now manipulate chemical data or instrumentation with the speed and capacity once requiring a roomful of electronic processors. These kinds of innovation have helped to transform soil chemistry from a qualitative science, rife with inadequate methodology and theory, to a highly technical subdiscipline that has begun to develop its own agenda for the study of natural systems.

As with any maturing field of investigation, controversy may attend the decision as to how university courses on soil chemistry should be taught in light of this rapid technical development. On the one hand, those still close to the traditional agricultural roots of soil science may favor full deferral of experimental and conceptual sophistication until the postgraduate curriculum is reached. This perspective would lead to a semiquantitative approach in the intermediate-level soil chemistry course, thereby respecting the less exact understanding of biological, as opposed to chemical, phenomena that exists at present. On the other hand, those who recognize soil science as a discipline basic not only to agriculture, but also to diverse fields such as environmental engineering, biogeochemistry, and hydrology, may wish to see an intermediate course on soil chemistry that evolves continually in sympathy with advances in pure chemistry. They would argue that the complexities of modern agricultural and environmental science demand that we produce a generation of soil chemists that attributes to spectroscopic data and thermodynamic or mechanistic modeling the same degree of intuitive recognition as its predecessors attributed to soil testing for plant nutrients and statistical regression analysis. If the soil chemistry that is taught is to maintain its traditional relevance to agriculture and environment, it must also keep stride with their changing needs.

This book is intended for use in one-semester or one-quarter courses on modern soil chemistry. A background in elementary soil chemistry and mineralogy, as found, for example, in *Fundamentals of Soil Science* by H. D. Foth or *Introduction to the Principles and Practice of Soil Science* by R. E. White, is assumed on the part of the student. An understanding of pure chemistry as presented in the usual two-year introductory sequence (general and organic chemistry) is also required. Exposure to the concepts of calculus is useful, but not necessary, since virtually all of the mathematics in this book involves only algebraic and numerical manipulations. Some familiarity with statistical methods will be helpful in the numerical aspects of the problem sets.

The general plan of the book is to introduce the principal chemical constituents of soils in the first four chapters, then to describe and apply important soil chemical processes in the following six chapters. The last three chapters present detailed applications of soil chemistry to soil acidity, salinity, and fertility. These chapters are *not* intended to review these extensive fields of applied soil science, but instead to provide the soil chemistry foundation for further, specialized courses on soil–plant relationships, fertilizers, and soil management.

An appendix on SI units and physical constants as used in soil chemistry is provided to summarize the scientific terminology found in the book. Students should read this appendix and work the problem set at the end of it *before* they begin to read the book itself. The 215 problems following the chapters in this book have been designed to reinforce or extend the main points discussed *and thus are regarded as an integral part of the text*. The more challenging problems, denoted with an asterisk, feature hints and/or answers that should help in finding a correct method of solution. No student should be satisfied with his or her understanding of soil chemistry without working at least part of these more difficult problems. Those students who wish to explore in greater depth the topics discussed will find a recommended reading list at the end of each chapter. It is expected that both the problem sets and the reading list will figure significantly in any course of lectures based on this book.

I should like to thank Louise DeHayes for her excellent typing of the manuscript and Linda Bobbitt for her creative preparation of the figures. The entire manuscript of this book was reviewed critically by Terry J. Logan, to whom I am most deeply grateful for keeping the errors and obscurities I have committed to a relative minimum. Finally, I must express my indebtedness to Stephen R. Judge of Oxford University Press, without whose initial encouragement this book would not have come to be written.

Riverside, Calif.   G. S.
February 1988

Contents

13 Soil Fertility 246

The Chemistry of Soils

"For the King, yes, of course. But which King?" The lad had one of those sudden serious moods which made him so mysterious and so endearing. "Unless we ourselves take a hand now, they'll foist a republic on us. If we want things to stay as they are, things will have to change. D'you understand?"

GIUSEPPE DI LAMPEDUSA
The Leopard

1

The Chemical Composition of Soils

1.1 Elemental Composition

Soils are porous media created at the land surface by weathering processes derived from biological, geological, and hydrologic phenomena. Soils differ from mere weathered rock because they show an approximately vertical stratification (the soil horizons) produced by the continual influence of percolating water and living organisms. From the point of view of chemistry, soils are multicomponent, open, biogeochemical systems containing solids, liquids, and gases. That they are open systems means soils exchange both matter and energy with the surrounding atmosphere, biosphere, and hydrosphere. These flows of matter and energy to or from soils are highly variable in time and space, but they are the essential fluxes that cause the development of soil profiles and govern the patterns of soil fertility.

Because they are open systems, soils undergo incessant biological and chemical transformations that link them physically with the atmosphere and hydrosphere, as illustrated in Fig. 1.1. Soils are represented in the figure by the storage component labeled "land," and the hydrosphere is represented by the storage components labeled "rivers" and "ocean." The transport of matter between the four storage components is represented in the figure by the labeled arrows: atmosphere-to-land transfers (AL), land-to-river transfers (LR), (usually equal to RO), and so on. The sizes of these transfer components on a global scale are estimated (a difficult task) in Table 1.1 for six important chemical elements found in soils. The data in the table give the mass of an element transferred each year from one storage component to another, worldwide.

In respect to oxygen, the chemical form of the element that is most significant to the quantity of global transport is water, and the data in Table

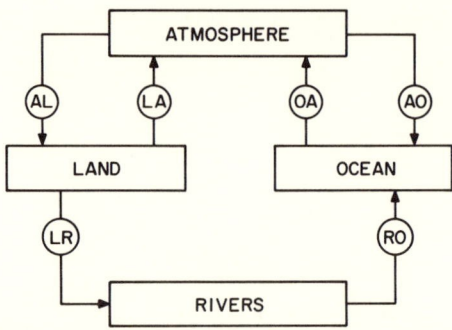

FIG. 1.1 Storage (rectangles) and transfer (circles) components in the global cycle of a chemical element. (Reprinted with permission from *Metal Ions in Biological Systems*, Vol. 20, *Concepts on Metal Ion Toxicity*, H. Sigel (ed.). Marcel Dekker, New York, 1986.)

1.1 refer to this liquid phase. Thus the transfer components for oxygen listed in the table actually are the same as the global hydrologic cycle. Note the major role played by transfer into and out of the ocean in this cycle (AO and OA) and that the cycle is balanced: AL = LA + LR = LA + RO, AO + RO = OA, etc.

In respect to total carbon, the subcycles involving land–atmosphere (AL and LA) and ocean–atmosphere (OA and AO) transfers tend to balance separately, because only a small relative portion of this carbon is discharged to the sea. About 6% of the LA transfer at present results from the burning of fossil fuels and global deforestation. The small relative difference between AO and OA is significant and reflects the accumulation of carbon in ocean floor sediments.

In respect to sulfur, there is a major component of the LA transfer (70%) produced by combustion and other anthropogenic processes, and an increasing AL transfer reflecting acid deposition. Weathering provides for a significant RO component, and the difference between OA and (AO + RO) indicates an accumulation of sulfur in ocean sediments.

The phosphorus cycle, like that of any chemical element in soil that is relatively insoluble, is dominated by the transport of particulate forms from rivers to the sea (RO) and from the land to the atmosphere (LA). Ocean sediments act as a major sink, as evidenced by the huge relative difference between (RO + AO) and OA. The lead cycle has similar characteristics, but with the added ominous feature that 90% of the LA component derives from human mining and industrial activity. This toxic element probably represents the most important environmental hazard on a global scale. Copper follows the same scenario as lead, with about 80% of the LA component being anthropogenic and with an enormous discharge of particulate forms to the sea (RO), where they accumulate in sediments.

TABLE 1.1 Global transfer components (Fig. 1.1) of some important chemical elements

Element	AL	LA	RO	OA	AO
Oxygen (10^{17} kg yr^{-1})[a]	1.0	0.70	0.30	3.7	3.4
Carbon (10^{14} kg yr^{-1})[b,c]	1.1	1.1	<0.01	0.90	0.93
Sulfur (10^{14} kg yr^{-1})[b,c]	0.7	1.6	2.1	1.6	2.6
Phosphorus (10^{9} kg yr^{-1})[b,c]	3.2	4.3	19	0.3	1.4
Lead (10^{8} kg yr^{-1})[d]	3.2	4.7	7.8	<0.01	1.4
Copper (10^{7} kg yr^{-1})[e]	6.2	7.1	632	<0.01	1.3

[a] As water, R. J. Chorley, *Introduction to Physical Hydrology.* Methuen, London, 1971.

[b] F. J. Stevenson, *Cycles of Soil.* Wiley, New York, 1986.

[c] B. Bolin and R. B. Cook, *The Major Biogeochemical Cycles and Their Interactions.* Wiley, New York, 1983.

[d] J. O. Nriagu, *The Biogeochemistry of Lead in the Environment,* Part A. Elsevier, Amsterdam, 1978.

[e] J. O. Nriagu, *Copper in the Environment,* Part I. Wiley, New York, 1979.

The role of soil as a dynamic reservoir in the global cycling of chemical elements is not reflected completely by the diagram in Fig. 1.1 because the relationship of soil to crustal rock, which often serves as soil parent material, has been omitted. Some idea of this relationship can be obtained from Table 1.2, which lists average mass concentrations of 50 chemical elements in soils and surficial rocks. The soil concentrations refer to samples taken approximately 0.2 m beneath the land surface from uncontaminated mineral soils in the conterminous United States. These concentration data are quite comparable with elemental composition figures for soils sampled worldwide. The average values in Table 1.2 often have large standard deviations, however, because of spatial heterogeneity in soils and rock masses. Figure 1.2 illustrates this point by vertical lines through filled circles that represent the average elemental concentrations in topsoils. (Note the logarithmic scale for concentration.)

Table 1.2 indicates that the 10 most abundant elements in soils are O > Si > Al > Fe > C > Ca > K > Na > Mg > Ti, whereas in crustal rocks they are O > Si > Al > Fe > Ca > Mg = Na > K > Ti > P. Aside from these most abundant elements, the entries in Table 1.2 often are classified as either macroelements or microelements depending on their typical mole concentrations in the *liquid* portion of soil (*the soil solution*). This classification will be discussed in Section 1.4. Suffice it to say here that the elements in the list from carbon to calcium—with the exception of F, Al, and P—qualify in this way as macroelements. The remaining 40 elements therefore are microelements. This latter designation is *not* necessarily the same as a *micronutrient*, which is an element essential to living organisms but absorbed by them only in trace amounts (see Problem 4). For example, phosphorus is a microelement but not a micronutrient, whereas iron is both.

TABLE 1.2 Mean elemental content (in mg kg^{-1}) of soil and crustal rocks, and the soil enrichment factor (EF)

Element	Soil[a,b]	Crust[b]	EF[c]	Element	Soil[a,b]	Crust[b]	EF[c]
Li	24	20	1.2	Zn	60	75	0.80
Be	0.92	2.6	0.35	Ga	17	18	0.94
B	33	10	3.3	Ge	1.2	1.8	0.67
C	25,000	480	52	As	7.2	1.5	4.8
N	2,000	25	80	Se	0.39	0.05	7.8
O	490,000	474,000	1.0	Br	0.85	0.37	2.3
F	950	430	2.2	Rb	67	90	0.74
Na	12,000	23,000	0.52	Sr	240	370	0.65
Mg	9,000	23,000	0.39	Y	25	30	0.83
Al	72,000	82,000	0.88	Zr	230	190	1.2
Si	310,000	277,000	1.1	Nb	11	20	0.55
P	430	1,000	0.43	Mo	0.97	1.5	0.65
S	1,600	260	6.2	Ag	0.05	0.07	0.71
Cl	100	130	0.77	Cd	0.35	0.11	3.2
K	15,000	21,000	0.71	Sn	1.3	2.2	0.59
Ca	24,000	41,000	0.59	Sb	0.66	0.20	3.3
Sc	8.9	16	0.56	I	1.2	0.14	8.6
Ti	2,900	5,600	0.52	Cs	4.0	3.0	1.3
V	80	160	0.50	Ba	580	500	1.2
Cr	54	100	0.54	La	37	32	1.2
Mn	550	950	0.58	Hg	0.09	0.05	1.8
Fe	26,000	41,000	0.63	Pb	19	14	1.4
Co	9.1	20	0.46	Nd	46	38	1.2
Ni	19	80	0.24	Th	9.4	12	0.78
Cu	25	50	0.50	U	2.7	2.4	1.1

[a]H. T. Schacklette and J. G. Boerngen, Element concentrations in soils and other surficial materials of the conterminous United States, *U.S. Geological Survey Prof. Paper* 1270 (1984).

[b]H. J. M. Bowen, *Environmental Chemistry of the Elements*. Academic Press, London, 1979.

[c]EF = soil content/crustal content.

Listed also in Table 1.2 are soil enrichment factors (EF), defined as the ratio of soil to crustal rock concentrations, with both expressed in the same units. The EF is a quantitative measure of the relative enrichment (or depletion) of a chemical element in soil as compared with rock. Given the variability of both soil and rock composition, noted above, values of EF in the range 0.5–2.0 about the value 1.0 probably should be interpreted as indicating no significant relative depletion or enrichment. Values of EF < 0.5 then would indicate significant depletion; values of EF between 2 and 10 would indicate some enrichment, and values > 10 would indicate strong enrichment. According to these criteria, Be, Mg, Ni, and perhaps P and Co would be depleted elements; B, F, S, As, Se, Br, Cd, Sb, and I would be somewhat enriched; and C and N would be strongly enriched elements in soils relative to crustal rocks. For most

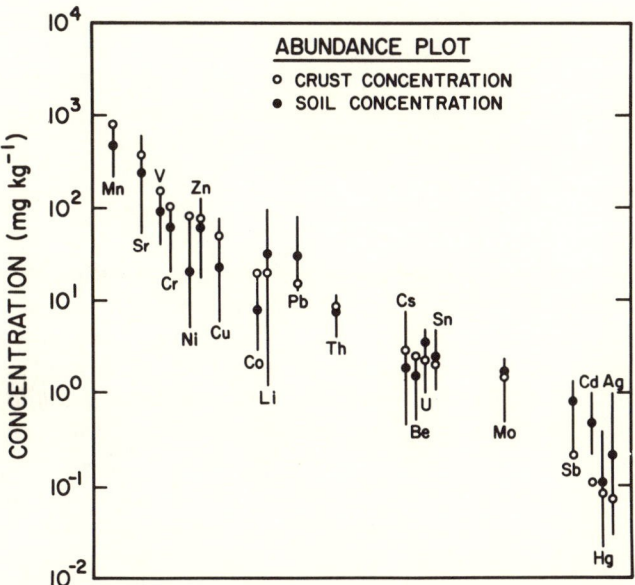

FIG. 1.2 Average concentrations (filled circles) and ranges (vertical lines) of trace metal concentrations in topsoils compared with average crustal concentrations. (Reprinted with permission from *Metal Ions in Biological Systems*, Vol. 20, *Concepts on Metal Ion Toxicity*, H. Sigel (ed.). Marcel Dekker, New York, 1986.)

of the microelements, as shown in Fig. 1.2, there is a close correspondence between soil and crustal rock concentrations, and this relationship can be explained by the principal chemical forms of these elements in soil or rock, a topic to be discussed in Section 1.3. The strong enrichment of carbon and nitrogen in soils is also a result of the principal chemical forms these elements assume, namely those associated with soil organic matter and soil organisms. The average C/N ratio of 13 in soils, indicated by the data in Table 1.2, is low and conducive to microbial mineralization, and further suggests the active biological milieu that distinguishes soil from crustal rock.

1.2 Solid Phases in Soils

About one-half to two-thirds of the soil volume is made up by solid matter. Of this material, typically more than 90% represents inorganic compounds, except for peat and muck soils wherein organic material accounts for $> 50\%$ of the solid matter. The inorganic solid phases in soils often do not have a simple stoichiometry (i.e., simple molar ratios of one element to another) because they are in a metastable state of transition from an inhomogeneous, irregular to a more homogeneous, regular atomic structure as a result of weathering processes. Nonetheless, a number of solid phases of relatively uniform composition (minerals) have been identified in soils worldwide. Table 1.3 lists the most common soil minerals along with their chemical formulas. Details of the atomic structures of these minerals are given in Chapter 2.

The two most abundant elements in soils are oxygen and silicon, according to Table 1.2, and these two elements combine chemically to form the 15 common silicates listed in Table 1.3. The first six silicates in the table are termed *primary minerals* because they are typically inherited from parent material, as opposed to precipitated through weathering processes. The key structural entity in these minerals is the Si—O bond, which is a more covalent, and therefore stronger, bond than typical metal–oxygen bonds (see Section 2.1). The relative resistance of any one of the minerals to decomposition by weathering can be correlated positively with the Si/O molar ratio of its fundamental silicate structural unit, since a larger ratio means a lesser need to incorporate metal cations into the mineral structure in order to neutralize the

TABLE 1.3   Common soil minerals

Name	Chemical formula	Importance
Quartz	SiO_2	Abundant in sand and silt
Feldspar	$(Na,K)AlO_2[SiO_2]_3$	Abundant in soil that is not leached
	$CaAl_2O_4[SiO_2]_2$	extensively
Mica	$K_2Al_2O_5[Si_2O_5]_3Al_4(OH)_4$	Source of K in most temperate-zone
	$K_2Al_2O_5[Si_2O_5]_3(Mg,Fe)_6(OH)_4$	soils
Amphibole	$(Ca,Na,K)_{2,3}(Mg,Fe,Al)_5(OH)_2$	Easily weathered to clay minerals
	$[Si,Al]_4O_{11}]_2$	and oxides
Pyroxene	$(Ca,Mg,Fe,Ti,Al)(Si,Al)O_3$	Easily weathered
Olivine	$(Mg,Fe)_2SiO_4$	Easily weathered
Epidote	$Ca_2(Al,Fe)_3(OH)Si_3O_{12}$	Highly resistant to chemical
Tourmaline	$NaMg_3Al_6B_3Si_6O_{27}(OH,F)_4$	weathering; used as "index mineral"
Zircon	$ZrSiO_4$	in pedologic studies
Rutile	TiO_2	
Kaolinite	$Si_4Al_4O_{10}(OH)_8$	Abundant in clay as products of
Smectite	$M_x(Si,Al)_8(Al,Fe,$	weathering; source of exchangeable
Vermiculite	$Mg)_4O_{20}(OH)_4$, where	cations in soils
Chlorite	M = interlayer cation	
Allophane	$Si_3Al_4O_{12} \cdot nH_2O$	Abundant in soils derived from
Imogolite	$Si_2Al_4O_{10} \cdot 5H_2O$	volcanic ash deposits
Gibbsite	$Al(OH)_3$	Abundant in leached soils
Goethite	$FeO(OH)$	Most abundant Fe oxide
Hematite	Fe_2O_3	Abundant in warm regions
Ferrihydrite	$Fe_{10}O_{15} \cdot 9H_2O$	Abundant in organic horizons
Birnessite	$(Na,Ca)Mn_7O_{14} \cdot 2.8H_2O$	Most abundant Mn oxide
Calcite	$CaCO_3$	Most abundant carbonate
Gypsum	$CaSO_4 \cdot 2H_2O$	Abundant in arid regions

oxygen anion charge. To the extent that metal cations are so excluded, the degree of covalency in the overall bonding arrangement will be greater and the mineral will be more resistant to decomposition in the soil environment. For the first six silicates in Table 1.3, the Si/O molar ratios of their fundamental structural units are as follows: 0.50 (quartz and feldspar, SiO_2); 0.40 (mica, Si_2O_5); 0.36 (amphibole, Si_4O_{11}); 0.33 (pyroxene, SiO_3); and 0.25 (olivine, SiO_4). The decreasing order of the Si/O molar ratio is the same as the observed decreasing order of resistance to chemical weathering in the sand or silt fractions of soils.

The minerals epidote, tourmaline, zircon, and rutile, listed in Table 1.3, are found to be highly resistant to weathering in the soil environment. Under the assumption of uniform parent material, the measured variation in the relative number of single-crystal grains of these minerals in the fine sand or coarse silt fractions of a soil profile can serve as a quantitative indicator of mass changes in the soil horizons produced by weathering. For example, if Z_p is the mass percentage of zircon grains in the fine-sand fraction of a soil parent material and Z_h is the mass percentage in the same size fraction of an overlying soil horizon, then Z_h/Z_p is equal to the mass of parent material that produced 1 kg of the weathered soil horizon.

The minerals listed from kaolinite to gypsum in Table 1.3 are termed *secondary minerals* because they nearly always result from the weathering transformations of primary silicates. Often these secondary minerals are of clay size and exhibit a relatively poorly ordered atomic structure. Variability in their composition through the substitution of ions into their structure (*isomorphic substitution*) also is noted frequently in soils. The secondary silicates, smectite and vermiculite, bear a net charge on their surfaces principally because of this variability in composition, as will be shown in Section 2.3. Kaolinite and the secondary metal oxides below it in the list also bear a net surface charge, but because of proton adsorption and desorption, not compositional variability. The metal oxides like gibbsite and goethite tend to persist in the soil environment longer than do the secondary silicates, because Si is more readily leached than Al, Fe, or Mn unless significant amounts of soluble organic matter are present to render the metals more mobile in the profile.

Organic matter is itself an important constituent of the solid portion of soils. The structural complexity of soil organic compounds has thus far precluded the making of a simple list of component solids like that in Table 1.3, but something can be said about the overall composition of *humic substances*, the dark, microbially transformed organic materials that persist in soils throughout profile development. The two most investigated humic substances are humic and fulvic acid; their chemical behavior is discussed in Section 3.2. Worldwide, the average composition of these two substances in soil is: $C_{187}H_{186}O_{89}N_9S$ (humic acid) and $C_{135}H_{182}O_{95}N_5S_2$ (fulvic acid). These average chemical formulas can be compared with the average C/N/P/S *molar* ratio of organic matter in soils, which is $278:17:1:1$, and with the famous "Redfield formula" for the average chemical composition of aquatic plants, which is $C_{106}H_{263}O_{110}N_{16}P$. Relative to soil organic matter as a whole, humic and fulvic acids are depleted in N but enriched in S. Their C/N molar ratio is about 50% larger than that of soil organic matter, indicating their greater resistance to net microbial mineralization. Relative to living organisms and biomolecules, as represented by the Redfield formula, humic and fulvic acids are enriched in C but depleted in H and N. The depletion of N again indicates the absence of constituents susceptible to net mineralization through biodegradation. The depletion of H, from a roughly $2:1$ H/C molar ratio in biological material to roughly $1:1$ in humic substances, suggests a greater degree of unsaturation in the latter, consistent with their more aromatic character that would tend to resist microbial attack.

1.3 Trace Elements in Soil Minerals

One of the most important aspects of the variability in composition of soil minerals is their content of trace elements. *A trace element is any chemical element whose mass concentration in a solid phase is $\leqslant 100\,mg\ kg^{-1}$.* Soil minerals bearing trace elements serve as reservoirs for the elements, releasing them slowly into the soil solution as weathering of the minerals continues. If a trace element is also a micronutrient, then the rate of mineral weathering becomes a critical factor in soil fertility. For example, the ability of soils to provide copper to plants depends on the rate at which this element is transformed from a solid phase to a soluble chemical form. Soil chemical and physical properties like pH, redox potential, and water content will affect the rate of this transformation and thus control Cu solubility. In a similar manner, the weathering rate of soil solids containing cadmium as a trace element will determine in part the potential hazard of this toxic element to plants.

The principal ways in which important trace elements occur in primary and secondary soil minerals are summarized in Tables 1.4 and 1.5. Table 1.5 also indicates the trace elements found typically in association with soil organic matter. The chemical phenomenon underlying the trace element occurrences described in these tables is called *coprecipitation*. Coprecipitation is the simultaneous precipitation of a chemical element with other elements by any mechanism and at any rate. The three broad types of coprecipitation are *inclusion, adsorption,* and *solid solution formation.*

If a pure solid phase that would be formed by a trace element has a very different atomic structure from that of the host mineral that coprecipitates with the trace element, then it is likely that the host mineral and the trace element will occur together only as morphologically distinct solids. This kind of

TABLE 1.4 Occurrence of trace elements in primary minerals

Element	Principal modes of occurrence in primary minerals
B	Tourmaline [$NaMg_3Al_6B_3Si_6O_{27}(OH,F)_4$]; isomorphic substitution for Si in micas
Ti	Rutile (TiO_2) and ilmenite ($FeTiO_3$); oxide inclusions in silicates
V	Isomorphic substitution for Fe in pyroxenes and amphiboles and for Al in micas; substitution for Fe in oxides
Cr	Chromite ($FeCr_2O_4$); isomorphic substitution for Fe or Al in other minerals of the spinel group
Co	Isomorphic substitution for Mn in oxides and for Fe in pyroxenes, amphiboles, and micas
Ni	Sulfide inclusions in silicates; isomorphic substitution for Fe in olivines, pyroxenes, amphiboles, micas, and spinels
Cu	Sulfide inclusions in silicates; isomorphic substitution for Fe and Mg in olivines, pyroxenes, amphiboles, and micas, and for Ca, K, or Na in feldspars
Zn	Sulfide inclusions in silicates; isomorphic substitution for Mg and Fe in olivines, pyroxenes, and amphiboles, and for Fe or Mn in oxides
As	Arsenate minerals: $FeAsO_4 \cdot 2H_2O$, $Mn_3(AsO_4)_2$, etc.
Se	Selenide minerals; isomorphic substitution for S in sulfides; iron selenite
Mo	Molybdenite (MoS_2); isomorphic substitution for Fe in oxides
Cd	Sulfide inclusions and isomorphic substitution for Cu, Zn, Hg, and Pb in sulfides
Pb	Sulfide and phosphate inclusions; isomorphic substitution for K in feldspars and micas, for Ca in feldspars, pyroxenes, and phosphates, and for Fe and Mn in oxides

association is termed *inclusion* with respect to the trace element. For example, CuS often occurs as an inclusion—a small separate phase—in primary silicates (Table 1.4).

If there is only limited structural compatibility between a trace element and the corresponding major element in a host mineral, then coprecipitation can produce a homogeneous mixture of the two elements at the host mineral–soil solution interface. This mechanism is termed *adsorption* because the mixed solid phase is restricted to the interfacial region and its composition can change as the host mineral continues to precipitate from the soil solution. Well-known examples of adsorption are the incorporation of oxyanions like borate, phosphate, or molybdate into secondary metal oxides (Table 1.5), and of transition metals like Fe or Ni into soil organic matter.

Finally, if structural compatibility is high and free diffusion of a trace element within the host mineral is possible, a major element in the host mineral can be replaced uniformly throughout by the trace element. This kind of homogeneous coprecipitation is *solid-solution formation*. It is enhanced if the

TABLE 1.5   Trace elements coprecipitated with secondary soil minerals and soil organic matter

Solid	Coprecipitated trace elements
Fe and Al oxides	B, P, V, Mn, Ni, Cu, Zn, Mo, As, Se
Mn oxides	P, Fe, Co, Ni, Zn, Mo, As, Se, Pb
Ca carbonates	P, V, Mn, Fe, Co, Cd
Illites	B, V, Ni, Co, Cr, Cu, Zn, Mo, As, Se, Pb
Smectites	B, Ti, V, Cr, Mn, Fe, Co, Ni, Cu, Zn, Pb
Vermiculites	Ti, Mn, Fe
Organic matter	Al, V, Cr, Mn, Fe, Ni, Cu, Zn, Cd, Pb

size and valence of the substituting element are comparable to those of the element replaced. Examples of solid-solution formation occur when secondary aluminosilicates precipitate and incorporate metals like Ni, Cu, and Zn to replace Al in their structures (Table 1.5) or when calcium carbonate precipitates with Cd replacing Ca in the structure. Isomorphic substitution of this kind is common also in primary silicates (Table 1.4). In this case, the trace element substitution occurs as the minerals crystallize from a silicate melt.

1.4 Soil Air and Soil Water

The fluid phases in soil constitute between one- and two-thirds of the soil volume. The gaseous fluid phase, *soil air*, typically is the same kind of mixture as atmospheric air. Because of biological activity in soil, however, the percentage composition of soil air can differ considerably from that of atmospheric air (781 mL N_2, 209 mL O_2, 9.3 mL Ar, and 0.31 mL CO_2 in 1 L of dry air). Well-aerated soil contains 180–205 mL O_2 per liter of soil air, but this figure can drop to 100 mL/L at 1 m below the soil surface, after inundation by rainfall or irrigation, or even to 20 mL/L in isolated soil microenvironments near plant roots. Similarly, the fractional volume of CO_2 in soil air is typically 3–30 mL/L, but can approach 100 mL/L at 1 m depth, in the vicinity of plant roots, or after the flooding of soil. This markedly higher CO_2 content of soil air relative to that of the atmosphere has a significant impact on both soil acidity and carbonate chemistry (Chapters 5 and 11). Soil air also contains variable but important contributions from NO, N_2O, NH_3, CH_4, and H_2S produced by microorganisms under conditions of low oxygen content (*anaerobic conditions*).

Soil water is found principally as a condensed phase in soil, although the content of water vapor in soil air can approach 30 mL/L in a wet soil. Soil water is a repository for dissolved solids and gases and for this reason is

TABLE 1.6 Values of the "Henry's law constant" at 25°C[a]

Gas	K_H (mol m^{-3} atm^{-1})	Gas	K_H (mol m^{-3} atm^{-1})
CO_2	34.06	NO	1.88
CH_4	1.50	O_2	1.26
NH_3	5.76×10^4	SO_2	1.24×10^3
N_2O	25.55	H_2S	1.02×10^2

[a]Based on data compiled in Section 2.16 of W. Stumm and J. J. Morgan, *Aquatic Chemistry*, Wiley, New York, 1981.

referred to commonly as the *soil solution*. In respect to dissolved solids, those that dissociate into ions (*electrolytes*) in the soil solution are most important to the chemistry of soils. The ion-forming chemical elements whose concentrations in uncontaminated soil solutions typically fall below 1.0 mmol m^{-3} are termed *microelements*. (For a discussion of concentration units, see the Appendix.) All others are *macroelements*. With reference to Table 1.2, this definition leads to C (HCO_3^-), N (NO_3^-), Na (Na^+), Mg (Mg^{2+}), Si ($Si(OH)_4^0$), S (SO_4^{2-}), Cl (Cl^-), K (K^+), Ca (Ca^{2+}), and oxygen being designated as macroelements. [The neutral species $Si(OH)_4^0$ is silicic acid.] The remaining 40 elements in the table are microelements.

The dissolution of gases from soil air into the soil solution is an important process contributing to the cycling of chemical elements in the soil environment. When equilibrium exists between soil air and soil water with respect to the partitioning of a gaseous species between the two phases, and if the concentration of the gas in the soil solution is low, the equilibrium can be described by a form of *Henry's law*:

$$K_H = [A(aq)]/P_A \tag{1.1}$$

where K_H is a parameter with the units mol m^{-3} atm^{-1}, known as the "Henry's law constant," [A] is the concentration of gas A in the soil solution (mol m^{-3}), and P_A is the partial pressure of A in soil air (atm). (The units used in Eq. 1.1 are discussed in the Appendix.) For example, $K_H = 34.06$ mol m^{-3} atm^{-1} at 25°C for CO_2. If the partial pressure of CO_2 in soil air is 0.03 atm, then $[CO_2(aq)]$ in the soil solution is 1.02 mol m^{-3}, according to Eq. 1.1. Table 1.6 lists values of K_H at 25°C for several gases found in soil air.

1.5 Soil Mineral Transformations

If soils were not open systems, soil minerals would not weather. It is the continual input and output of percolating water, biomass, and solar energy in soils that makes them change with the passage of time. These changes are perhaps reflected most dramatically in the morphological development of soil horizons, but they are also very apparent in the mineralogy of the soil clay fraction.

Table 1.7 is a summary of the changes in clay fraction mineralogy observed during the course of soil profile development. These changes, known collectively as the *Jackson–Sherman weathering stages*, can be classified as "early stage," "intermediate stage," or "advanced stage." Early-stage weathering is recognized through the importance of sulfates, carbonates, and primary silicates, other than quartz and muscovite, in the soil clay fraction, These minerals can survive only if soils remain very dry, or very cold, or very wet, most of the time—that is, if they lack the throughputs of water, air, and thermal energy

TABLE 1.7   Jackson–Sherman weathering stages[a,b]

Characteristic minerals in soil clay fraction	Characteristic soil chemical and physical conditions
Early stage	
Gypsum	Very low content of water and organic matter, very limited leaching
Carbonates	
Olivine/pyroxene/amphibole	Reducing environments
Fe(II)-bearing micas	Limited amount of time for weathering
Feldspars	
Intermediate stage	
Quartz	Retention of Na, K, Ca, Mg, Fe(II), and silica:
Dioctahedral mica/illite	Ineffective leaching and alkalinity
Vermiculite/chlorite	Igneous rock rich in Ca, Mg, Fe(II), but no Fe(II) oxides
Smectites	Silicates easily hydrolyzed
	Flocculation of silica, transport of silica into the weathering zone
Advanced stage	
Kaolinite	Removal of Na, K, Ca, Mg, Fe(II), and silica:
Gibbsite	Effective leaching, fresh water
Iron oxides	Oxidation of Fe(II)
(goethite, hematite)	Acidic compounds, low pH
Titanium oxides	Dispersion of silica
(anatase, rutile, ilmenite)	Al-hydroxy polymers

[a]M. L. Jackson and G. D. Sherman, Chemical weathering of minerals in soils, *Adv. Agron.* **5**:219–318 (1953).

[b]M. L. Jackson, Clay transformations in soil genesis during the quaternary, *Soil Sci.* **99**:15–22 (1965).

that characterize open systems in nature. Intermediate-stage weathering features quartz, muscovite, and secondary aluminosilicates prominently in the clay fraction. These minerals can survive under leaching conditions that do not deplete silica and the macroelements, and do not result in the complete oxidation of ferrous iron [Fe(II)], which is incorporated into illite and smectite. Advanced-stage weathering, on the other hand, is associated with intensive leaching and strongly oxidizing conditions, such that only hydrous oxides of aluminum, ferric iron [Fe(III)], and titanium persist ultimately. Kaolinite will be an important clay mineral only if the removal of silica by leaching is not complete, or if there is an invasion of silica-rich waters as can occur, for example, when leachate from the upper part of a soil toposequence moves laterally into the profile of a lower part.

The order of increasing persistence of the soil minerals listed in Table 1.7 is downward, both among and within the three stages of weathering. The primary minerals, therefore, tend to occur higher in the list than the secondary minerals, and the former can be linked with the latter by a variety of chemical reactions. Of these reactions, the most important type is *hydrolysis*, which may be illustrated by the chemical equations:

$NaAlSi_3O_8(s) + 8H_2O(\ell) =$
(albite)

$$Na^+(aq) + Al(OH)_2^+(aq) + 3Si(OH)_4^0(aq) + 2OH^-(aq) \qquad (1.2)$$

$NaAlSi_3O_8(s) + 8H_2O(\ell) =$
(albite)

$$Al(OH)_3(s) + Na^+(aq) + 3Si(OH)_4^0(aq) + OH^-(aq) \qquad (1.3)$$
$$\text{(gibbsite)}$$

In both of these reactions, which are taken to proceed from left to right, the dissolution of the feldspar, albite, occurs through chemical reaction with water to form dissolved species (denoted "aq"). Equation 1.2 is termed a *congruent dissolution* because only dissolved species make up the products, whereas Eq. 1.3 is termed an *incongruent dissolution* because a solid-phase product— gibbsite—is formed as well. The basic chemical principles underlying the development of these equations are discussed in Special Topic 1, at the end of this chapter.

Another important weathering reaction is *complexation* (often inappropriately called "chelation"), which describes the reaction of complexing anions with soil minerals:

$K_2[Si_6Al_2]Al_4O_{20}(OH)_4(s) + 6C_2O_4H_2(aq) + 4H_2O(\ell) = 2K^+(aq)$
(muscovite)

$$+ 6C_2O_4Al^+(aq) + 6Si(OH)_4^0(aq) + 8OH^-(aq) \qquad (1.4)$$

The organic compound on the left side of Eq. 1.4, oxalic acid (ethanedioic acid), dissociates and releases an anion, $C_2O_4^{2-}$, that forms a soluble complex with Al^{3+} (see Section 4.2) to enhance the congruent dissolution of muscovite by hydrolysis. This soluble complex, $C_2O_4Al^+(aq)$, helps to prevent the hydrolysis of Al^{3+} that otherwise could lead to gibbsite precipitation, as in Eq. 1.3.

   Cation exchange, on the other hand, is a weathering reaction associated with the *incongruent* dissolution of muscovite to form vermiculite in soils that retain both Ca^{2+} and $Si(OH)_4^0$:

$$K_2[Si_6Al_2]Al_4O_{20}(OH)_4(s) + 0.8Ca^{2+}(aq) + 1.3Si(OH)_4^0(aq)$$
(muscovite)

$$= 1.1Ca_{0.7}[Si_{6.6}Al_{1.4}]Al_4O_{20}(OH)_4(s) + 2K^+(aq) + 0.4OH^-(aq) + 1.6H_2O(\ell)$$
(vermiculite)
$$(1.5)$$

The Ca^{2+} ion exchanges with K^+ to occupy the interlayer position in vermiculite (see Section 2.3). This kind of exchange reaction would be favored in a poorly drained, calcareous soil having abundant dissolved calcium ions and silicic acid in the soil solution.

   Incongruent dissolution also is accompanied often by *oxidation–reduction* if iron or some other "redox element" is involved in a weathering reaction. An example is the incongruent dissolution of biotite, which contains ferrous iron, to form vermiculite, which contains both ferrous and ferric iron, and goethite, which contains only ferric ion. Oxidation–reduction (or *redox*) reactions will be discussed in Chapter 6.

   Finally, *hydration–dehydration* can be added to the list of the five significant weathering reactions recognized generally. Examples include the transformation of hematite to ferrihydrite and anhydrite to gypsum:

$$5Fe_2O_3(s) + 9H_2O(\ell) = Fe_{10}O_{15} \cdot 9H_2O(s) \qquad (1.6)$$
(hematite)   (ferrihydrite)

$$CaSO_4(s) + 2H_2O(\ell) = CaSO_4 \cdot 2H_2O(s) \qquad (1.7)$$
(anhydrite)   (gypsum)

Both of these mineral hydration reactions are favored as the relative humidity of soil water approaches 100%.

   The weathering reaction types surveyed very briefly in this section provide a chemical basis for the transformations of soil minerals both within and between the Jackson–Sherman weathering stages. In respect to silicates, a "master variable" controlling these transformations is the concentration of silicic acid in the soil solution. As the concentration of $Si(OH)_4^0$ decreases through leaching, the mineralogy of the soil clay fraction passes from the primary minerals of the early stage to the secondary minerals of the intermediate and advanced stages. Should the $Si(OH)_4^0$ concentration increase through an influx of silica, on the other hand, the clay mineralogy can be expected to shift upward in Table 1.7. This possible behavior is in fact implied by the equals signs in the chemical reactions in Eqs. 1.2–1.7.

FOR FURTHER READING

J. B. Dixon and S. B. Weed (eds.), *Minerals in Soil Environments*. Soil Science Society of America, Madison, WI, 1977. The standard reference work on soil mineral structures and chemistry.

A. Kabata-Pendias and H. Pendias, *Trace Elements in Soils and Plants*. CRC Press, Boca Raton, FL, 1984. A compendium of analytical data on trace elements in the lithosphere and biosphere, organized according to the Periodic Table.

J. A. Kittrick (ed.), *Soil Mineral Weathering*. van Nostrand Reinhold, New York, 1986. A carefully edited collection of fundamental scientific papers on chemical weathering in soils.

F. J. Stevenson, *Cycles of Soil*. Wiley, New York, 1986. An in-depth discussion of the biogeochemical cycles of C, N, P, S, and some microelements, addressed to the interests of soil chemists.

PROBLEMS

The more difficult problems are indicated by an asterisk.

1. The annual atmospheric deposition rate of cadmium onto land is 5.7×10^6 kg yr^{-1} worldwide, and that onto the oceans is 2.4×10^6 kg yr^{-1}. The emission rate of Cd into the atmosphere from land is 8.1×10^6 kg yr^{-1}, whereas that from the oceans is 10^3 kg yr^{-1}. Cadmium is transported by rivers (in suspended particles) to the oceans at the rate of 2.3×10^7 kg yr^{-1}. Use these data to construct a global transfer cycle for Cd following the format in Fig. 1.1 and Table 1.1. To which elemental cycles in Table 1.1 is that of Cd similar?

2. The concept of the *residence time* can be applied to any one of the storage components in Fig. 1.1 if the annual rates of input and output for the component are equal (steady-state conditions). The residence time in years is then the ratio of the mass of an element in the storage component to the output rate for the element. Calculate the residence times of water in the atmosphere, rivers, and oceans, given that the masses of water in these three storage components are: 1.3×10^{16} kg, 1.08×10^{15} kg, and 1.35×10^{21} kg, respectively.

3. The combined mass of terrestrial carbon stored in soils and the biota is estimated to be 2.55×10^{15} kg. Use this result and the data in Table 1.1 to calculate the overall residence time of terrestrial carbon. For soil C (humus), the steady-state output rate is about 3×10^{12} kg yr^{-1} from a stored mass of 1.5×10^{15} kg. What is the residence time of soil C?

4. Consult a basic soil or plant science textbook to prepare a list of the 16 chemical elements essential to plant growth. Compare the macronutrients and micronutrients in your list with the macroelements and microelements in Table 1.2. The elements in this table are listed in order of increasing atomic number. Is there a relation between the macronutrients and atomic number?

5. The table given here lists the average chemical composition of terrestrial higher plants. Calculate enrichment factors for these elements (relative to crustal rocks) and compare them when possible with the EF values in Table 1.2.

Element	Content (g kg^{-1})	Element	Content (g kg^{-1})
H	54	Mg	4.2
C	412	P	3.0
N	33	K	8.0
O	463	Ca	21.0

*6. Calculate an average chemical formula for land plants based on the data for H, C, N, O, and P given in Problem 5. Compare your result with the chemical formulas for humic and fulvic acid, and the "Redfield formula." (*Answer:* $C_{354}H_{552}O_{298}N_{24}P$)

7. Calculate the corresponding concentrations of CO_2 dissolved in soil water as the CO_2 partial pressure in soil air increases in the order: 3.02×10^{-4} (atmospheric CO_2), 0.003, 0.01, 0.05, 0.10 atm (flooded soil).

*8. The ideal gas law, $PV = nRT$, can be applied to the constituents of soil air to a good approximation. (*P* is pressure, *V* is volume, *n* is the number of moles, *R* is the molar gas constant, and *T* is absolute temperature, as described in the Appendix.) Use the ideal gas law to show that Eq. 1.1 can be rewritten in the useful form:

$$H = [A(g)]/[A(aq)]$$

where [ ] is a concentration in moles per cubic meter, and $H = 10^3/K_H RT$ is a *dimensionless* constant based on $R = 0.08205$ atm $\cdot$ L $\cdot$ mol$^{-1} \cdot$ K^{-1} and *T* in units of Kelvins (K). Prepare a table of *H* values based on Table 1.6.
(*Answer:* See the reference cited in Table 1.6.)

*9. The CO_2/N_2O molar ratio in soil air was found to be 6.0 under reducing conditions that led to denitrification. Calculate the corresponding molar ratio in soil water.
(*Answer:* 4.5)

10. Write a balanced chemical reaction for the congruent dissolution of olivine (Mg_2SiO_4) by hydrolysis.

11. Write a balanced chemical reaction for the incongruent dissolution of K-feldspar ($KAlSi_3O_8$) to produce kaolinite by hydrolysis.

*12. Write a balanced chemical reaction for the weathering of muscovite to the smectite, $Na_{0.9}[Si_{7.6}Al_{0.4}]Al_{3.5}Mg_{0.5}O_{20}(OH)_4$. This formula is an example of the clay mineral, *montmorillonite*.
(*Hint:* Refer to Eq. 1.5.)

13. Consult a textbook on inorganic chemistry (or a good dictionary) and prepare a brief paragraph that defines the terms, "soluble complex" and "chelate." Be sure to give concrete examples of each term relevant to soil solutions.

14. When carbon dioxide dissolves in the soil solution, it solvates to form the species $CO_2 \cdot H_2O$, which is the same as H_2CO_3, carbonic acid. Carbonic acid, in turn, dissociates a proton to leave the species HCO_3^-, bicarbonate ion. Write a series of balanced chemical reactions that shows the formation of $H_2CO_3(aq)$ from $CO_2(g)$, the formation of bicarbonate from carbonic acid, and the reaction of bicarbonate with Ca-feldspar ($CaSi_2Al_2O_8$) to form calcite and aqueous silicic acid.

15. In respect to its soil solution chemistry, fulvic acid can be represented by the formula HL, where L^- represents 1 mol of dissolved fulvic acid charge, regardless of the nature of the organic ligands involved. Use this convention to write a balanced chemical reaction for the congruent dissolution of kaolinite, with the complex, $Al(OH)_2L^0$, as one of the products.

Special Topic 1

Balancing Chemical Reactions

Chemical reactions like those in Eqs. 1.2–1.7 must fulfill two general conditions: *mass balance* and *charge balance*. Mass balance requires that the number of moles of each chemical element be the same on both sides of the reaction when written as a chemical equation. Charge balance requires that the net total free charge be the same on both sides of the reaction. These constraints can be applied to develop the correct form of a chemical reaction when only the principal product and reactant are given.

As a first example, consider the incongruent dissolution of albite, $NaAlSi_3O_8$, to produce gibbsite, $Al(OH)_3$. Since 1 mol of the reactant albite contains 1 mol Na, 1 mol Al, and 3 mol Si, these amounts, by mass balance, must appear in the products as well. Therefore,

$$NaAlSi_3O_8(s) \rightarrow Al(OH)_3(s) + Na^+(aq) + 3Si(OH)_4^0(aq) \qquad (s1.1)$$

can be written down as a first step in developing the complete reaction. Missing still are mass balance for O and H and overall charge balance. The mass balance problem can be remedied by considering a *mechanism* for the dissolution reaction—hydrolysis:

$$NaAlSi_3O_8(s) + H_2O(\ell) \rightarrow$$
$$Al(OH)_3(s) + Na^+(aq) + 3Si(OH)_4^0(aq) \quad (s1.2)$$

The charge balance problem requires adding either a monovalent cation to the left side of the reaction or a monovalent anion to the right side. If the cation is chosen, it should be a proton to reflect the acidity of liquid water [denoted $H_2O(\ell)$]:

$$NaAlSi_3O_8(s) + H_2O(\ell) + H^+(aq) \rightarrow$$
$$Al(OH)_3(s) + Na^+(aq) + 3Si(OH)_4^0(aq) \quad (s1.3)$$

Charge balance is achieved with just 1 mol of protons, so the stoichiometric coefficient of H_2O in Eq. s1.3 can be determined from proton mass balance. Since there are 15 mol of H among the products (3 mol from gibbsite and $4 \times 3 = 12$ mol from silicic acid), 7 mol H_2O must react with albite:

$$NaAlSi_3O_8(s) + 7H_2O(\ell) + H^+(aq) =$$
$$Al(OH)_3(s) + Na^+(aq) + 3Si(OH)_4^0(aq) \quad (s1.4)$$

Note that 15 mol O now appear on both sides of Eq. s1.4 to give oxygen mass balance. This reaction can be compared with Eq. 1.3, which has 8 mol H_2O and OH^- instead of 7 mol H_2O and H^+. The difference is accounted for by adding the ionization reaction of water to Eq. s1.4,

$$H_2O(\ell) = H^+(aq) + OH^-(aq) \quad (s1.5)$$

and canceling $H^+(aq)$ from both sides of the result.

More complicated dissolution reactions can be developed in the same way. Consider the weathering of muscovite via cation exchange and silicic acid addition to form vermiculite, as in Eq. 1.5. The fundamental reaction is:

$$K_2[Si_6Al_2]Al_4O_{20}(OH)_4(s) + Ca^{2+}(aq) + Si(OH)^0(aq) \rightarrow$$
$$Ca_{0.7}[Si_{6.6}Al_{1.4}]Al_4O_{20}(OH)_4(s) + K^+(aq) \quad (s1.6)$$

Mass balance for Al requires 6 mol on both sides of the reaction; thus the stoichiometric coefficient for vermiculite must be $6/5.4 = 1.1$. Mass balance for Ca then requires $1.1 \times 0.7 = 0.8$ mol Ca on both sides and that for K requires 2 mol K on both sides.

$$K_2[Si_6Al_2]Al_4O_{20}(OH)_4(s) + 0.8Ca^{2+}(aq) + Si(OH)_4^0(aq)$$
$$\rightarrow 1.1Ca_{0.7}[Si_{6.6}Al_{1.4}]Al_4O_{20}(OH)_4(s) + 2K^+(aq) \quad (s1.7)$$

Mass balance for Si now can be imposed. The right side of Eq. s1.7 has $1.1 \times 6.6 = 7.3$ mol Si vs. 6 mol in muscovite; thus the coefficient for silicic acid must be 1.3. Charge balance requires an additional 0.4 mol cation charge on the right side of Eq. s1.7 (1.6 mol Ca charge against 2 mol K charge). This deficit can be made up by adding $0.4H^+$ to the left side:

$$K_2[Si_6Al_2]Al_4O_{20}(OH)_4(s) + 0.8Ca^{2+}(aq) + 0.4H^+(aq) + 1.3Si(OH_4^0(aq)$$

$$\rightarrow 1.1Ca_{0.7}[Si_{6.6}Al_{1.4}]Al_4O_{20}(OH)_4(s) + 2K^+(aq) \qquad (s1.8)$$

Finally, mass balance for H $(4 + 0.4 + 1.3 \times 4 = 9.6$ mol H on the left side of Eq. s1.8) leads to the addition of 2.6 mol H_2O as one of the products:

$$K_2[Si_6Al_2]Al_4O_{20}(OH)_4(s) + 0.8Ca^{2+}(aq) + 0.4H^+(aq) + 1.3Si(OH)^0(aq)$$

$$= 1.1Ca_{0.7}[Si_{6.6}Al_{1.4}]Al_4O_{20}(OH)_4(s) + 2K^+(aq) + 2.6H_2O(\ell) \qquad (s1.9)$$

Equation s1.9 shows 29.2 mol O on the left side vs. 29.0 mol O on the right side, a small discrepancy (0.7%) caused by round-off errors in the stoichiometric coefficients. Note that Eq. s1.9 can be made the same as Eq. 1.5 by adding the ionization reaction of 0.4 mol H_2O.

2

Soil Minerals

2.1 Ionic Solids

The chemical elements making up soil minerals occur typically as ionic species whose electron configuration is unique and remains the same regardless of whatever other ions may occur in the mineral structure. The interaction between one ion and another of opposite charge in the structure nonetheless is strong enough to form a chemical bond, termed an *ionic bond*. Ionic bonds differ from covalent bonds, which involve distortion of the electron configurations of the bonding atoms that results in the sharing of electrons. The electron sharing mixes up the detailed structures of the atoms, so it is not possible to assign to each a unique electron configuration that is the same regardless of the partner with which the covalent bond forms. This loss of electronic identity in the atoms leads to a more coherent fusion of their atomic structures and makes covalent bonds stronger than ionic bonds.

Ionic and covalent bonds are conceptual idealizations that real chemical bonds only approximate. In general, a chemical bond shows some degree of ionic character and some degree of electron sharing. The Si—O bond, for example, is said to be an even partition between ionic and covalent character, and the Al—O bond is thought to be about 40% covalent, 60% ionic. Aluminum, however, is exceptional in this respect, for almost all of the metal–oxygen bonds that occur in soil minerals are strongly ionic. Covalence thus plays a minor role in the structure of most soil minerals, aside from the important feature that Si—O bonds impart particular stability against weathering, discussed in Section 1.2. In respect to metal–oxygen bonds, covalence figures prominently only in the short-range molecular environment of the metal, and this environment does not change much from one mineral to another. The overall structure and relative stability of a mineral are determined by the longer-range ionic interaction whose behavior can be understood on the basis of classical electrostatics and the coulomb law.

Given this point of view, the two most important atomic properties of the metal elements in soil minerals are ionic valence and radius. The valence is simply the ratio of the electric charge of an ionic species to the charge of the proton. The ionic radius is a less direct concept because the radius of a single ion in a solid cannot be measured. The ionic radius of a metal thus is a *defined* quantity whose calculation is based on the following assumptions: (1) the radius of the oxygen ion in all minerals is 0.140 nm, (2) the sum of cation and anion radii equals the measured interatomic distance between the two ions, and (3) the ionic radius depends on the coordination number, but otherwise is independent of the type of mineral structure containing the ion. (The coordination number is the number of anions that are nearest neighbors of a cation in a mineral structure.) Table 2.1 lists standard ionic radii calculated under these assumptions from crystallographic data. Note that the radii depend on the valence (Z) as well as the coordination number (CN) of the metal cation. The radius decreases as the valence increases, but it increases with increasing coordination number for a constant valence. The coordination

Table 2.1 Ionic radius (IR), coordination number (CN), and valence (Z) of metal cations[a]

Metal	Z	CN	IR (nm)	Metal	Z	CN	IR (nm)
Li	1	4	0.059	Ni	2	6	0.069
	1	6	0.076	Cu	2	4	0.057
Na	1	6	0.102		2	6	0.073
Mg	2	6	0.072	Zn	2	6	0.074
Al	3	4	0.039	Rb	1	6	0.152
	3	6	0.054		1	12	0.172
K	1	6	0.138	Sr	2	6	0.118
	1	8	0.151	Zr	4	8	0.084
	1	12	0.164	Ag	1	6	0.115
Ca	2	6	0.100	Cd	2	6	0.075
	2	8	0.112	Cs	1	6	0.167
Ti	4	6	0.061			12	0.188
Mn	2	6	0.083	Ba	2	6	0.135
	3	6	0.065	Hg	1	6	0.119
	4	6	0.053		2	6	0.102
Fe	2	6	0.078	Pb	2	6	0.119
	3	6	0.065		4	6	0.078
Co	2	6	0.075				
	3	6	0.061				

[a]R. D. Shannon, Revised effective ionic radii and systematic studies of interatomic distances in halides and chalcogenides, *Acta Cryst.* **A32**:751–767 (1976).

numbers found typically for cations in soil minerals are 4, 6, 8, and 12. The geometric arrangements of anions these numbers represent are illustrated in Fig. 2.1. Each of the arrangements corresponds to a regular geometric solid (*polyhedron*). It is evident that the magnitude of the anionic electrostatic field acting on a cation can be larger as its coordination number increases. This larger field deforms (*polarizes*) the cation such that it expands more into the void between the surrounding anions, thereby causing its radius to increase with its coordination number.

The electrostatic picture of ionic solids has significant implications for what kinds of atomic structures these solids can have in nature. The structures of most of the minerals in soils can be rationalized on the physical grounds that the atomic configuration observed is that which tends to minimize the total electrostatic energy of the crystal. This concept has been formulated in a most useful fashion through a set of descriptive statements known as the *Pauling Rules*:

Rule 1. A polyhedron of anions is formed about each cation. The cation–anion distance is determined by the sum of the respective radii, and the coordination number is determined by the radius ratio of cation to anion.

Minimum radius ratio	Coordination number
1.00	12
0.732	8
0.414	6
0.225	4

Rule 2. In a stable crystal structure, the sum of the strengths of the bonds that reach an anion from adjacent cations is equal to the absolute value of the anion valence.

Rule 3. The cations maintain as large a separation as possible and have anions interspersed between them so as to screen their charges. In geometric terms, this means that anion polyhedra tend *not* to share edges or especially faces. If edges are shared, they are shortened.

Rule 4. In a structure comprising different kinds of cation, those of high valence and small coordination number tend *not* to share polyhedron elements with one another.

Rule 5. The number of essentially different kinds of ions in a crystal structure tends to be as small as possible. Thus the number of types of coordination polyhedra in a close-packed array of anions tends to be a minimum.

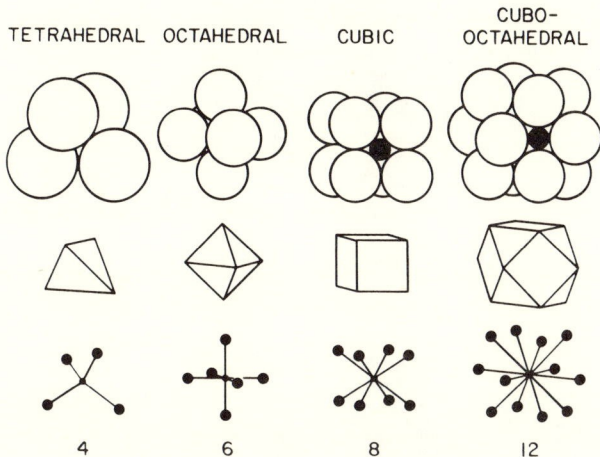

FIG. 2.1 Anion configurations for the principal cation coordination numbers found in soil minerals. For each coordination number, the anion cluster (upper row), corresponding regular polyhedron (middle row), and chemical bond orientations (bottom row) are shown.

Pauling Rule 1 is a statement having the same physical meaning as Fig. 2.1. The anion polyhedra mentioned in the rule are shown in the middle of the figure, and the bottom row of "ball-and-stick" drawings shows the cation–anion bonds whose lengths are determined by the ionic radii. The radius of the smallest sphere that can reside in the central void created by packing anions in the four ways shown at the top of the figure can be calculated with the methods of Euclidean geometry. It turns out that this radius is always proportional to the radius of the coordinating anion. For example, in the case of tetrahedral coordination, the smallest sphere that can fit inside the four coordinating anions has a radius that is 22.5% of the anion radius, and for six coordinating anions it is 41.4% of the anion radius. These minimum radii are listed in the table that accompanies Pauling Rule 1. Specific examples of the cation–oxygen radius ratio can be calculated with the *IR* data in Table 2.1 and the assumed O^{2-} radius of 0.140 nm. Any cation with a coordination number of 6, for example, should have an ionic radius $\geqslant 0.058$ nm ($= 0.414 \times 1.40$). This is the case for all but two of the *IR* values in the table for $CN = 6$, illustrating the point that the Pauling Rules are good *approximations* based on a strictly electrostatic/geometric viewpoint.

Pauling Rule 2 is related closely to local charge balance in a crystal. The *strength* of an ionic bond from a cation to an anion is defined by the equation:

$$s = \frac{Z}{CN} \tag{2.1}$$

TETRAHEDRAL SHEET

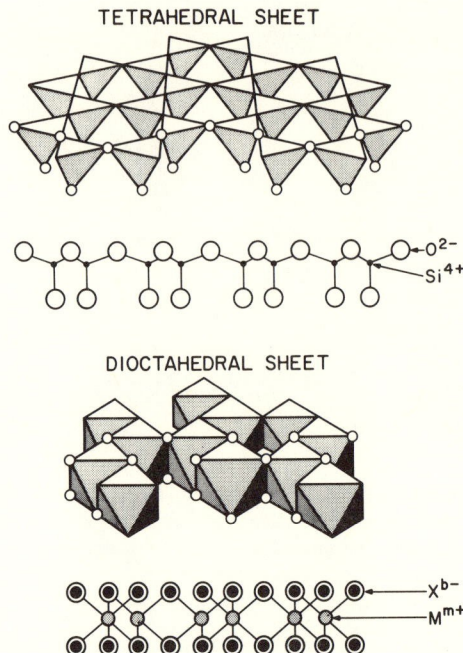

DIOCTAHEDRAL SHEET

FIG. 2.2   Sheet structures of SiO_4 tetrahedra and MX_6 octahedra (e.g., $Al(OH)_6$ octahedra). The open circles at the corners of the tetrahedra and octahedra appear in a section view below each sheet structure. (Reprinted with permission from G. Sposito, *The Surface Chemistry of Soils*, Oxford Univ. Press, New York, 1984.)

where Z and CN are set out in Table 2.1. Rule 2 states that the sum of all s values for the cations bonded to a given anion equals the magnitude of the anion valence. As an example, consider the oxygen ions in quartz (SiO_2), which are coordinated to Si^{4+} ions. The radius of Si^{4+} is 0.026 nm, and its coordination number is 4. It follows from Eq. 2.1 that $s = 1.0$ for Si^{4+}. Since the absolute value of the valence of O^{2-} is 2, Pauling Rule 2 permits only two Si^{4+} to bond to an O^{2-} in SiO_2. This means that each O^{2-} in quartz must serve as the corner of no more than two silica tetrahedra like the one pictured on the left in Fig. 2.1. Hypothetical atomic structures for quartz that involve, say, O^{2-} at the corners of an isolated tetrahedron alternating with three tetrahedra linked together are ruled out, even though they would satisfy the chemical formula, SiO_2.

Pauling Rule 3 is based on the coulomb law applied to cations. The repulsive electrostatic interaction between the cations in a crystal is weakened effectively, or "screened," by the negatively charged anions in the respective coordination polyhedra of the cations. If the cations have a large valence as does, for example, Si^{4+}, then the polyhedra can do no more than share corners if the cations are to be kept as far apart as possible in a structural arrangement that achieves the lowest possible total electrostatic energy. An example of a *sheet* of silica tetrahedra sharing corners is shown in Fig. 2.2. If the cation valence is somewhat smaller, as it is for Al^{3+}, the sharing of polyhedron edges becomes possible. Figure 2.2 also shows this kind of sharing for a sheet of octahedra comprising six anions X^{b-} (e.g., OH^-) bound to the metal cation M^{m+} (e.g., Al^{3+}). Edge sharing brings the cations closer together than corner sharing, however, so the task of charge screening by the anions is more difficult. They respond to this difficulty by approaching one another slightly along the shared edge to enhance screening. Doing so, they shorten the edge relative to unshared edges of the polyhedra.

Pauling Rules 4 and 5 continue in the spirit of Rule 3. They reflect the fact that stable ionic crystals containing different kinds of cations cannot tolerate much sharing of the coordination polyhedra or much variability in the type of coordination environment. These and the other three Pauling Rules serve as useful guides to the atomic structural interpretation of the chemical formulas for soil minerals.

2.2 Primary Silicates

Primary silicates appear in soils from the physical disintegration of parent rock material. They are to be found mainly in the sand and silt fractions, except for soils at the early to intermediate stages of the Jackson–Sherman weathering sequence (Table 1.7), wherein they survive in the clay fraction as well. The chemical weathering of primary silicates contributes to the native fertility and electrolyte content of soils. Among the major decomposition products of these minerals are the soluble chemical species Na^+, Mg^{2+}, K^+, Ca^{2+}, Mn^{2+}, and Fe^{2+} in the soil solution. The metal cations Co^{2+}, Cu^{2+}, and Zn^{2+} occur as trace elements in primary silicates (Table 1.4) and are released to the soil solution by weathering. All of these free-cation soluble species are readily bioavailable and, except for Na^+, are essential to the nutrition of green plants. The cations Na^+, Mg^{2+}, and Ca^{2+}, released through the weathering of primary silicates, also provide a major input to the electrolyte concentration in arid-zone soil solutions. Often the bivalent cations coming from primary silicates exceed Na^+ in concentration and, therefore, lessen the sodicity hazard of irrigation waters that leach these soils (see Chapter 12).

The names and chemical formulas of primary silicate minerals important to soils are listed in Table 2.2. The fundamental building block in the atomic structure of these minerals is the silica tetrahedron, SiO_4^{4-}. Silica tetrahedra can occur as isolated units, in single or double chains linked together by shared corners (Pauling Rules 2 and 3), in sheets (Fig. 2.2), or in fully three-dimensional frameworks. Each mode of occurrence defines a class of primary silicates, as summarized in Fig. 2.3.

Table 2.2 Names and chemical formulas of primary silicates

Name	Chemical formula	Mineral group
Forsterite	Mg_2SiO_4	Olivine
Fayalite	Fe_2SiO_4	Olivine
Chrysolite	$Mg_{1.8}Fe_{0.2}SiO_4$	Olivine
Enstatite	$MgSiO_3$	Pyroxene
Orthoferrosilite	$FeSiO_3$	Pyroxene
Diopside	$CaMgSi_2O_6$	Pyroxene
Tremolite	$Ca_2Mg_5Si_8O_{22}(OH)_2$	Amphibole
Actinolite	$Ca\ Mg_4FeSi_8O_{22}(OH)_2$	Amphibole
Hornblende	$NaCa_2Mg_5Fe_2AlSi_7O_{22}(OH)$	Amphibole
Muscovite	$K_2[Si_6Al_2]Al_4O_{20}(OH)_4$	Mica
Biotite	$K_2[Si_6Al_2]Mg_4Fe_2O_{20}(OH)_4$	Mica
Phlogopite	$K_2[Si_6Al_2]Mg_6O_{20}(OH)_4$	Mica
Orthoclase	$KAlSi_3O_8$	Feldspar
Albite	$NaAlSi_3O_8$	Feldspar
Anorthite	$CaAl_2Si_2O_8$	Feldspar

The *olivines* comprise individual silica tetrahedra in a structure held together with bivalent metal cations like Mg^{2+}, Fe^{2+}, Ca^{2+}, and Mn^{2+} in octahedral coordination (Fig. 2.1). Solid solution occurs with the minerals forsterite and fayalite (Table 2.2) to produce a series of mixtures with specific names, like chrysolite, which contains 10–30 mol % of fayalite. As discussed in Section 1.2, the olivines have the smallest Si/O molar ratio among the primary silicates, and, therefore, they feature the least amount of overall covalence in their chemical bonds. Their weathering in the soil environment is relatively

FIG. 2.3 The principal classes of silicate structures in primary soil minerals.

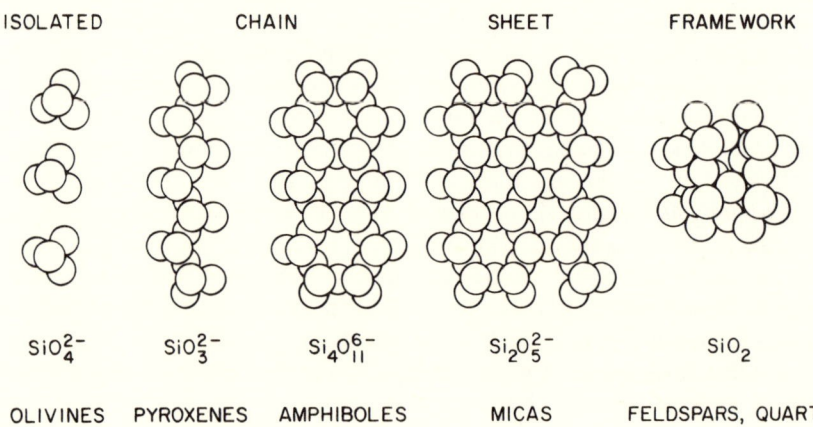

ISOLATED	CHAIN	SHEET		FRAMEWORK
SiO_4^{2-}	SiO_3^{2-}	$Si_4O_{11}^{6-}$	$Si_2O_5^{2-}$	SiO_2
OLIVINES	PYROXENES	AMPHIBOLES	MICAS	FELDSPARS, QUARTZ

rapid, beginning along cracks and at the crystal surface to form altered rinds containing oxidized-iron solid phases and smectite (Table 1.3). A chemical reaction illustrating this process is:

$$12.4Mg_{1.63}Fe(II)_{0.37}SiO_4(s) + 0.8AlOH^{2+}(aq) + 25.4H^+(aq)$$
(olivine)

$$+1.6H_2O(\ell) + O_2(g) = Mg_{0.40}[Si_{7.2}Al_{0.8}]Mg_6O_{20}(OH)_4(s)$$
(saponite/smectite)

$$+4.59FeO(OH)(s) + 5.2Si(OH)_4^0(aq) + 13.5Mg^{2+}(aq) \qquad (2.2)$$
(goethite)

The principal reaction mechanisms contributing to this incongruent dissolution are hydrolysis and oxidation [of Fe(II) to Fe(III)]. Note the consumption of protons and the production of silicic acid and Mg^{2+} in the soil solution.

The *pyroxenes* and *amphiboles* contain single and double chains of silica tetrahedra forming the repeating unit $Si_2O_6^{4-}$ or $Si_4O_{11}^{6-}$, with Si/O ratios near 0.33–0.36. The amphiboles feature isomorphic substitution of Al^{3+} for Si^{4+} (Table 2.2), and both mineral groups have a variety of bivalent metal cations, as well as Na^+ and Fe^{3+}, in octahedral coordination with O^{2-} to link the silica chains together. The weathering reactions of these silicates are similar to that in Eq. 2.2: Mg-rich smectite, with Al and Si in tetrahedral coordination and oxidized Fe in octahedral coordination, is produced along with iron oxides and soluble silica; Na^+, Ca^{2+}, and Mg^{2+} are released to the soil solution.

The *micas* are built up from two sheets of silica tetrahedra ($Si_2O_5^{2-}$ repeating unit) fused to each planar side of a sheet of metal cation octahedra (Fig. 2.2). The octahedral sheet typically contains Al, Mg, and Fe ions coordinated to O^{2-} and OH^-. If the metal cation is trivalent, only two of the three possible cationic sites in the octahedral sheet can be filled to achieve charge balance and the sheet is termed *dioctahedral*. If the metal cation is bivalent, all three possible sites are filled and the sheet is *trioctahedral*. Isomorphic substitution of Al for Si, Fe(III) for Al, and Fe or Al for Mg occurs typically in the micas along with the many trace element substitutions mentioned in Table 1.4.

Muscovite and biotite are the common soil micas, the former being dioctahedral, the latter trioctahedral (Table 2.2). In both minerals, Al^{3+} substitutes for Si^{4+}. The resulting charge deficit is balanced by K^+ that coordinates to 12 oxygen ions in the cavities of two opposing tetrahedral sheets belonging to a pair of mica layers stacked on top of one another. (More details of this arrangement will be given in Sections 2.3 and 7.1.) Thus the K^+ ions link adjacent mica layers together. The initial weathering reaction of muscovite to vermiculite is presented in Eq. 1.5. An important feature of this reaction is the reduction in the amount of interlayer cation charge ($2K^+ \rightarrow 0.7Ca^{2+}$) caused by a reduction in the amount of substituted Al in tetrahedral coordination ($2Al \rightarrow 1.4Al$, shown in square brackets) in the silica sheet. This reduction continues as the vermiculite weathers further to smectite:

$$Ca_{0.7}[Si_{6.6}Al_{1.4}]Al_4O_{20}(OH)_4(s) + 0.42Mg^{2+}(aq)$$
$$+ 0.86Si(OH)_4^0(aq) + 0.56H^+(aq)$$
$$= 1.05AlOH_{0.65}^{2+}[Si_{7.1}Al_{0.9}]Al_{3.6}Mg_{0.4}O_{20}(OH)_4(s)$$
$$+ 0.7Ca^{2+}(aq) + 1.56H_2O(\ell) \qquad (2.3)$$

where cations located in the tetrahedral sheet are enclosed in square brackets (see also Tables 1.3 and 2.2).

Similar processes occur with biotite, except that the reduction of interlayer cation charge can be accomplished by the oxidation of ferrous iron as well as by the loss of tetrahedrally coordinated aluminum:

$$K_2[Si_6Al_2]Mg_4Fe(II)_2O_{20}(OH)_4(s) + 3Mg^{2+}(aq) + 2Si(OH)_4^0(aq)$$
$$\text{(biotite)}$$

$$= 1.25hr\ Mg_{0.4}[Si_{6.4}Al_{1.6}]Mg_{5.2}Fe(III)_{0.8}O_{20}(OH)_4(s)$$
$$\text{(vermiculite)}$$

$$+ FeO(OH)(s) + 2K^+(aq) + 4H^+(aq) \qquad (2.4)$$
$$\text{(goethite)}$$

Note that the interlayer cation charge was decreased by 0.8 proton charges simply by the oxidation of Fe(II) to Fe(III) in the transition from biotite to vermiculite. The remainder of the octahedrally coordinated Fe was expelled from the sheet silicate to precipitate as goethite.

The atomic structure of the *feldspars* is a continuous, three-dimensional framework of tetrahedra sharing corners, like quartz, except that some of the tetrahedra contain Al instead of Si, thus requiring either monovalent or bivalent metal cations to occupy cavities in the framework for charge balance. These minerals, the most abundant in soils, have repeating units of either $AlSi_3O_8^-$, with Na^+ or K^+ for charge balance, or $Al_2Si_2O_8^{2-}$, with Ca^{2+} for charge balance (Table 2.2). Solid solution among the three minerals thus formed is extensive. Feldspars may weather eventually to kaolinite and gibbsite (Eq. 1.3 and Problem 11 in Chapter 1), but their initial decomposition appears to produce allophane and smectite:

$$4KAlSi_3O_8(s) + 4H^+(aq) + (n + 16)H_2O(\ell)$$
$$\text{(orthoclase)}$$

$$= Si_3Al_4O_{12} \cdot nH_2O(s) + 9Si(OH)_4^0(aq) + 4K^+(aq) \qquad (2.5a)$$
$$\text{(allophane)}$$

$$4KAlSi_3O_8(s) + 0.5Mg^{2+}(aq) + 2H^+(aq) + 10H_2O(\ell)$$
(orthoclase)

$$= K[Si_{7.5}Al_{0.5}]Al_{3.5}Mg_{0.5}O_{20}(OH)_4(s) + 4.5Si(OH)_4^0(aq) + 3K^+(aq)$$
(montmorillonite/smectite)

$$(2.5b)$$

Note again the consumption of protons and the production of silicic acid and soluble cations, as in Eq. 2.3.

The general characteristics of primary silicate weathering illustrated by Eqs. 1.2–1.5 and 2.2–2.5 are as follows:

1. Loss of tetrahedrally coordinated Al
2. Oxidation of Fe(II)
3. Consumption of protons
4. Release of silica and the metal cations Na^+, K^+, Mg^{2+}, and Ca^{2+}

In the case of the sheet silicates (micas), there is also an important reduction of interlayer charge accompanying characteristics 1 and 2 just given. From the soil weathering sequence in Table 1.7, one can conclude that soil development renders tetrahedral Al and ferrous iron unstable in response to continual throughputs of oxygenated fresh water (i.e., rainwater), which provides protons and receives soluble macroelements. If the elements received by the soil solution are not leached, the secondary silicates that characterize the inter-mediate stage of weathering will precipitate, as in Eqs. 1.5 and 2.2–2.5 If leaching is extensive, the desilicated minerals characteristic of the advanced stage of weathering will begin to predominate in the clay fraction instead. These concepts of metastable mineral sequences and the rates of dissolution involved with mineral weathering are discussed fully in Chapter 5.

2.3 Clay Minerals

Clay minerals are aluminosilicates that predominate in the clay fractions of soils at the intermediate to advanced stages of weathering. These minerals, like the micas, are sandwiches of tetrahedral and octahedral sheet structures like those in Fig. 2.2. This bonding together of the tetrahedral and octahedral sheets occurs through the apical oxygen ions in the former structure and always produces a significant distortion of the anion arrangement in the final layer structure formed. The distortion occurs primarily because the apical oxygen ions in the tetrahedral sheet cannot be fit to the corners of the octahedra to form a layer while preserving the ideal hexagonal pattern of the tetrahedra. In order to fuse the two sheets, pairs of adjacent tetrahedra must rotate and thereby distort the symmetry of the hexagonal cavities in the basal plane of the tetrahedral sheet (Fig. 2.2). Besides this distortion, the sharing of edges in the octahedral sheet shortens them (Pauling Rule 3), and isomorphic substitution of the cations in both sheets tends to make the thickness of the layer structure smaller and its basal surfaces slightly corrugated. These effects occur in both the micas and the clay minerals.

The clay minerals usually are classified into three *layer types*, distinguished by the number of tetrahedral and octahedral sheets combined, and further into five *groups*, differentiated by the kinds of isomorphic cation substitutions that occur. The layer types are shown in Fig. 2.4, and the groups are described in Table 2.3. The 1:1 layer type consists of one tetrahedral and one octahedral sheet. In soil clays, it is represented by the kaolinite group, with the general chemical formula $[Si_4]Al_4O_{10}(OH)_8 \cdot nH_2O$, where the cation enclosed in square brackets is in tetrahedral coordination and n is the number of moles of hydration water. Normally there is no significant isomorphic substitution for Si or Al in this clay mineral (substitution of Fe(III) for Al up to 3 mol % is observed in Oxisols). As is common with soil clay minerals, the octahedral

FIG. 2.4 The three layer types of clay minerals (Table 2.3). (Reprinted with permission from G. Sposito, *The Surface Chemistry of Soils*, Oxford Univ. Press, New York, 1984.)

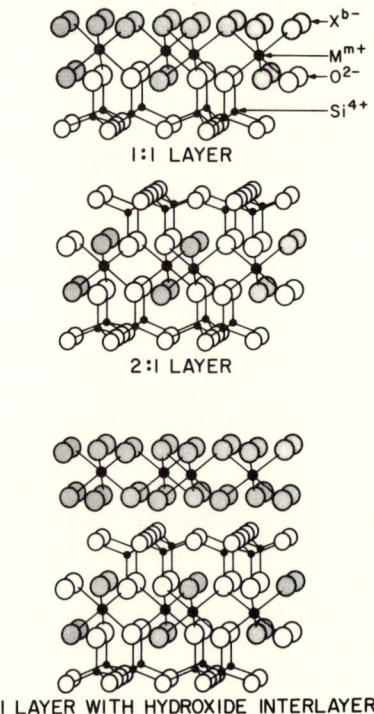

Table 2.3 Clay mineral groups (Fig. 2.4)

Group	Layer type	Layer charge (x)	Typical chemical formula[a]
Kaolinite	1:1	<0.01	$[Si_4]Al_4O_{10}(OH)_8 \cdot nH_2O (n = 0$ or 4)
Illite	2:1	1.4–2.0	$M_x[Si_{6.8}Al_{1.2}]Al_3Fe_{0.25}Mg_{0.75}O_{20}(OH)_4$
Vermiculite	2:1	1.2–1.8	$M_x[Si_7Al]Al_3Fe_{0.5}Mg_{0.5}O_{20}(OH)_4$
Smectite[b]	2:1	0.5–1.2	$M_x[Si_8]Al_{3.2}Fe_{0.2}Mg_{0.6}O_{20}(OH)_4$
Chlorite	2:1 with hydroxide interlayer	Variable	$(Al(OH)_{2.55})_4 \cdot [Si_{6.8}Al_{1.2}]Al_{3.4}Mg_{0.6}O_{20}(OH)_4$

[a] $n = 0$ is kaolinite and $n = 4$ is halloysite; M = monovalent interlayer cation.

[b] Principally montmorillonite in soils.

sheet has two-thirds cation site occupancy (dioctahedral sheet). The 2:1 layer type has two tetrahedral sheets that sandwich an octahedral sheet (Fig. 2.4). The three soil clay mineral groups with this structure are illite, vermiculite, and smectite. If a, b, and c are the stoichiometric coefficients of Si, octahedral Al, and Fe(III), respectively, in the chemical formulas of these groups, then

$$x = 12 - a - b - c \tag{2.6}$$

is the *layer charge*, the number of moles of excess electron charge per chemical formula that is produced by isomorphic substitution. As indicated in Table 2.3, the three 2:1 groups differ from one another in two principal ways. The layer charge decreases in the order illite > vermiculite > smectite, and the vermiculite group is further distinguished from the smectite group by the extent of isomorphic substitution in the tetrahedral sheet. Among the smectites, those in which the substitution of Al for Si exceeds that of Fe(II) or Mg for Al are called *beidellite* (see Eq. 2.3), and those in which the reverse is true are called *montmorillonite* (see Eq. 2.5b). The sample chemical formula in Table 2.3 for smectite represents montmorillonite. In any of these 2:1 minerals, the layer charge is balanced by cations that reside on or in the cavities of the basal plane of the oxygen atoms of the tetrahedral sheet. These interlayer cations are represented by M in the chemical formula (Table 2.3). The 2:1 layer type with hydroxide interlayer is represented by dioctahedral chlorite in soil clays (Fig. 2.4). The octahedrally coordinated cations in chlorite reside in two sheets: one comprising $M(OH)_2O_4^{m-10}$ octahedra (with $M^{m+} = Al^{3+}$, Fe^{3+}, or Mg^{2+}) sandwiched in a 2:1 layer, and one comprising principally $Al(OH)_6^{3-}$ octahedra situated on a basal surface of the 2:1 layer. To preserve the electroneutrality of the whole structure, the octahedral occupancy often is larger than the expected value of 8 per chemical formula for the two dioctahedral sheets, such that the excess positive charge balances the excess negative charge created by isomorphic substitution in the tetrahedral sheets.

Structural discorder in all of the clay minerals listed in Table 2.3 is created through isomorphic substitutions for their principal cation components. The range of these substitutions in the Periodic Table is very broad, as indicated in Table 1.5. Even more pronounced structural disorder exists in silica and in aluminosilicates freshly precipitated in soils, since these compounds typically are amorphous. (If the repeating structure based on the chemical formula of a solid phase persists throughout a molecular region whose diameter is at least as large as 3 nm, the solid phase is said to be *crystalline*. If structural regularity does not exist over molecular distances this large, the solid phase is termed *amorphous*. This distinction concerning structural regularity is intended principally as a general guide.) Among the crystalline soil clay minerals, there is wide variation in molecular order, with disorder, in the absence of isomorphic substitution, created by dislocations (microcrevices between offset rows of atoms) and irregular stacking of crystalline layers. This kind of disorder exists, for example, in the kaolinite group minerals.

Structurally disordered aluminosilicates, known collectively as *allophane* and *imogolite* (Table 1.3), are common in the clay fractions of soils formed on volcanic ash parent material. The atomic structure of allophane is not well understood, but is thought to consist of a $1:1$ aluminosilicate layer framework riddled with defects (vacant ion sites) and containing Al in both the tetrahedral and the octahedral sheets. These many defects promote a curling of the layer into the form of a hollow spherule ~ 5 nm in diameter whose outer boundary would contain many apertures through which small molecules or ions from the soil solution could enter. As this structural concept suggests, allophane often is found in association with clay minerals of the kaolinite group, especially the hydrated species, halloysite (Table 2.3). Imogolite, with the general empirical formula, $Si_2Al_4O_{10} \cdot 5H_2O$, exhibits a tubular morphology. The tube unit in the structure contains Al only in octahedral coordination and exposes a defective, gibbsitelike surface.

There are three principal kinds of weathering reactions involving the $2:1$ layer-type clay minerals:

$$Ca_{0.7}[Si_{6.6}Al_{1.4}]Al_4O_{20}(OH)_4(s) + 3.6Al(OH)_{2.6}^{0.4+}(aq)$$
$$\text{(vermiculite)}$$

$$= (Al(OH)_{2.6})_{3.6}[Si_{6.6}Al_{1.4}]Al_4O_{20}(OH)_4(s) + 0.7Ca^{2+}(aq)$$
$$\text{(hydroxy-interlayer vermiculite)}$$

$$(2.7a)$$

$$Ca_{0.7}[Si_{6.6}Al_{1.4}]Al_4O_{20}(OH)_4(s) + 0.5Mg^{2+}(aq) + 0.2H^+(aq) + 2.7Si(OH)_4^0(aq)$$
$$\text{(vermiculite)}$$

$$= 1.24Al_{0.24}Ca_{0.09}[Si_{7.5}Al_{0.5}]Al_{3.6}Mg_{0.4}O_{20}(OH)_4(s)$$
$$\text{(beidellite/smectite)}$$

$$+ 0.6Ca^{2+}(aq) + 5H_2O(\ell) \quad (2.7b)$$

$Al_{0.3}[Si_{7.5}Al_{0.5}]Al_{3.6}Mg_{0.4}O_{20}(OH)_4(s) + 0.8H^+(aq) + 8.2H_2O(\ell)$
(beidellite/smectite)

$$= 1.1[Si_4]Al_4O_{10}(OH)_8(s) + 3.1Si(OH)_4^0(aq) + 0.4Mg^{2+}(aq) \quad (2.7c)$$
(kaolinite)

The reaction in Eq. 2.7a is cation exchange between Ca^{2+} on vermiculite and an Al-hydroxy complex to form hydroxy-interlayer vermiculite (*chloritized vermiculite*). This reaction is common under acidic conditions wherein dissolved Al produced by mineral weathering is abundant. The product clay mineral resembles chlorite except that the amount of Al-hydroxy material in the interlayer is less (see Table 2.3). The same kind of reaction is possible for smectite. Equation 2.7b represents the weathering of vermiculite to the smectite, beidellite (see Table 1.7), which is also favored by acidic conditions. This reaction is of the same kind as in Eq. 2.3. Finally, Eq. 2.7c represents the proton attack and hydrolysis of smectite (beidellite) that produces kaolinite (Table 1.7). This reaction is favored by fresh water and good drainage.

2.4 Oxides and Hydroxides

Because of their great abundance in the lithosphere and their low solubilities in the normal range of soil pH values, aluminum, iron, and manganese form the most important oxide, oxyhydroxide, and hydroxide minerals in soils. These minerals are listed in Table 2.4, and representative octahedral structures are shown in Fig. 2.5.

Among the iron compounds listed in Table 2.4, goethite is the one most often found in soils regardless of climatic region. However, under oxic conditions and in the presence of iron-complexing ligands that inhibit crystallization (e.g., organic ligands or silicate anions), ferrihydrite ($Fe_{10}O_{15} \cdot 9H_2O$) may precipitate from a soil solution. This poorly crystalline solid comprises sheets of octahedra with Fe(III) coordinated to O, OH, and OH_2 in a defect-sprinkled arrangement that is similar to that in hematite. Ferrihydrite can transform either to hematite, which ultimately will transform to goethite, or to goethite directly.

Goethite is the most thermodynamically stable of the iron oxides and therefore is the solid phase expected finally in soil clays. As indicated in Fig. 2.5, the oxygen ions in goethite lie in planes, and the Fe^{3+} cations are coordinated in distorted octahedra that share edges. Some of the octahedral vertex ions are hydroxyl groups that form hydrogen bonds with neighboring oxygen ions. Isomorphic substitution of Al for Fe in goethite is common, especially in highly weathered soils having abundant dissolved Al (see Chapter 11).

Gibbsite is the most important of the aluminum minerals listed in Table 2.4. Its molecular structure is illustrated on the right in Fig. 2.5. The dioctahedral

Table 2.4   Metal oxides, oxyhydroxides, and hydroxides found commonly in soils

Name	Chemical formula[a]	Name	Chemical formula[a]
Anatase	TiO_2	Hematite	$\alpha\text{-}Fe_2O_3$
Birnessite	$Na_{0.7}Ca_{0.3}Mn_7O_{14}\cdot 2H_2O$	Ilmenite	$FeTiO_3$
Boehmite	$\gamma\text{-}AlOOH$	Lepidocrocite	$\gamma\text{-}FeOOH$
Ferrihydrite	$Fe_{10}O_{15}\cdot 9H_2O$	Lithiophorite	$(Al,Li)MnO_2(OH)_2$
Gibbsite	$\gamma\text{-}Al(OH)_3$	Maghemite[b]	$\gamma\text{-}Fe_2O_3$
Goethite	$\alpha\text{-}FeOOH$	Magnetite[b]	$FeFe_2O_4$

[a]γ denotes cubic close-packing of anions, whereas α denotes hexagonal close-packing.
[b]Some of the Fe(III) are in tetrahedral coordination.

sheets included in the structure are bound together by hydrogen bonds between opposed hydroxyl groups. Hydrogen bonding also occurs between hydroxyl groups along the edges of unfilled octahedra within a sheet, thereby producing additional distortion of the aluminum octahedra beyond the distortion caused by the sharing of edges.

The most commonly found manganese mineral in soils is birnessite, with lithiophorite, the other manganese oxyhydroxide listed in Table 2.4, restricted

FIG. 2.5 The atomic structures of goethite and gibbsite, showing $FeO_3(OH)_3$ and $Al(OH)_6$ octahedra in sheets. The dashed lines in the goethite structure indicate hydrogen bonds between OH and O ions. (Reprinted with permission from G. Sposito, *The Surface Chemistry of Soils*, Oxford Univ. Press, New York, 1984.)

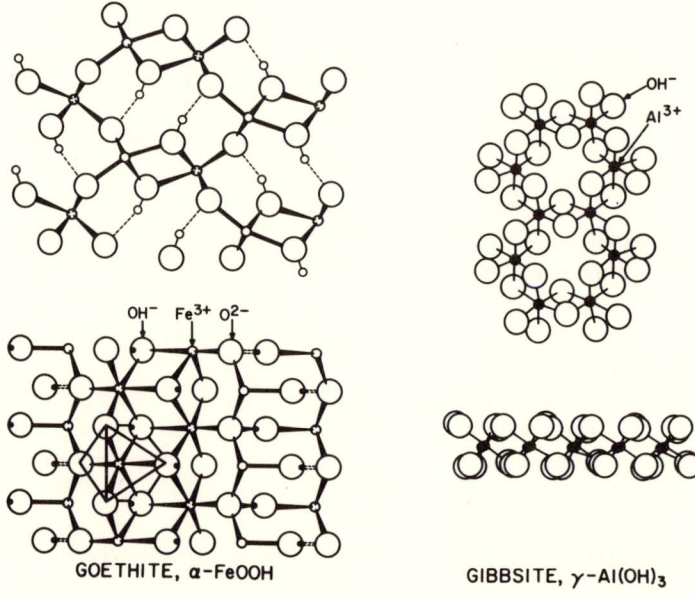

GOETHITE, α–FeOOH   GIBBSITE, γ–Al(OH)$_3$

largely to acid soils. Birnessite contains sheets of MnO_6^{8-} octahedra linked in some fashion with $Mn(III)$, $Mn(II)$, $Na(I)$, and $Ca(II)$ ions coordinated to both hydroxyl groups and water molecules. In lithiophorite, sheets of MnO_6^{8-} octahedra alternate with sheets containing $Al_{0.67}Li_{0.33}$ octahedra.

The metal oxides and hydroxides represent the "climax mineralogy" of soils, as indicated in Table 1.7. They can form directly from the weathering of primary silicates (Eqs. 1.3, 2.2, and 2.4) or from the hydrolysis and desilication of clay minerals like smectite and kaolinite. The 2:1 layer silicates can react collectively with positively charged metal hydroxy polymers (Eq. 2.7a) to form interlayer surface coatings of the metal hydroxides. In a similar fashion, metal oxide or hydroxide surfaces bearing hydroxyl groups can react with negatively charged polymers (e.g., silica) in the soil solution and become coated. Chemical weathering in soil brings about the dissolution of minerals that release oxidized elements into the aqueous phase of soil, and it is these elements that hydrolyze readily to form hydroxy-polymers, the most important containing Al, Fe(III), and Si. Aluminum hydroxy-polymers are metastable, dissolved species whose formation and complexation by the interlayer surfaces of vermiculite and smectite are favored by pH values below 6.0, low concentrations of organic compounds, and frequently varying water content. Their presence on an interlayer surface can be interpreted as an intermediate step in the formation of dioctahedral chlorite (Table 2.3). Conversely, the gradual stripping of the hydroxide interlayer in chlorite observed in some soils can be viewed as a weathering regression toward vermiculite or smectite. It is possible for iron hydroxy-polymers to precipitate in a similar fashion on the basal surfaces of 1:1 clay minerals; the incidence of these precipitates on 2:1 clay minerals in soils is very low. In the case of silica polymers, the more probable surface precipitates would be found on the exposed hydroxyl surfaces of kaolinite, gibbsite, and, to some extent, goethite.

2.5 Carbonates and Sulfates

The important carbonate minerals in soils include calcite ($CaCO_3$), dolomite [$CaMg(CO_3)_2$], nahcolite ($NaHCO_3$), trona [$Na_3H(CO_3)_2 \cdot 2H_2O$], and soda ($Na_2CO_3 \cdot 10H_2O$). Calcite may be, and dolomite appears always to be, a primary mineral in soils. Secondary calcite that precipitates from soil solutions enriched in soluble Mg often coprecipitates with $MgCO_3$ to form a *magnesian calcite*, $Ca_{1-y}Mg_yCO_3$, with the mole fraction y of $MgCO_3$ ranging typically up to about 0.05. This mode of formation accounts for nearly all the secondary Mg carbonate found in soils. Like the secondary metal oxides and hydroxides, secondary Ca/Mg carbonates can occur as coatings on other minerals.

Pedogenic calcite is a normal weathering product of Ca-bearing primary silicates (pyroxenes, amphiboles, and feldspars) as well as primary carbonates.

For example, the dissolution of anorthite can produce both smectite and calcite:

$$2CaAl_2Si_2O_8(s) + 0.5Mg^{2+}(aq) + 3.5Si(OH)_4^0(aq) + CO_2(g)$$
$$(anorthite)$$

$$= Ca_{0.5}[Si_{7.5}Al_{0.5}]Al_{3.5}Mg_{0.5}O_{20}(OH)_4(s) + CaCO_3(s)$$
$$(montmorillonite/smectite) \qquad (calcite)$$

$$+ 0.5Ca^{2+}(aq) + 5H_2O(\ell) \tag{2.8}$$

This incongruent dissolution takes advantage of soluble magnesium and silica available from the weathering of primary silicates and the ubiquitous presence of CO_2 in soils. Note that the reaction products are favored by abundant CO_2, since it is a reactant, and are inhibited by abundant H_2O, since it is one of the products. The formation of calcite from primary carbonates also is favored by abundant CO_2, but not as a source of dissolved carbonate ions. Instead, the carbonic acid formed when CO_2 dissolves in the soil solution serves as a source of protons to aid in the dissolution of calcite or dolomite:

$$CO_2(g) + H_2O(\ell) = H_2CO_3^*(aq) = H^+(aq) + HCO_3^-(aq) \tag{2.9a}$$

$$CaCO_3(s) + H^+(aq) = Ca^{2+}(aq) + HCO_3^-(aq) \tag{2.9b}$$

where $H_2CO_3^*$ designates conventionally the sum of undissociated carbonic acid ($H_2CO_3^0$) and solvated $CO_2(CO_2 \cdot H_2O)$, since these two dissolved species are very difficult to distinguish by chemical analysis. The transport of the two ions on the right side of Eq. 2.9b is facilitated by the formation of the soluble complex, $CaHCO_3^+(aq)$, which has lower charge than Ca^{2+}. If leaching is moderate and followed by drying of the soil, the reaction in Eq. 2.9b is reversed and secondary calcite forms. Calcium coprecipitation with Mn, Fe(II), Co(II), or Cd by adsorption onto calcite is not uncommon (Table 1.5). The metals Zn and Cu also may coprecipitate with calcite by inclusion as the hydroxycarbonate minerals, hydrozincite $[Zn_5(OH)_6(CO_3)_2]$, malachite $[Cu_2(OH)_2CO_3]$, or azurite $[Cu_3(OH)_2(CO_3)_2]$.

Like the secondary carbonates, the Ca, Mg, and Na sulfates tend to accumulate as weathering products in soils that develop under arid conditions (Table 1.7). The principal minerals in this group are gypsum ($CaSO_4 \cdot 2H_2O$), anhydrite ($CaSO_4$), epsomite ($MgSO_4 \cdot 7H_2O$), mirabilite ($Na_2SO_4 \cdot 10H_2O$), and thenardite (Na_2SO_4). Gypsum, like calcite, can dissolve and reprecipitate in a soil profile leached by rainwater or irrigation water [the downward transport facilitated by formation of the uncharged soluble complex, $CaSO_4^0(aq)$] and can occur as a coating on soil minerals, including calcite. The Na sulfates, like the Na carbonates, form at the top of the soil profile as it dries through evaporation.

In acid soils, sulfate produced through sulfur oxidation or introduced by amendments of gypsum can react with the abundant Fe and Al in the soil solution to precipitate the minerals jarosite $[KFe_3(OH)_6(SO_4)_2]$, alunite $[KAl_3(OH)_6(SO_4)_2]$, basaluminite $[Al_4(OH)_{10}SO_4 \cdot 5H_2O]$, or jurbanite $(AlOHSO_4 \cdot 5H_2O)$ (see Section 5.4). These minerals may dissolve incongruently to form ferrihydrite or gibbsite upon further attack by protons in a percolating soil solution.

FOR FURTHER READING

J. B. Dixon and S. B. Weed (eds.), *Minerals in Soil Environments*. Soil Science Society of America, Madison, WI, 1977. Details of the crystallographic properties and chemical weathering reactions of soil minerals are to be found in this standard reference.

H. Megaw, *Crystal Structures: A Working Approach*. W. B. Saunders, Philadelphia, 1973. Chapter 2 of this excellent crystallography textbook gives a complete introduction to the concepts of ionic radius, coordination number, and bond strength in relation to ionic crystals.

L. Pauling, *The Nature of the Chemical Bond*. Cornell Univ. Press, Ithaca, NY, 1960. Chapter 13 of this classic monograph, particularly Section 13.6, discusses the origin and applications of the Pauling Rules.

The following three articles can be studied (in the order listed) to gain a deeper understanding of the relationship between the atomic structure of primary silicates and their weathering reactions.

M. J. Wilson, Chemical weathering of some primary rock-forming minerals, *Soil Sci.* **119**:349–355 (1975).

M. L. Jackson, Clay transformations in soil genesis during the quaternary, *Soil Sci.* **99**:15–22 (1965).

R. A. Eggleton, The relations between crystal structure and silicate weathering rates, in S. M. Colman and D. P. Dethier (eds.), *Rates of Chemical Weathering of Rocks and Minerals*, pp. 21–40. Academic Press, Orlando, FL, 1986.

PROBLEMS

The more difficult problems are indicated by an asterisk.

1. Isomorphic substitution of Na^+ for Ca^+ is much more common than Na^+ for K^+ in mineral structures. Similarly, Li^+ replaces octahedrally coordinated Mg^{2+} much more often than it replaces Na^+. Use Table 2.1 to explain these trends and to discuss the relative importance of ionic valence and radius in isomorphic substitution.

2. Alumino-goethite $[Fe_{1-y}Al_yO(OH)]$, ferri-kaolinite $[Si_4(Al_{1-y}Fe_y)_4-O_{10}(OH)_8]$, and magnesian calcite $[Ca_{1-y}Mg_yCO_3]$ are examples of coprecipitated soil minerals, with the metal having the stoichiometric

coefficient y being in the minor component. For each of these solids, rewrite the chemical formula so as to indicate $1 - y$ moles of the major component mineral combined with y moles of the minor component mineral. (The minor component $AlO(OH)$ in aluminogoethite is known as diaspore when it occurs as a pure solid phase).

3. Show that the OH^- in the gibbsite structure (Fig. 2.4) satisfies Pauling Rule 2.

*4. Use Eq. 2.1 and Pauling Rule 2 to show that a corner of a Si—O tetrahedron can be linked to one other Si—O tetrahedron, but not solely to one other Al—O tetrahedron. In the latter case, show that either two monovalent cations or one bivalent cation with $CN = 8$ are needed to satisfy the Pauling Rule. (The minerals orthoclase and anorthite are. examples.)

5. Derive Eq. 2.6 and calculate the specific values of the layer charge x that corresponds to the chemical formulas of illite, vermiculite, and smectite given in Table 2.3. Assume that Fe in the formulas is Fe(III).

*6. The cation exchange capacity (CEC) of a vermiculite or smectite comes largely from its layer charge. In units of cmol kg^{-1},

$$CEC = (x/M_r) \cdot 10^5$$

where x is the layer charge, and M_r is the relative molecular mass of the clay mineral. Use the data in the table below to calculate M_r (ignore the interlayer cation) and CEC for vermiculite and smectite in Table 2.3. (*Answer:* $x = 1.5$, $M_r = 732.6$, $CEC = 205$ cmol kg^{-1} for vermiculite; $x = 0.6$, $M_r = 724.8$, $CEC = 83$ cmol kg^{-1} for smectite)

Element	M_r	Element	M_r
H	1.008	Al	26.982
O	15.999	Si	28.086
Mg	24.312	Fe	55.847

7. Orthoclase can weather to kaolinite and gibbsite under humid tropical conditions. Write a balanced chemical reaction for this transformation.

8. The weathering of biotite as shown in Eq. 2.4 is typical of temperate humid regions. In tropical, humid regions, the clay mineral product is typically kaolinite, not vermiculite. Develop a balanced chemical reaction for the weathering of biotite to kaolinite and goethite.

9. Combine Eq. 2.5a and 2.5b to derive a chemical reaction for the conversion of allophane to montmorillonite. Is this reaction favored or inhibited by acidic conditions?

*10. Write a balanced cation exchange reaction for the transformation of the montmorillonite in Eq. 2.5b to a hydroxy-interlayer montmorillonite, analogous to the vermiculite in Eq. 2.7a.
(*Hint:* The stoichiometric coefficient of $Al(OH)_{2.6}^{0.4+}(aq)$ is 2.5 in this reaction.)

11. Develop a chemical equation that describes a reaction involving trona, nahcolite, and CO_2. Which of the two Na carbonates would be favored by increasing the CO_2 partial pressure in soil?

*12. Develop a series of chemical equations that describes a balanced reaction involving the dissolution of gypsum and dolomite to form magnesian calcite with 3 mole % $MgCO_3$ (i.e., $y = 0.03$ in the chemical formula of magnesian calcite).
(*Hint:* Dissolve $1-2y$ moles of gypsum and y moles of dolomite.)

13. Write a balanced chemical reaction for the transformation of jurbanite to gibbsite.

*14. Gypsum is added to an acid soil containing the Al-saturated beidellite in Eq. 2.7c. Develop a chemical equation that describes the resulting formation of Ca-saturated beidellite and jurbanite. This reaction could provide exchangeable Ca and soluble SO_4, as well as reduce the bioavailability of Al.
(*Hint:* 0.45 mol gypsum is required in this reaction.)

*15. Using the notation from Problem 2, write a balanced chemical reaction for the incongruent dissolution of ferri-kaolinite, containing 2 mol % Fe(III) substituted for Al, into goethite, gibbsite, and silicic acid.
(*Hint:* $y = 0.02$ in the kaolinite.)

3

Soil Organic Matter

3.1 Biomolecules in Soils

Soils are biological milieux teeming with microorganisms. Ten grams of fertile soil may contain a population of bacteria alone that equals the world population of human beings. One kilogram of soil may contain 500 billion bacteria, 10 billion actinomycetes, and nearly 1 billion fungi. Even the fauna population can approach 500 million in a kilogram of soil. To this microbial biomass can be added the contribution of plant roots, whose combined length even for a single plant can exceed 600 km in the top meter of a soil profile.

Soil microorganisms play an essential role in the catalysis of oxidation–reduction reactions (Chapter 6). The exudates released by microorganisms and healthy plant roots are important to soil acidity (Chapter 11) and to the cycling of trace elements in soil (Chapters 4 and 13). Of the exudates, organic acids are among the best characterized. Table 3.1 lists five aliphatic organic acids that are found commonly in association with microbiological activity or rhizosphere chemistry. (The *rhizosphere* is the local soil environment influenced significantly by plant roots.) These acids have the general structural property of acids have the general structural property of containing the molecular organic unit COOH, which is the *carboxyl group*. The carboxyl group can dissociate its proton easily in the normal range of soil pH (see the third column of Table 3.1) and so is an example of an acid. The dissociated proton can attack soil minerals to provoke their decomposition (see Eqs. 2.2, 2.3, and 2.5), and the remaining *carboxylate* anion (COO^-) can form soluble complexes with metal cations released by mineral weathering (see Eq. 1.4). The total concentration of organic acids in the soil solution ranges from 0.01 to 5 mol m^{-3}, which is quite large relative to trace metal concentrations ($\leqslant 1$ mmol m^{-3}). These acids have

TABLE 3.1 Common aliphatic organic acids in soils

Name	Chemical formula	pH_{dis}[a]
Formic acid	HCOOH	3.8
Acetic acid	CH_3COOH	4.8
Oxalic acid	HOOCCOOH	1.3
Tartaric acid	H O | H HOOCC CCOOH H | O H	3.0
Citric acid	COOH H |H HOOCCCCCOOH H |H O H	3.1

[a]pH value at which the most acidic carboxyl group has a 50% probability to be dissociated in aqueous solution.

very short lifetimes in soil (perhaps hours), but they are produced continually throughout the life cycles of microorganisms.

Formic acid (methanoic acid), the first entry in Table 3.1, is a monocarboxylic acid produced by bacteria and found in the root exudate of corn. Acetic acid (ethanoic acid) also is produced microbially—especially under anaerobic conditions—and is found in the root exudates of grasses and herbs. Formic and acetic acid concentrations in the soil solution range from 2 to 5 mol m^{-3}. Oxalic acid (ethanedioic acid) and tartaric acid (D-2,3-dihydroxybutanedioic acid) are dicarboxylic acids, excreted by the roots of cereals; their soil solution concentrations range from 0.05 to 1 mol m^{-3}. The tricarboxylic citric acid (2-hydroxypropane-1,2,3-tricarboxylic acid) is produced by fungi and excreted by plant roots. Its soil solution concentration is <0.05 mol m^{-3}. Besides these *aliphatic* organic acids, soil solutions contain *aromatic* acids whose fundamental structural unit is a benzene ring. To this ring, carboxyl (benzene carboxylic acids) or hydroxyl (phenolic acids) groups can be bonded in a variety of arrangements. The soil solution concentration of these acids is in the range 0.05–0.3 mol m^{-3}.

Organic acids with the general chemical formula

$$\begin{array}{c} H \\ R\!-\!C\!-\!COOH \\ NH_2 \end{array}$$

TABLE 3.2　Common amino acids in soils

Name	Chemical formula
Glycine	$\overset{\displaystyle NH_2}{\underset{\displaystyle H}{\overset{\mid}{HC}}}-COOH$
Alanine	$CH_3-\overset{\displaystyle NH_2}{\underset{\displaystyle H}{\overset{\mid}{C}}}-COOH$
Aspartic acid	$HOOC-CH_2-\overset{\displaystyle NH_2}{\overset{\mid}{CH}}-COOH$
Glutamic acid	$HOOC-CH_2-CH_2-\overset{\displaystyle NH_2}{\underset{\displaystyle H}{\overset{\mid}{C}}}-COOH$
Arginine	$NH_2-\underset{\displaystyle NH}{\overset{\displaystyle \parallel}{C}}-NH-CH_2-CH_2-CH_2-\overset{\displaystyle NH_2}{\overset{\mid}{CH}}-COOH$
Lysine	$NH_2-CH_2-CH_2-CH_2-CH_2-\overset{\displaystyle NH_2}{\overset{\mid}{CH}}-COOH$

are *amino acids*, where R represents an organic unit like CH_3. These acids, whose concentration in the soil solution is typically in the range $0.05-0.6$ mol m^{-3}, can account for as much as one-half the N in soil humus. Several of the most abundant amino acids in soils are listed in Table 3.2. Glycine and alanine are examples of *neutral* amino acids, for which the side-chain unit R contains neither the carboxyl group nor the *amino group*, NH_2. The name is apt because the COOH group contributes negative charge by dissociating a proton, while the NH_2 group contributes positive charge by accepting a proton to become NH_3^+. Neutral amino acids account for about two-thirds of the soil amino acids. *Acidic* amino acids, for which R includes a carboxyl group (aspartic and glutamic acids), and *basic* amino acids, for which R includes an amino group (arginine and lysine), account for about equal portions of the remaining one-third. Amino acids can combine according to the general reaction:

$$R-\underset{\displaystyle NH_2}{\overset{\displaystyle H}{C}}-COOH + R'-\underset{\displaystyle NH_2}{\overset{\displaystyle H}{C}}-COOH = R-\underset{\displaystyle NH_2}{\overset{\displaystyle H}{C}}-CONH-\overset{\displaystyle R'}{\overset{\mid}{CH}}-COOH$$

$$+ H_2O \qquad\qquad (3.1)$$

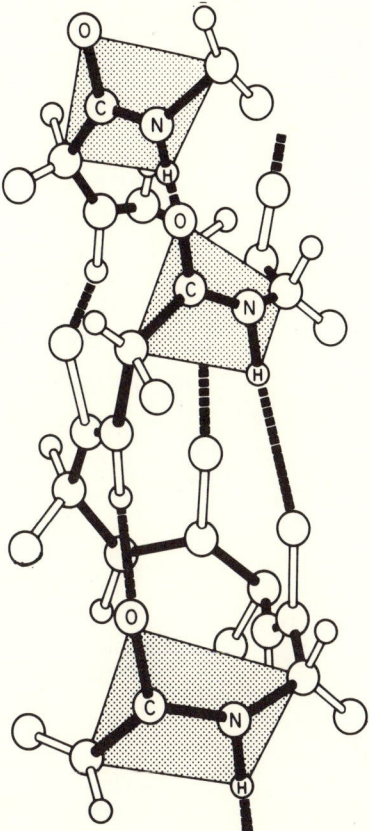

FIG. 3.1 Structure of a right-hand helix containing peptide repeating units. The shaded planes highlight the amide group, CONH, and the broken lines denote hydrogen bonds. (Reprinted with permission from G. Sposito, *The Surface Chemistry of Soils*, Oxford Univ. Press, New York, 1984.)

to form the *peptide*,

$$\underset{NH_2}{\overset{H}{R-C-CONH}}-\overset{R'}{\underset{}{CH}}$$

(Fig. 3.1). This group is the fundamental repeating unit in proteins. Since the peptide group is repeated, proteins are *polymers*; since water is a product in peptide formation (Eq. 3.1), proteins are specifically *condensation polymers* of amino acids. Peptides of varying composition and structure are the dominant form of amino acids in soils. They accumulate typically as complexes with organic and inorganic colloids (Sections 3.4 and 3.5).

TABLE 3.3 Common monosaccharides in soils

Name	Chemical formula
Glucose	
Galactose	
Mannose	
Xylose	
Glucuronic acid	
Glucosamine	

Another important class of biopolymers in soil is the *carbohydrates*. These compounds, which may account for up to one-half of the organic carbon in soils, include the *monosaccharides* listed in Table 3.3. The monosaccharides have a ring structure with a characteristic substituent group and arrangement of hydroxyls. In glucose, galactose, and mannose, the substituent group is CH_2OH, whereas in xylose it is H, in glucuronic acid it is COOH, and in glucosamine it is NH_2. (Note the close structural relationship among glucose, glucuronic acid, and glucosamine in Table 3.3.) Monosaccharides polymerize to form *polysaccharides*. For example, two glucose units can link together through oxygen at the site of HOH in each to form a repeating unit of *cellulose* after eliminating water (Fig. 3.2). Thus cellulose is a condensation polymer of glucose. It can account for up to one-sixth of the organic carbon in soil. The monosaccharides are technically polyhydroxyl alcohols. An important class of weakly acidic alcohols in soils in the *phenols*, which, as noted above, feature OH groups on benzene rings. Coniferyl alcohol is an example of a phenol that polymerizes to form *lignin* (Fig. 3.2), which, along with cellulose, is an important precursor of soil humic substances.

FIG. 3.2 Structural units in lignin and cellulose, two important biopolymers in soils. (Reprinted with permission from G. Sposito, *The Surface Chemistry of Soils*, Oxford Univ. Press, New York, 1984.)

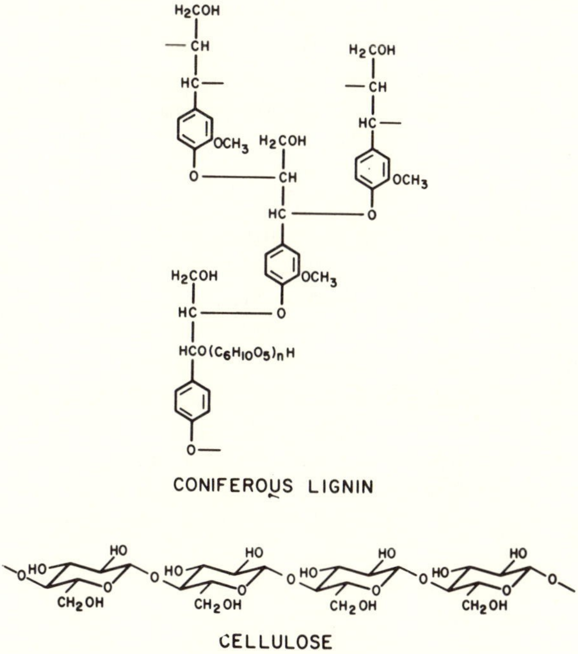

CONIFEROUS LIGNIN

CELLULOSE

The biomolecules just described are the most abundant ones in soils, but they by no means exhaust the long list of organic compounds produced by living organisms in the soil environment. Organic phosphorus compounds, which can account for up to 80% of soil P, occur principally in the form of inositol phosphates (benzene rings with H_2PO_4 bound to the ring carbon atoms), and organic sulfur compounds, which also can account for nearly all the soil S, occur principally as S-containing amino acids, phenols, and polysaccharides. The chemistry of the biomolecules of low relative molecular mass, such as those listed in Tables 3.1–3.3, has a strong influence on acid–base and metal complexation reactions in soils, whereas the chemistry of the biopolymers influences the surface and colloid chemistry of soils through adsorption reactions both with the constituents of the soil solution and with the solid phases in soil.

3.2 Humic Substances

The totality of organic matter in soil, except for materials identifiable as unaltered or partially altered biomass (plant parts and microorganisms), is called *humus*. This dark-colored constituent of the solid matter in soil plays a significant role in the formation of aggregates, in the control of soil acidity, in the cycling of nutrient elements, and in the detoxification of hazardous compounds. It comprises biomolecules, like those discussed in Section 3.1, as well as the *humic substances*. In simple terms, humic substances are compounds in humus that are not synthesized directly to sustain the life cycles of the soil biomass. More specifically, they are polymeric compounds produced through microbial action that differ from biopolymers because of their molecular structure and their long-term persistence in soil. This definition of humic substances implies no particular set of organic compounds, relative molecular mass, or chemical reactivity. What is essential is a dissimilarity to biomolecular structure and an evolved, biologically refractory nature.

The biochemical processes by which humus forms are not fully understood, but there is agreement that four stages of development occur in the transformation of soil biomass to humus: (1) decomposition of the biomass components, including lignin, into simple organic compounds; (2) microbial metabolism of the simple compounds; (3) cycling of C, H, N, and O between soil organic matter and the microbial biomass; (4) microbially mediated polymerization of the cycled organic compounds. The principal humus-forming compounds involved in stages 3 and 4 are believed to be phenolic polymers, derived from stages 1 and 2, which are converted to a reactive class of compounds containing oxygenated benzene rings (quinones) that polymerize readily.

The chemical properties of humic substances are often investigated after fractionation of soil organic matter based on solubility characteristics. Organic material that has been solubilized by mixing soil with a 500 mol m^{-3} NaOH solution is separated from the insoluble material (termed *humin*) and brought to pH 1 with concentrated HCl. The precipitate that forms after this acidification is called *humic acid*, whereas the remaining, soluble organic material is called *fulvic acid*. Repeated extractions of this type are often done on the humin and humic acid fractions to enhance separation. The humic and fulvic acids recovered also are subjected to centrifugation and ion exchange resin treatments to remove inorganic constituents and biomolecules.

The average chemical composition of soil humic and fulvic acids worldwide is summarized in Table 3.4. As a general rule, there is relatively more C and less O in humic acid than in fulvic acid. The average chemical formulas for these two fractions given in Section 1.2 were based on the composition data in Table 3.4. On the basis of a unit containing 1 mol of S, these formulas indicate that the average relative molecular mass of humic acid is larger than that of fulvic acid, which typically is < 2000. (For a discussion of relative molecular mass, see the Appendix.) Thus humic acid is relatively more polymerized than fulvic acid and therefore is at a more advanced stage of humification.

Table 3.4 also shows that fulvic acid contains more carboxyl groups per unit mass than humic acid. The total functional group acidity of humic substances is usually calculated as the sum of carboxyl and phenolic OH groups. Fulvic acid has the larger total functional group acidity, but both humic and fulvic acid possess a much larger dissociable proton charge per unit mass (6.7 and 11.2 mol kg^{-1}, respectively) than the typical cation exchange capacity of 2:1 clay minerals (< 2 mol kg^{-1}). The implication of this fact for cation exchange in soils is discussed in Section 9.1.

The investigation of molecular structures in humic substances is a difficult area of current research. Although it is not currently possible to describe the molecular configuration of fractionated humic substances in any but the most general terms, the functional groups in humic substances—especially those most reactive with protons and metal cations—have been characterized reasonably well. These include, in decreasing order of typical content: carboxyl, phenolic, and alcoholic OH, quinone and ketonic carbonyl (C=O), amino, and sulfhydryl (SH) groups (Table 3.5). The prominence of the carboxyl and phenolic OH groups in this list underscores the significant acidity of humic substances. As shown in Table 3.4, total functional group acidities (moles of dissociable protons per unit mass) ranging from 3 to 17 mol kg^{-1} have been reported for soil humic substances across all terrestrial climatic zones. Since most of the total functional group acidity of humic substances dissociates between pH 5 and 7, humic and fulvic acid molecules are expected to bear a net negative charge in soils.

Spectroscopic and other physicochemical methods applied to humic substances have shown that there are four principal structural characteristics of humic and fulvic acids influencing their chemical reactivity.

TABLE 3.4 Average chemical composition of soil humic substances worldwide[a]

Substance	C Mean	C Range	H Mean	H Range	N Mean	N Range	S Mean	S Range	O Mean	O Range	COOH Mean	COOH Range	Phenolic OH Mean	Phenolic OH Range
	- g kg^{-1} -										- - - - - - - - mol kg^{-1} - - - - - - - -			
Humic acid	560	530–570	47	30–65	32	8–55	8	1–15	355	320–385	3.6	1.5–6.0	3.1	2.1–5.7
Fulvic acid	457	407–506	54	38–70	21	9–33	19	1–36	448	390–500	8.2	5.2–11.2	3.0	0.3–5.7

[a]Data compiled in M. Schnitzer and S. U. Khan, *Soil Organic Matter* (Elsevier, Amsterdam, 1978), Chap. 1.

TABLE 3.5 The important functional groups in soil humus

Functional group	Structural formula
Carboxyl	$-\overset{\overset{\displaystyle O}{\|\|}}{C}-OH$
Carbonyl	$-\overset{\overset{\displaystyle O}{\|\|}}{C}-$
Amino	$-NH_2$
Imidazole	Aromatic ring NH
Phenolic OH	Aromatic ring OH
Alcoholic OH	$-OH$
Sulfhydryl	$-SH$

1. *Polyfunctionality:* the existence of a variety of functional groups and a broad range of functional group reactivity, representative of a heterogeneous mixture of interacting polymers.

2. *Macromolecular charge:* the development of anionic character on a macromolecular framework, with the resultant effects on functional group reactivity and molecular conformation.

3. *Hydrophilicity:* the tendency to form strong hydrogen bonds with water molecules solvating polar functional groups like COOH and OH.

4. *Structural lability:* the capacity to associate intermolecularly and to change molecular conformation in response to changes in pH value, redox conditions, electrolyte concentration, and functional group binding.

These properties of humic substances are common as well to biopolymers like proteins and polysaccharides, but in humic substances they reflect the behavior of a heterogeneous mixture of interacting polymeric molecules instead of the behavior of a structurally well-defined, single type of macromolecule. Thus the degree of complexity associated with these four properties is much larger for humic substances than for biomolecules.

3.3 Cation Exchange Reactions

Soil organic matter plays a major role in the buffering of both proton and metal cation concentrations in the soil solution. The chemical basis for this buffer capacity is cation exchange. By analogy with Eqs. 1.5 and 2.7a, a cation exchange reaction involving the dissociable protons on soil humus and a cation like Ca^{2+} in the soil solution can be written:

$$SH_2(s) + Ca^{2+}(aq) = SCa(s) + 2H^+(aq) \qquad (3.2)$$

where SH_2 represents an amount of humus (S) bearing 2 mol of dissociable protons, and SCa is the same amount of humus bearing 1 mol of exchangeable Ca^{2+}. The symbol S^{2-} then would represent an amount of humus bearing 2 mol of negative charge that can be neutralized by cations drawn from the soil solution.

The prospect of interpreting S in Eq. 3.2 at the level of molecular detail typical for clay minerals like vermiculite or biopolymers like protein, however, is dimmed by the need to consider, in the case of humus, many competing exchange reactions involving polymer fragments. Even if the molecular structure of each possible cation–humus association were worked out, the use of Eq. 3.2 for each would still entail the determination of a large number of molecular parameters—too many for the set of data usually available from a cation exchange experiment to provide. For this reason, and because of the complicated way the four structural characteristics mentioned previously influence humic substance reactivity, the modeling of cation exchange reactions involving soil humus has always interpreted Eq. 3.2 in some *average* sense. The humus reactant and product in the equation are taken to represent averages over a heterogeneous mixture whose cation exchange reactions can be described *by analogy* with what is known to apply for well-defined polymers like clay minerals or proteins.

This perspective is emphasized by expressing the H^+–Ca^{2+} exchange reaction in an alternate form:

$$2SH(s) + Ca^{2+}(aq) = S_2Ca(s) + 2H^+(aq) \qquad (3.3)$$

In this case, SH represents an amount of humus bearing 1 mol of dissociable protons, and S_2Ca is twice the amount. Equations 3.2 and 3.3 are equivalent ways to represent the *same* cation exchange process, and neither has any molecular structure implication for the humus material investigated. Suspensions containing known concentrations of total humus colloid charge can be prepared without prior information about the structures of the acidic functional groups producing the charge. Equations 3.2 and 3.3 do *not* imply, for example, that a "humus anion" has either the valence -1 or -2. The choice of which equation to use is a matter of personal preference, because both satisfy the general requirements of mass and charge balance, as discussed in Special Topic 1 (Chapter 1).

The *cation exchange capacity* (CEC) of soil humus is the maximum number of moles of proton charge dissociable from unit mass of humus under given conditions of temperature, pressure, soil solution composition, and humus concentration. The method used most widely to measure CEC for humus

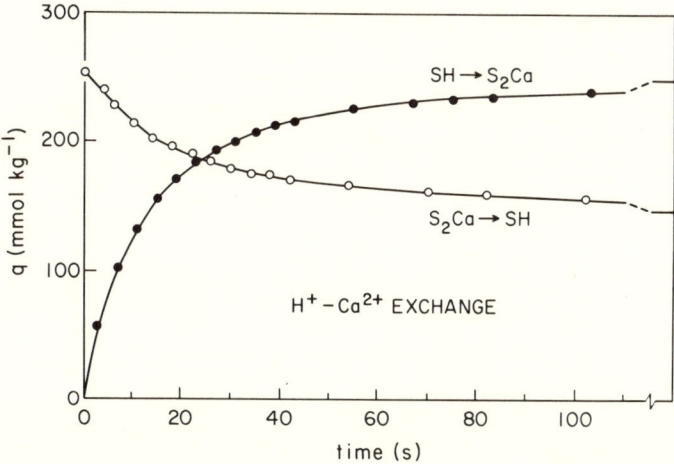

FIG. 3.3 Time dependence of the moles of charge of exchangeable Ca ($=2S_2Ca$) on sphagnum peat during H $\rightarrow$ Ca exchange (filled circles) and Ca $\rightarrow$ H exchange (open circles). [Data from K. Bunzl, W. Schmidt, and B. Sansoni, Kinetics of ion exchange in soil organic matter. IV. Adsorption and desorption of Pb^{2+}, Cu^{2+}, Zn^{2+}, and Ca^{2+} by peat, *J. Soil Sci.* **27**:32–41 (1976).]

involves determining the moles of protons exchanged in the reaction:

$$2SH(s) + Ba^{2+}(aq) = S_2Ba(s) + 2H^+(aq) \tag{3.4}$$

where the Ba^{2+} ions are supplied in a 100 mol m^{-3} Ba(OH)$_2$ solution. Measurements of this kind indicate that the CEC of colloidal humic acids ranges typically between 4 and 9 mol$_c$ kg^{-1} whereas for peat materials (soil humus plus woody plant parts) it ranges from 1 to 4 mol$_c$ kg^{-1}. (For a discussion of moles of charge, mol$_c$, see the Appendix.) These values are comparable to the total acidities calculated from acidic functional group analysis (Table 3.4).

The kinetics of H^+–Ca^{2+} exchange, as expressed in Eq. 3.3, are illustrated in Fig. 3.3 for a suspension of sphagnum peat. The data show the time development of the formation of S_2Ca (filled circles) after the addition of 50 μmol of Ca^{2+} charge and the depletion of S_2Ca (open circles) after the addition of 50 μmol of H^+ charge to a suspension containing 0.1 g peat. It is apparent that the exchange process is relatively rapid. Note that the reaction in Eq. 3.3 proceeds to the right more readily than to the left from comparable initial conditions. Additional experiments and data analysis showed that the exponential time dependencies in Fig. 3.3 can be described by a *film diffusion mechanism*. The basic concept in this mechanism is that the rate of cation exchange is controlled by diffusion of the exchanging ions through a thin (2– 50 μm), immobile film of solution surrounding the exchanger particle in suspension. Film diffusion, which is discussed in Special Topic 2 at the end of this chapter, is thought to be a common process leading to the observed rates of cation exchange on soil colloids (see Section 9.3).

When the metal cation replacing a proton on soil humus is monovalent, it is often considered a "background ion" in the analysis of proton exchange data. This is done on the hypothesis that all monovalent metal cations have a much lower affinity for humus than the proton. Attention is then focused on the species SH. Experimental measurements of the number of moles of strong acid or strong base added to a suspension (or solution) of humus material are combined with pH measurements to calculate the relative *formation function* for SH(s) [or SH(aq)]:

$$\delta n_H = \frac{(n_A - [H^+]V) - (n_B - [OH^-]V)}{m_s} \tag{3.5}$$

where n_A is the number of moles of strong acid (like HCl) added, and n_B is the number of moles of strong base (like NaOH) added to bring a suspension (or solution) to the volume V with a free aqueous proton concentration equal to $[H^+]$ moles per unit volume. The concentration of $[H^+]$ can be determined through a pH measurement, as can that of $[OH^-]$. (Usually, $[OH^-] \approx 10^{-14}/[H^+]$ in dilute solutions if concentrations are in moles per cubic decimeter.) The numerator in Eq. 3.5 is the difference between H^+ bound and OH^- bound by the humus sample, with each bound ion calculated as the difference between moles of added ion and moles of free ion. After division by m_s, the mass of humus sample, one has computed the *relative* formation function δn_H for proton binding. To convert δn_H into n_H, the absolute formation function, one chooses a value of $[H^+]$ so small (e.g., that at pH 11) that all dissociable protons are considered stripped from the sample. The calculated value of δn_H at this $[H^+]$ then is subtracted from each δn_H at all larger $[H^+]$ to give n_H. The resulting plot of n_H against $[H^+]$ (or $\log[H^+]$) is the true formation function for SH.

Figure 3.4 shows n_H calculated in this way for a solution of fulvic acid extracted from humus sampled in a subsurface horizon of a poorly drained Spodosol. An addition of KOH brought the solution to pH 11, after which HCl was added incrementally to decrease the pH and produce the exchange reaction:

$$SK(aq) + H^+(aq) = SH(aq) + K^+(aq) \tag{3.6}$$

The pH meter used was calibrated so as to give $-\log[H^+]$ directly in the 100 mol m^{-3} KCl solution used as "ionic background," where $[H^+]$ is in moles per cubic decimeter. Equation 3.5 was used to compute δn_H at each measured value of $-\log[H^+]$ after blank correction and the assumption was made that $n_H = \delta n_H - (\delta n_H)_{pH=11}$.

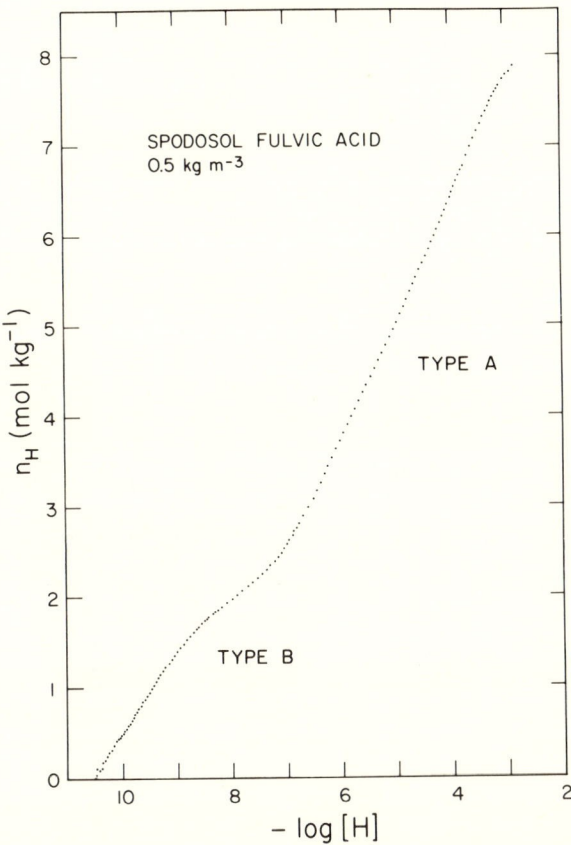

FIG. 3.4 Formation function for proton binding by a soil fulvic acid dissolved in 100 mol m^{-3} KCl solution at 25°C. In this system, pH = $-\log[H^+] + 0.11$, where $[H^+]$ is in moles per cubic decimeter. [Data from G. Sposito, K. M. Holtzclaw, C. S. LeVesque, and C. T. Johnston, Trace metal chemistry in arid-zone field soils amended with sewage sludge: II. Comparative study of the fulvic acid fraction, *Soil Sci. Soc. Am. J.* **46**:265–270 (1982).]

The portions of the formation function curve labeled "Type A" and "Type B" are thought to correspond primarily to COOH and phenolic OH functional groups, respectively. The plateau in n_H around $-\log[H^+] = 8$ suggests that the Type B groups are fully protonated at that point and that they contribute about 2.0 mol kg^{-1} to n_H. The Type A groups contribute about 6.0 mol kg^{-1} to give an estimated 8.0 mol kg^{-1} total functional group acidity. Direct measurements of the COOH content and total functional group acidity gave 4.7 mol kg^{-1} and 8 mol$_c$ kg^{-1}, respectively. Thus about 80% of the Type A groups are carboxyls; the other 20% may be acidic OH not associated with carbonyl groups.

The *acid-neutralizing capacity* (ANC) of a solution containing humus is the sum of concentrations of dissociated humus and free OH^-, less the concentration of free protons:

$$ANC = (TFA - n_H)c_s + [OH^-] - [H^+] \qquad (3.7)$$

where TFA is the total functional group acidity in moles of charge per kilogram, and c_s is the humus concentration in kilograms per cubic decimeter. Clearly, ANC will increase with pH. In the case of the fulvic acid in Fig. 3.4, $ANC \approx (8-4) \times (5 \times 10^{-4}) + 10^{-8} - 10^{-6} = 2 \times 10^{-3}$ mol dm^{-3} at $-\log[H^+] = 6$. The change in ANC with pH (strictly, the derivative $dANC/dpH$) is called the *buffer intensity*, β_H. If the ANC increases greatly as pH increases, then the solution constituents have a large increase in their capacity to bind and thus neutralize protons; this corresponds to a large buffer intensity. Speaking generally, one can estimate the buffer intensity to be greatest where the formation function changes most rapidly with $-\log[H^+]$ in Fig. 3.4. This occurs in the range $4 < -\log[H^+] < 6$, which is typical for soil humus materials. It is for this reason that soil humus is so important in the buffering of acid soils (Chapter 11).

3.4 Reactions with Organic Molecules

The organic compounds that react with soil organic matter are derived mainly from pesticides, fertilizers, green manures, and the degradation products of these soil amendments. Humus in solid form, either as a separate colloid or as a coating on mineral surfaces, can immobilize these compounds and, in some instances, detoxify them significantly. Soluble humus, like the fulvic acid fraction, can form complexes with organic compounds that then may travel freely with percolating water down into the soil profile. Pesticides that otherwise might be localized near the soil surface can be transported by this mechanism.

As discussed in Sections 3.2 and 3.3, soil humus bears a net negative charge even in the acidic pH range. This property causes it to react as a cation exchanger with organic compounds that contain N atoms bonded in tetrahedral coordination (*quaternized*) and bearing a positive charge. These structures occur in both chain (aliphatic) and benzene ring (aromatic) compounds, the latter being common in pesticide preparations. The general reaction scheme is analogous to Eq. 3.6:

$$SH + R-\overset{+}{\underset{R_3}{\overset{R_1}{N}}}-R_2 = \overset{R_1\ R_2\ R_3}{\underset{R}{SN}} + H^+ \qquad (3.8)$$

where the Rs represent organic units bonded to the N atom. Spectroscopic studies of this reaction indicate that some electron transfer from the humus functional groups to the quaternized N compound takes place, thereby enhancing the stability of the humus–organic complex.

Organic compounds that contain functional groups that become positively charged when in protonated form can also react with soil humus by cation exchange, as in Eq. 3.8. Basic amino acids like arginine (Table 3.2), which contains two protonatable NH_2 groups, are good examples of these positively charged compounds, as are the *s*-triazine pesticides, which contain protonatable NH substituents on an aromatic ring. Even without the development of net positive charge, protonated functional groups like COOH and NH can form *hydrogen bonds* with electronegative atoms such as O, N, and F. As an example, the C=O group in the phenylcarbamate and substituted urea pesticides can form a hydrogen bond (···) with NH in soil organic matter: C=O ··· HN. (Hydrogen bonds of this type are also illustrated in Fig. 3.1.) On the other hand, NH groups in the *s*-triazine pesticides can form hydrogen bonds with C=O groups in humus. Humus contains carboxyl, hydroxyl, carbonyl, and amino groups in a broad variety of molecular environments that lead to a spectrum of possibilities for hydrogen bonding with organic compounds. The additive effect of these possibilities makes hydrogen bonding an important reaction mechanism despite its relatively low bonding energy ($\sim 5\%$ of an O—H bond energy).

Much of the molecular framework of soil organic matter is not electrically charged. This nonionic structure can nevertheless react strongly with the uncharged part of an organic compound through *van der Waals interactions*. The van der Waals interaction involves weak bonding between polar units, either permanent (like OH and C=O) or induced by the presence of neighboring molecules. The induced van der Waals interaction is the result of correlations between fluctuating polarizations created in the electron configurations of two nearby nonpolar molecules. Although the time-averaged polarization induced in each molecule is zero (otherwise it would not be a nonpolar molecule), the correlations between the two induced polarizations do not average to zero. These correlations produce a net attractive interaction between the two molecules at very small intermolecular distances (around 0.1 nm). The van der Waals interaction between just two molecules is very weak, but when many molecules in a polymeric structure like humus interact simultaneously, the van der Waals component is additive and strong. These characteristic features of the van der Waals interaction will be discussed further in Sections 3.5 and 10.3.

The van der Waals interaction between nonionic compounds, or nonionic portions of compounds, and soil organic matter is often stronger than the interactions between these compounds and soil water. Water molecules in the vicinity of large nonpolar molecules are not attracted and so cannot orient their very polar OH bonds in ways that are compatible with the normal structure of liquid water (*hydrophobic effect*). The resultant disorder of this situation produces a low water solubility of the nonpolar molecule and a propensity for it to react with soil organic matter through van der Waals interactions. This reaction usually can be described by a *distribution coefficient*, K_d:

$$K_d = n/[A(aq)] \tag{3.9}$$

where n is the number of moles of a compound A that reacts with unit mass of *soil* in equilibrium with the soil solution concentration $[A(aq)]$ in moles per cubic meter or moles per kilogram of solution. Thus the units of K_d are cubic meters per kilogram or kilograms per kilogram. It is customary to express K_d as the product of another parameter, K_{om}, and the organic matter mass fraction of soil, f_{om} (kg_{om} kg^{-1}):

$$K_d \equiv K_{om} f_{om} \tag{3.10}$$

Then Eq. 3.9 can be rewritten in the form:

$$K_{om} = (n/f_{om})/[A(aq)] \tag{3.11}$$

where n/f_{om} is the number of moles of A that reacts with unit mass of *soil organic matter*. Equation 3.11 is analogous to Henry's law in Eq. 1.1, in that it represents a partition of a compound between two phases, here soil humus and the soil solution. This partitioning, in the case of nonionic molecules like the halogenated, aromatic rings in polychlorinated biphenyl (PCB) polymers or the organophosphates, is expected to favor soil humus when van der Waals interactions and the hydrophobic effect are significant. Since the latter is inversely related to water solubility, it is reasonable to expect also that there will be an inverse relationship between K_{om} and water solubility. Such a negative correlation has often been observed statistically and is of the general form:

$$\log K_{om} = a - b \log S_w \tag{3.12}$$

where S_w is the water solubility (mol m^{-3}) of a compound whose partition coefficient is K_{om}, and a and b are empirical parameters. For example, suppose that, for nonionic organic compounds reacting with a silt loam containing 19 g organic matter per kilogram, $a = 2.188$, $b = 0.729$ based on K_{om} values in the range 10 to 10^4 kg kg^{-1} correlated to S_w values in the range 10^{-3} to 10 mol m^{-3}. With these two parameter values, Eq. 3.12 predicts that 2-chlorobiphenyl (PCB), with a water solubility of 2.69×10^{-2} mol m^{-3}, will have a K_{om} of 2.15×10^3 kg kg^{-1}. On the other hand, benzene, for which $S_w = 22.9$ mol m^{-3}, will have a K_{om} of only 15.7 kg kg^{-1}. *Increasing water solubility corresponds to decreasing partitioning into soil organic matter.* Consistent with this view, it has been observed that the rates of reaction between nonionic organic compounds and soil humus can be described with a mechanism involving the diffusion of the organic compound into the humus colloid (*particle diffusion*) instead of the film diffusion mechanism that governs cation exchange.

3.5 Reactions with Soil Minerals

The coating of soil minerals by humus plays a major role in the cycling of chemical elements and in the formation of soil aggregates. Humus bound to clay minerals appears to be relatively stable against biodegradation and so presents a reactive surface to dissolved solutes in the soil solution. But the partial dissolution of humus coatings and the release of soluble organic compounds by soil fauna and flora also provide a suite of reactive anions in the soil solution that may bind to uncoated clay minerals and help to dissolve them. Thus soil minerals serve as both a substrate for humus and a source of metal ions with which to form soluble humus complexes (see Section 4.3).

The association between the solid phases in soil and humus is not understood well because the molecular structures in soil organic matter have not been worked out completely. In Section 3.2, it is pointed out that humic substances (the microbially transformed, polymeric organic materials that persist in soils) are of such structural complexity as to defy, for the present, any well-defined conceptualization except as to their reactive functional groups (Table 3.5). Therefore, insofar as soil mineral reactions are concerned, the most accessible characteristic of humus is its functional group behavior. Stereochemistry is also of great importance, but this characteristic is relatively poorly defined in the absence of more complete structural information. The catalog of functional groups in Table 3.5 thus provides a basis for understanding how humus reacts with soil particles.

Fortunately, many studies have been carried out wherein model organic compounds of known molecular structure relevant to soil humus were reacted with specimen soil minerals. These studies have provided definitive information about the mechanisms through which the surfaces of clay minerals and metal hydrous oxides associate with a specific kind of organic functional group bound to a small molecule or to a soluble polymer. A summary of the principal mechanisms that have been deduced in this way is given in Table 3.6. These mechanisms should apply to the functional groups in humus when it reacts with soil minerals.

The *cation exchange* mechanism in Table 3.6 can be represented by the reaction:

$$B^+(aq) + M^+\!\!\equiv\; = \;B^+\!\!\equiv\; + \;M^+(aq) \tag{3.13}$$

where B refers typically to a molecular unit comprising a quaternized nitrogen atom in an aliphatic chain or a heterocyclic aromatic ring, and M^+ is a

TABLE 3.6 Mechanisms of association between organic functional groups and soil minerals

Mechanism	Organic functional groups involved
Cation exchange	Amino, ring NH, heterocyclic N (aromatic ring)
Protonation	Amino, heterocyclic N, carbonyl, carboxylate
Anion exchange	Carboxylate
Water bridging	Amino, carboxylate, carbonyl, alcoholic OH
Cation bridging	Carboxylate, amines, carbonyl, alcoholic OH
Ligand exchange	Carboxylate
Hydrogen bonding	Amino, carbonyl, carboxyl, phenolic OH
van der Waals interactions	Uncharged organic units

monovalent exchangeable metal cation initially bound to a clay mineral surface depicted symbolically by $\equiv$. The most important specific example of B^+ is a unit containing the protonated amino group, NH_3^+.

The *protonation* mechanism refers to the association between an organic functional group and a surface-bound proton or a proton in an acidic water molecule solvating a bivalent exchangeable metal cation. The mineral surfaces in soils can develop acidity in a variety of ways (proton exchange, dissociation of hydroxyls, hydrolysis of solvated metal cations, etc.). This surface acidity offers the possibility that proton-selective organic functional groups like NH_2 and $C{=}O$ can be bound through a protonation reaction. This mechanism is expected to be most important under conditions of low pH or low water content in soils, when the acidity of mineral surfaces is greatest (see Sections 7.3 and 7.5.)

The *anion exchange* mechanism is the analog of the reaction in Eq. 3.13, wherein the signs of the valences are reversed, B symbolizes a carboxylate group (COO^-), and M is replaced by a univalent, exchangeable anion (e.g., Cl^- or NO_3^-) bound to a protonated surface hydroxyl (OH_2^+). This mechanism is not observed often, possibly because of the weakness of the electrostatic bonds involved, but it should be prominent in acid soils whose clay fraction comprises primarily metal oxides.

Another weak adsorption mechanism for either anionic or polar functional groups (e.g., carboxylate or carbonyl) is *water bridging*, which involves complexation with a water molecule solvating an exchangeable cation:

$$B + (H_2O)M^{m+}\equiv\ =\ B(H_2O)M^{m+}\equiv \qquad (3.14)$$

where B is an anionic or polar molecular unit, and M^{m+} is an exchangeable cation. Water bridging is expected to occur when M is strongly solvated (e.g., Mg^{2+}), since, in that case, B is less likely to displace the water molecule. If displacement of the water molecule does occur, a direct bond is formed between B and M^{m+}, and the bonding mechanism is termed *cation bridging*. Clearly, whether water bridging or cation bridging takes place will depend on the nature of the functional group and the relative ease with which it can desolvate the exchangeable cation. For example, the carboxylate groups in humic substances associate with montmorillonite through cation bridging when monovalent exchangeable cations are present and through water bridging when bivalent exchangeable cations are present (see Sections 7.1 and 7.2).

Ligand exchange refers specifically to direct bond formation between a carboxylate group and either Al or Fe(III) in a soil mineral bearing hydroxyl groups (see Section 8.4). This mechanism involves stronger chemical bonds than those involved in anion exchange or in the two bridging mechanisms. Evidence for ligand exchange in carboxylate reactions with metal oxides and the edges of clay minerals is abundant though indirect. The general reaction scheme can be expressed by the equations:

$$—MOH(s) + H^+(aq) = —MOH_2^+(s) \tag{3.15}$$

$$—MOH_2(s) + S—COO^- = —MOOC—S(s) + H_2O(\ell) \tag{3.16}$$

where —MOH represents 1 mol of reactive hydroxyls bound to a metal M (M = Al or Fe) in a soil mineral, and S—COO$^-$ is 1 mol of humus carboxylate groups. The protonation step in Eq. 3.15 is thought to make the ligand exchange (OH$_2$ for COO$^-$) in Eq. 3.16 more favorable. Thus ligand exchange is enhanced at low pH. This condition also favors chemical weathering, and ligand exchange is considered to be a first step in the dissolution of hydroxylated soil minerals by humus carboxyls. A subsequent step would involve the detachment of the complex S—COOM from the mineral surface and translocation in the soil solution (see also Section 5.1).

Hydrogen bonding between organic functional groups and either clay minerals or metal hydrous oxides does not appear to be significant in humus reactions with soil minerals, probably because the electronegativity of mineral surface oxygens usually is not large enough to form strong H bonds. On the other hand, *van der Waals interactions* are just as important to humus–soil mineral associations as they are to organic compound–humus reactions. For polymeric humus material, the van der Waals interaction with the atoms in a mineral surface can be quite strong and relatively long-ranged. The influence of this interaction on biopolymers, such as proteins and carbohydrates, is believed to be the reason for the formation of very stable complexes when these large molecules react with soil minerals. The effects of van der Waals interactions are especially apparent when conditions in the soil solution suppress the ionization of acidic functional groups on large organic molecules—for example, when the pH value is such as to make the net charge on them vanish (see Section 7.4).

FOR FURTHER READING

S. Burchill, D. J. Greenland, and M. H. B. Hayes, Adsorption of organic molecules, in D. J. Greenland and M. H. B. Hayes (eds.), *The Chemistry of Soil Processes*, pp. 221–400. Wiley, Chichester, U.K., 1981. An elegant, encyclopedic review of the reactions of organic compounds—both natural and synthetic—with soil minerals and humus.

P. M. Huang and M. Schnitzer (eds.), *Interactions of Soil Minerals with Natural Organics and Microbes*, SSSA Spec. Publ. No. 17, Soil Science Society of America, Madison, WI, 1986. The 15 chapters of this outstanding reference work provide advanced reviews of the biochemistry of humus and its reactions with aqueous cations, enzymes, and soil minerals.

R. A. Saar and J. H. Weber, Fulvic acid: Modifier of metal-ion chemistry, *Environ. Sci. Technol.* **16**:510A–517A (1982). A useful review of fulvic acid reactions with aqueous cations that emphasizes methodologies for isolating and characterizing complexes.

F. J. Stevenson, *Humus Chemistry*, Wiley, New York, 1982. Highly recommended as a supplement to the present chapter for in-depth discussions on all aspects of soil organic matter.

PROBLEMS

The more difficult problems are indicated by an asterisk.

1. Develop a reaction analogous to Eq. 3.1 to show that cellulose (Fig. 3.2) is a condensation polymer of glucose (Table 3.3).

2. Develop a reaction like Eq. 3.1 to show that kaolinite (Table 2.3) is a condensation polymer of hydrated silica $[SiO_2 \cdot 2H_2O]$ and gibbsite $[Al(OH)_3]$. How can this concept be extended to the other clay minerals in Table 2.3?

*3. Organic acids like citric and oxalic are known to inhibit the precipitation of crystalline hydrous oxides of Al and Fe(III) like gibbsite and goethite (Table 2.4). Develop a mechanism for this inhibition by adapting Eqs. 3.15 and 3.16 to MOH in aqueous solution.
   (*Hint:* Combine the two equations so that the species MOH_2^+ does not appear.)

4. Polysaccharides appear to be more effective than humic substances in binding clay minerals into stable aggregates. Give an explanation for this observation based on structural concepts discussed in Sections 2.3, 3.1, 3.2, and 3.5.

5. Humic substances do not associate with 2:1 clay minerals in the interlayer region unless pH < 3. Give two reasons why this behavior should be expected.

6. Combine Eqs. 3.3 and 3.6 to derive a cation exchange reaction for $K^+ \rightarrow Ca^{2+}$ on colloidal humus.

*7. Show that for pure water, $ANC = [OH^-] - [H^+]$ and the buffer intensity $\beta_H = dANC/dpH \approx 2.303 (10^{-pH} + 10^{pH-14})$.
(*Hint:* The derivative of 10^{-x} with respect to x is $-2.303 \cdot 10^{-x}$.)

*8. The formation function for proton binding by a sample of dissolved humus can be described mathematically by the equation:

$$n_H = \frac{b_1 K_1 10^{-pH}}{1 + K_1 10^{-pH}} + \frac{b_2 K_2 10^{-pH}}{1 + K_2 10^{-pH}}$$

where $b_1 = 6$ mol kg^{-1}, $b_2 = 2$ mol kg^{-1}, $K_1 = 10^{4.7}$, and $K_2 = 10^{9.2}$.
(*a*) Prepare a graph of n_H as a function of pH in the range 11–3 (see Fig. 3.4).
(*b*) Given that the concentration of dissolved humus is 0.1 kg m^{-3}, calculate the ANC as a function of pH in the range 4–7.
(*Answer:* $ANC \approx b_1 c_s/(1 + K_1 10^{-pH}) + 10^{pH-14} - 10^{-pH}$)

*9. Use the equation and data in Problem 8 to calculate the buffer intensity of the dissolved humus as a function of pH in the range 3–7.
(*Hint:* The derivative of 10^{-x} with respect to x is $-2.303 \cdot 10^{-x}$. Then $\beta_H = dANC/dpH \approx 2.303[b_1 c_s K_1 10^{-pH}/(1 + K_1 10^{-pH})^2 + 10^{pH-14} + 10^{-pH}]$.)

10. Tetrachloroethylene (PCE) is a common organic solvent that may contaminate groundwater if leached through soil from a disposal site. Given its water solubility of about 5 mol m^{-3}, estimate the distribution coefficient for PCE in a soil containing 20 g humus per kilogram. Is PCE immobilized well by the soil?

11. Develop a chemical reaction to describe the association of a protonated *s*-triazine molecule TH^+ (T = *s*-triazine) with a humus carboxylate group through hydrogen bonding.

12. Apply Eq. 3.13 to the bonding of a humus NH_3^+ group to Na-saturated montmorillonite, then rewrite it to describe the bonding of a humus COO^- group to Cl-saturated goethite.

13. Apply Eq. 3.14 to humus carboxylate bonding to Ca-saturated montmorillonite.

*14. Rewrite Eq. 3.14 so as to describe carboxylate bonding to Na-montmorillonite through cation bridging.

*15. Combine Eqs. 3.15 and 3.16 with a carboxyl dissociation reaction to derive a chemical equation showing the formation of —MOOC—S from —MOH and S—COOH.

Special Topic 2

Film Diffusion Kinetics in Cation Exchange

The film diffusion mechanism in cation exchange is based on three assumptions: (1) that a thin film of inhomogeneous solution separates an exchanger surface from a homogeneous bulk solution, (2) that the exchanging cations diffuse across the film much more rapidly than the concentrations of these ions change in the bulk solution, and (3) that equilibrium between the ions in the exchanger and those at the film–exchanger interface is established on a time scale that is very small relative to the time required for ions to diffuse across the film. Thus if τ_{bulk} is the time scale for changes in the bulk concentration of a cation, τ_{dif} is the time scale for it to diffuse across the film, and τ_{ex} is the time scale for it to equilibrate at the exchanger surface, then assumptions (2) and (3) stipulate: $\tau_{bulk} \gg \tau_{dif} \gg \tau_{ex}$. A simplified mathematical description of film diffusion can be developed as follows.

Consider as an example the $H^+ - Ca^{2+}$ exchange reaction in Eq. 3.3 whose kinetics were illustrated in Fig. 3.3. If c is the concentration of Ca^{2+} in the bulk solution and c' is its concentration at the film–exchanger interface, then the rate at which the ion diffuses across a film of thickness δ is described mathematically by a form of *Fick's law*:

$$j = \frac{D}{\delta} (c - c') \tag{s2.1}$$

where j is the rate of diffusion from bulk solution to exchanger surface per unit area of exchanger surface (mol m^{-2} s^{-1}), and D is the *diffusion coefficient* of the ion (m^2 s^{-1}). Mass balance requires that j in Eq. s2.1 be proportional to the rate of accumulation of Ca^{2+} charge by the exchanger, dq/dt:

$$\frac{dq}{dt} = 2Sj = 2 \frac{SD}{\delta} (c - c') \tag{s2.2}$$

where q is the moles of Ca^{2+} *charge* per unit mass accumulated by the exchanger (as in Fig. 3.3), and S is the specific surface area of the exchanger. The factor of 2 appears in Eq. s2.2 because q is equal to twice the number of moles of S_2Ca per unit mass in the exchanger.

Given that $\tau_{bulk} \gg \tau_{dif}$, the concentration c will be effectively constant as q and c' change according to Eq. s2.2. The numerical value of c must be the same as the equilibrium value of c' attained when diffusion has ceased: $c = c'_\infty =$ the value of c' at "infinite" elapsed time. Therefore, Eq. s2.2 can be written in the form:

$$\frac{dq}{dt} = 2 \left(\frac{SD}{\delta} \right) (c'_\infty - c') \tag{s2.3}$$

Since $\tau_{dif} \gg \tau_{ex}$, the variables q and c' can be related through a *distribution coefficient* K_d that characterizes Ca^{2+} equilibrium at the exchanger surface:

$$K_d = n/c' = (q/2)/c' \qquad (s2.4)$$

where n is the moles of S_2Ca per unit mass of exchanger. The parameter K_d will in general vary with q and c'. The combination of Eqs. s2.3 and s2.4 permits the rate of Ca^{2+} accumulation by the exchanger to be expressed:

$$\frac{dq}{dt} = \frac{SD}{\delta K_d}(q_\infty - q) \qquad (s2.5)$$

where q_∞ is the equilibrium ("infinite time") value of q.

The kinetics expression of Eq. s2.5 is of the type that leads to an exponential time dependence in q. If Ca^{2+} is replacing H^+ on the exchanger, q will increase from zero to q_∞ as time passes (see Fig. 3.3). If H^+ is replacing Ca^{2+} on the exchanger and were to do so completely, such that $q_\infty = 0$, Eq. s2.5 would reduce to the expression:

$$\frac{dq}{dt} = -\left(\frac{SD}{\delta K_d}\right)q \qquad (s2.6)$$

which is a standard equation that leads to an exponential decrease of q with time.

The parameter

$$k = \frac{SD}{\delta K_d} \qquad (s2.7)$$

is called the film diffusion *rate constant*. It is related to the *half-life* for the exponential decline of q with time through the conventional expression:

$$t_{1/2} = \frac{0.693}{k} = 0.693\left(\frac{\delta K_d}{SD}\right) \qquad (s2.8)$$

Typical values for the parameters on the right side of Eq. s2.8 are: $S = 10^4$ m^2 kg^{-1}, $D = 10^{-9}$ m^2 s^{-1}, $\delta = 10^{-5}$ m, and $K_d = 10$ m^3 kg^{-1}. These values lead to $t_{1/2}$ on the order of 10 s, in qualitative agreement with the time dependence in Fig. 3.3.

4

The Soil Solution

4.1 Sampling the Soil Solution

The soil solution was described in Section 1.4 as a mixture of liquid water and dissolved substances like sodium chloride or carbon dioxide gas. Speaking more precisely, one can define the soil solution as the aqueous liquid phase in soil whose composition is influenced by flows of matter and energy between it and its surroundings and by the gravitational field of the earth. This more careful definition shows that the soil solution is an *open system* (see Section 1.1). That it is designated also as a *phase* means two things: (1) it has uniform macroscopic properties, like temperature and electrolyte concentration, and (2) it can be isolated from the soil and investigated in the laboratory.

The uniformity of soil solution properties applies to some small element of volume in a soil profile (e.g., a soil ped). Both along the profile and laterally in a field-scale environment, the chemical properties of the soil solution often show great variability. To this spatial variability may be added the temporal variability of soil solution properties brought on by diurnal and seasonal changes in the soil environment as well as by human or microbiological activity. Temporal variations in the soil solution also result from the kinetics of chemical reactions themselves, as will be discussed in Section 4.2.

The problem of isolating the "true" soil solution has not yet been solved, but several preferred methods of removing aqueous phases from soils for laboratory study have been established as operational compromises between chemical accuracy and analytical convenience. Among these methods, the most used are collection of drainage water, displacement by a fluid immiscible with water, and direct extraction by vacuum, applied pressure, or centrifugation.

The collection of drainage water from a soil horizon or a whole profile has the advantage of sampling the soil solution *in situ*, but suffers from uncertainties related to the disruption of natural drainage patterns and reactions between dissolved constituents and the collection device. There is also the limitation that the soil must be water-saturated near the point of sample collection. This latter constraint often leads to great variations in sample volume that affect the composition of the solution collected.

The displacement method involves the use of a dense, unreactive organic liquid of low solubility (e.g., trichlorotrifluoroethane), which is centrifuged with a soil sample to expel its aqueous phase and, through the greater density of the displacement fluid, to float the aqueous phase to the top of the centrifuge tube. Yield percentages as large as 60% of the soil solution are possible, depending on soil texture, and contamination is potentially negligible. The displacement method also has the advantage of not requiring a water-saturated soil sample, but it does necessitate the disturbance of soil structure and the use of specialized laboratory equipment.

In the vacuum extraction method, the aqueous phase of a soil *in situ*, or a "disturbed" soil sample saturated previously with water in the laboratory, is withdrawn through a filter by vacuum. This method suffers from both negative and positive interferences caused by the filter (principally from adsorption–desorption reactions with dissolved constituents) as the extracted solution passes through it. There are also uncertainties associated with the effect of vacuum extraction on soil solution flow patterns and on chemical reactions between dissolved constituents and soil solid phases. Finally, if the soil sample has been saturated with water prior to the extraction, the composition of the extract may differ considerably from that of any "true" soil solution at ambient water contents. Despite these difficulties—which are shared with the pressure extraction and centrifugation methods as well—the vacuum extraction technique, once standardized, is convenient for routine analyses. It usually provides aqueous solutions whose composition reflects something of the true reactions between the soil solution and solid soil constituents.

For any of the common methods of obtaining soil solutions, however, there is still the problem of the inherent *microscopic* heterogeneity in soil aqueous phases caused by the electrical charge on soil particles, discussed in Sections 1.2, 2.3, and 3.3 (see also Sections 7.2 and 8.1). This charge will create poorly defined zones of accumulation or depletion of ions in the soil solution near soil particle surfaces, with accumulation occurring for ions whose charge is opposite that of the neighboring surface and depletion for the reverse situation. Because of this phenomenon, successive increments of extracted soil solution that represent different regions near soil particle surfaces should not have the same composition, and the uniformity of composition of these extracted solution increments cannot be used as a means to verify that the "true" soil solution has been sampled.

Standard laboratory procedures have been compiled in *Methods of Soil*

Analysis (see the "For Further Reading" section at the end of this chapter) for the determination of the composition of extracted soil solutions. These composition data, which provide total concentrations of dissolved (i.e., filterable under designated conditions) constituents, pH, electrolytic conductivity, and so on, make up the primary information needed for the description of soil solutions, at known temperature and pressure, according to the principles of chemical kinetics and thermodynamics.

4.2 Soluble Complexes

A *complex* is said to form whenever a molecular unit, such as an ion, acts as a central group to attract and form a close association with other atoms or molecules. Examples of complexes in the soil solution were provided abundantly throughout Chapters 1 and 2. The aqueous species $Si(OH)_4^0$, $Al(OH)_2^+$, and HCO_3^- are complexes, with Si^{4+}, Al^{3+}, and CO_3^{2-}, respectively, acting as the central group. The associated ions, OH^- or H^+, in these complexes are termed *ligands*. Usually this term is applied to anions or neutral molecules coordinated to a metal cation in a complex, but it can be applied also to cations coordinated to an anion, as in bicarbonate or the phosphate complex, $H_2PO_4^-$. If two or more functional groups of a single ligand are coordinated to a metal cation in a complex, the complex is termed a *chelate*. The complex formed between Al^{3+} and citric acid (Table 3.1), $[Al(COO)_2-COH(CH)_2COOH]^+$, in which two COO^- groups and one COH group are coordinated to Al^{3+}, is an example of a chelate.

If the central group and ligands in a complex are in direct contact, the complex is called *inner-sphere*. If one or more water molecules is interposed between the central group and a ligand, the complex is *outer-sphere*. If the ligands in a complex are water molecules (as, for example, in $Ca(H_2O)_6^{2+}$), these two terms do not apply and the unit is called instead a *solvation complex*. The "free" cations and anions represented in Chapters 1 and 2 by, for example, $Na^+(aq)$ or $OH^-(aq)$ actually are solvation complexes, since ion charge serves to attract and bind water dipoles in any aqueous solution.

Inner-sphere complexes usually are much more stable than outer-sphere complexes because the latter cannot easily involve ionic or covalent bonding between the central group and ligand, whereas the former can. Thus the "driving force" for inner-sphere complexes is the heat evolved through strong bond formation between the central group and ligand. For outer-sphere complexes, the heat of bond formation is not so large and the "driving force" involves favorable stereochemical aspects of coordination about the central group, such as in the adjustment that occurs when an anion coordinates to a cation solvation complex to form an electrostatic bond with the cation.

Table 4.1 lists the principal complexes found in well-aerated soil solutions. The ordering of free-ion and complex species in each row from left to right is

TABLE 4.1 Representative chemical species in soil solutions

Cation	Principal species	
	Acid soils	Alkaline soils
Na^+	Na^+	Na^+, $NaHCO_3^0$, $NaSO_4^-$
Mg^{2+}	Mg^{2+}, $MgSO_4^0$, org[a]	Mg^{2+}, $MgSO_4^0$, $MgCO_3^0$
Al^{3+}	org, AlF^{2+}, $AlOH^{2+}$	$Al(OH)_4^-$, org
Si^{4+}	$Si(OH)_4^0$	$Si(OH)_4^0$
K^+	K^+	K^+, KSO_4^-
Ca^{2+}	Ca^{2+}, $CaSO_4^0$, org	Ca^{2+}, $CaSO_4^0$, $CaHCO_3^+$
Cr^{3+}	$CrOH^{2+}$	$Cr(OH)_4^-$
Cr^{6+}	CrO_4^{2-}	CrO_4^{2-}
Mn^{2+}	Mn^{2+}, $MnSO_4^0$, org	Mn^{2+}, $MnSO_4^0$, $MnCO_3^0$, $MnHCO_3^+$, $MnB(OH)_4^+$
Fe^{2+}	Fe^{2+}, $FeSO_4^0$, $FeH_2PO_4^+$	$FeCO_3^0$, Fe^{2+}, $FeHCO_3^+$, $FeSO_4^0$
Fe^{3+}	$FeOH^{2+}$, $Fe(OH)_3^0$, org	$Fe(OH)_3^0$, org
Ni^{2+}	Ni^{2+}, $NiSO_4^0$, $NiHCO_3^+$, org	$NiCO_3^0$, $NiHCO_3^+$, Ni^{2+}, $NiB(OH)_4^+$
Cu^{2+}	org, Cu^{2+}	$CuCO_3^0$, org, $CuB(OH)_4^+$, $Cu[B(OH)_4]_4^0$
Zn^{2+}	Zn^{2+}, $ZnSO_4^0$, org	$ZnHCO_3^+$, $ZnCO_3^0$, org, Zn^{2+}, $ZnSO_4^0$, $ZnB(OH)_4^+$
Mo^{5+}	$H_2MoO_4^0$, $HMoO_4^-$	$HMoO_4^-$, MoO_4^{2-}
Cd^{2+}	Cd^{2+}, $CdSO_4^0$, $CdCl^+$	Cd^{2+}, $CdCl^+$, $CdSO_4^0$, $CdHCO_3^-$
Pb^{2+}	Pb^{2+}, org, $PbSO_4^0$, $PbHCO_3^+$	$PbCO_3^0$, $PbHCO_3^+$, org, $Pb(CO_3)_2^{2-}$, $PbOH^+$

[a]Organic complexes (e.g., fulvic acid complexes).

roughly according to decreasing concentration typical for either acid or alkaline soils. A normal soil solution will easily contain 100–200 different soluble complexes, many of them involving metal cations and the organic ligands discussed in Chapter 3. The main effect of pH on these complexes, as is evident in Table 4.1, is to favor free metal cations and protonated anions at low pH and carbonate or hydroxyl complexes at high pH.

Most of the complexes listed in Table 4.1 are metal–ligand complexes, which may be either inner-sphere or outer-sphere. As an example, consider the formation of a neutral sulfate complex with a bivalent metal cation as the central group:

$$M^{2+}(aq) + SO_4^{2-}(aq) = MSO_4^0(aq) \qquad (4.1)$$

where the metal M can be Ca, Mg, Mn, Cu, etc. Detailed spectroscopic investigation shows that $MSO_4^0(aq)$ can be either inner- or outer-sphere, with the latter type dominant. The rate at which MSO_4^0 forms is quite high, as is usual for metal–ligand complexes. In mathematical terms, this rate of formation can be expressed by minus the time derivative of $d[M^{2+}]/dt$, where the square brackets represent a concentration in moles per liter (cubic decimeter). The rate of soluble-complex formation can be measured by a variety of spectroscopic and electrochemical techniques. It is common to *assume* that the observed rate can be represented mathematically by the difference of two terms:

$$-\left(\frac{d[M^{2+}]}{dt}\right) = R_f - R_b \tag{4.2}$$

where R_f and R_b each are functions of the composition of the solution in which the reaction in Eq. 4.1 takes place, as well as of the temperature and pressure. Equation 4.2 need not have any direct relationship to the mechanism by which MSO_4^0 forms. For example, there may be intermediate species that do not appear in the overall reaction, but that help to determine the observed rate and prevent it from being a simple difference expression. Whenever Eq. 4.2 is appropriate, however, R_f and R_b are interpreted as the respective rates of formation ("forward reaction") and dissociation ("backward reaction") of MSO_4^0. It is common to *assume* that these two rates depend on powers of the concentrations of the reactants and products:

$$-\left(\frac{d[M^{2+}]}{dt}\right) = k_f[M^{2+}]^\alpha[SO_4^{2-}]^\beta - k_b[MSO_4^0]^\delta \tag{4.3}$$

where k_f, k_b, α, β, and δ are parameters. The parameters α, β, and δ are the *order* of the reaction with respect to the associated species [e.g., αth order with respect to $M^{2+}(aq)$]. The parameters k_f and k_b are the *rate constants* of the formation ("forward") and dissociation ("backward") reactions, respectively. Each of the five parameters in Eq. 4.3 may depend on the solution composition, temperature, and pressure.

Usually, experimental conditions can be arranged to make one of the terms on the right side of Eq. 4.2 negligible. When this is possible, the observed reaction rate and concentration data as a function of time can be analyzed graphically or by statistical curve-fitting computer programs to estimate values for the parameters in Eq. 4.3. Table 4.2 indicates the kind of graphic analysis that will result in straight-line plots of concentration data for zero-,

TABLE 4.2   Graphic analysis of Eq. 4.4

Reaction order (b)	Plotting variables	Slope	Intercept	Half-life[a]
Zero	[A] vs. time	$-K$	$[A]_0$	$[A]_0/2K$
One	ln[A] vs. time	$-K$	$\ln[A]_0$	$0.693/K$
Two	1/[A] vs. time	$+K$	$1/[A]_0$	$1/K[A]_0$

[a]Valid only for positive K, with $[A]_0$ equal to the initial concentration of species A.

first-, and second-order kinetics expressions of the general form:

$$-\left(\frac{d[A]}{dt}\right) = K[A]^b \tag{4.4}$$

where A is some chemical species. Note that the parameter K in Eq. 4.4 may equal the product of a rate constant with a concentration (maintained constant during an experiment) raised to a power. The parameter b, like α, β, or δ in Eq. 4.3, need *not* be the same as the stoichiometric coefficient of species A in the chemical reaction investigated, since expressions like Eq. 4.3 are strictly empirical. Table 4.2 also lists the *half-life* of a reaction according to its order. This parameter (*applicable only when K is positive*) is equal to the time required for the concentration of species A to decrease to one-half its initial value.

Table 4.3 lists measured half-life parameters for a number of complex formation and dissociation reactions that are representative of soil solutions. In each case, A has been identified as the first species on the left side of the reaction, and the rate of reaction has been written as in Eq. 4.4 with $K = k_f$. First-order reactions occur in the table when there is only one reactant other than H_2O, whose very large concentration is assumed to be constant. Second-order reactions occur in the table when there are two reactants, whose concentrations are assumed equal. For example, in the formation reaction for

TABLE 4.3 Kinetics parameters for complex formation and dissociation reactions at $25°C$[a]

Reaction	Order[b]	Half-life(s)[c]
$CO_2 + H_2O = H_2CO_3^*$	1	10
$Fe^{3+} + H_2O = FeOH^{2+} + H^+$	1	10^{-7}
$FeOH^{2+} + H^+ = Fe^{3+} + H_2O$	2	10^{-6}
$Mn^{2+} + SO_4^{2-} = MnSO_4^0$	2	10^{-5}
$MnSO_4^0 = Mn^{2+} + SO_4^{2-}$	1	10^{-9}
$Ni^{2+} + C_2O_4^{2-} = NiC_2O_4^0$	2	1
$NiC_2O_4^0 = Ni^{2+} + C_2O_4^{2-}$	1	10^{-1}
$Al^{3+} + F^- = AlF^{2+}$	2	10^{3} [d]
$CO_2 + OH^- = HCO_3^-$	2	10
$HCO_3^- = CO_2 + OH^-$	1	10^3
$Ca(H_2O)_6^{2+} + H_2O' = Ca^{2+}(H_2O)_5 H_2O' + H_2O$	1	10^{-8} [e]

[a]Mostly from data compiled in J. F. Pankow and J. J. Morgan, Kinetics for the aquatic environment, *Environ. Sci. Technol.* **15**:1155–1164 (1981).

[b]Sum of the orders with respect to individual species other than water.

[c]Initial concentrations of 10 mmol m^{-3} assumed for second-order reactions.

[d]Data at pH 3 from B. J. Pankey, H. H. Patterson, and C. S. Cronan, Kinetics of Aluminum fluoride complexation in acidic waters, *Environ. Sci. Technol.* **20**:160–165 (1986).

[e]Data from Chapter 1 in A. E. Martell (ed.), *Coordination Chemistry*, Vol. 2, ACS Monograph 174, American Chemical Society, Washington, D.C., 1978.

$MnSO_4^0$ (row 4 in Table 4.3), the rate of reaction is expressed by the equation:

$$\frac{d[Mn^{2+}]}{dt} = -k_f[Mn^{2+}][SO_4^{2-}] = -k_f[Mn^{2+}]^2$$

under the assumptions that (a) the rate of dissociation is negligible, (b) the reaction orders with respect to Mn^{2+} and SO_4^{2-} are the same as the stoichiometric coefficients of these two species in Eq. 4.1, and (c) $[Mn^{2+}] = [SO_4^{2-}]$. The experimental value of k_f in dilute solutions is about 2×10^{10} mol^{-1} dm^3 s^{-1} and the corresponding half-life (row 3 in Table 4.2) is the order of 10 μs, if $[Mn^{2+}]_0 = 10$ mmol m^{-3}. Note that, by definition, the total order of the reaction in Eq. 4.1 is $\alpha + \beta = 2$, regardless of whether the concentrations of the reactants are assumed equal.

None of the half-life values in Table 4.3 is > 1 h (3.6×10^3 s), indicating that reactions involving soluble complexes are relatively rapid. The condition of equilibrium is defined by $R_f = R_b$, if the reaction rate can be expressed as in Eq. 4.2. If R_f and R_b are also proportional to species concentrations raised to powers equal to stoichiometric coefficients (e.g., as in Eq. 4.3 if $\alpha = \beta = \delta = 1$), then the condition of equilibrium leads to an expression like:

$$\frac{k_f}{k_b} = \frac{[MSO_4^0]_e}{[M^{2+}]_e[SO_4^{2-}]_e} \equiv {}^cK_s \tag{4.5}$$

as applied to the reaction in Eq. 4.1, where $[MSO_4^0]_e$ is the concentration of MSO_4^0 *at equilibrium*, etc. The parameter ${}^cK_s$ defined by the right side of Eq. 4.5 is called the *conditional stability constant* for the complex MSO_4^0. It is "conditional" because it is equal numerically to k_f/k_b, which is generally a function of solution composition, temperature, and pressure. As a numerical example, at 298 K, ${}^cK_s \approx 2 \times 10^{10}$ mol^{-1} dm^3 s$^{-1}/2 \times 10^9$ s$^{-1} = 10$ mol^{-1} dm^3 in the case of $MnSO_4^0$. Thus the ratio, $[MnSO_4^0]/[Mn^{2+}][SO_4^{2-}]$, equals 10 mol^{-1} dm^3 at equilibrium. Equation 4.5 shows that ${}^cK_s$ can be calculated either with kinetics data (k_f and k_b) or with equilibrium data (the $[\ ]_e$). The units of concentration used in this example are discussed in the Appendix.

4.3 Speciation Equilibria

The total concentrations of dissolved constituents in a soil solution represent the sum of "free" (i.e., solvation complex) and complexed forms of the constituents that are stable enough to be considered definite chemical species. The distribution of a given constituent among its possible chemical forms can be described with conditional stability constants, like that in Eq. 4.5, if complex formation and dissociation reactions either are at equilibrium or are so unfavorable kinetically that the reactants can be assumed to be perfectly stable species to a high degree of approximation. This requirement of stable states is often met in natural soils: both ion exchange (Sections 3.3 and 9.3) and soluble-complex formation are usually fast reactions, and certain oxidation–reduction or precipitation–dissolution reactions (Chapters 5 and 6) are very slow on the time scale of a laboratory or a field experiment. But these generalizations can fail in important special cases. As suggested by the data in Table 4.3, the half-lives for metal complex formation and dissociation reactions in aqueous solution at concentrations typical for soils range over about 15 orders of magnitude, from 10^{-9} s for the dissociation of $MnSO_4^0$ to 10^6 s for the formation of $FeCl^{2+}$. The two extremes of this spectrum of time scales present no practical limitations on the applicability of conditional stability constants to soil solutions, whereas the upper midrange of 10^2–10^4 s (e.g., the formation of the complex AlF^{2+}) requires careful consideration of equilibrium times.

The way in which conditional stability constants are used to calculate the distribution of chemical species can be illustrated conveniently by consideration of the forms of dissolved Al in an acidic soil solution. Suppose that the pH of the soil solution is 4.6 and that the total concentration of Al is 10 mmol m^{-3}. The concentrations of the complex-forming ligands, sulfate, fluoride, and fulvic acid have the values 50, 2, and 10 mmol m^{-3}, respectively. The complexes between these ligands and Al are $AlSO_4^+$, AlF^{2+}, and AlL^{2+}, where L refers to "fulvic acid ligands" (see Section 3.3). These complexes are not the only ones formed with Al, SO_4, F, or L, nor are the three ligands the only ones that form Al complexes in the soil solution, but they will serve to introduce speciation calculations in a relatively simple manner.

According to the speciation concept, the total concentration of Al (as determined, e.g., by the 8-hydroxyquinoline method to exclude polymeric species) is the sum of free and complexed forms:

$$Al_T = [Al^{3+}] + [AlOH^{2+}] + [AlSO_4^+] + [AlF^{2+}] + [AlL^{2+}] \tag{4.6}$$

where the square brackets denote species concentrations in moles per cubic decimeter conventionally. (The hydroxy species, $AlOH^{2+}$, is a major one at pH <5.) Each of the complexes in Eq. 4.6 can be described by a conditional stability constant:

$$^cK_1 = \frac{[AlOH^{2+}]}{[Al^{3+}][OH^-]} \approx 10^{9.0} \ \text{mol}^{-1} \ \text{dm}^3 \tag{4.7a}$$

$$^cK_2 = \frac{[AlSO_4^+]}{[Al^{3+}][SO_4^{2-}]} \approx 10^{3.2} \ \text{mol}^{-1} \ \text{dm}^{-3} \tag{4.7b}$$

$$^cK_3 = \frac{[AlF^{2+}]}{[Al^{3+}][F^-]} \approx 10^{7.0} \ \text{mol}^{-1} \ \text{dm}^3 \tag{4.7c}$$

$$^cK_4 = \frac{[AlL^{2+}]}{[Al^{3+}][L^-]} \approx 10^{8.6} \ \text{mol}^{-1} \ \text{dm}^3$$

Common to each of the stability constant expressions is the concentration of the free-ion species, Al^{3+}. Therefore, Eq. 4.6 can be rewritten in the form:

$$Al_T = [Al^{3+}]\left\{1 + \frac{[AlOH^{2+}]}{[Al^{3+}]} + \frac{[AlSO_4^+]}{[Al^{3+}]} + \frac{[AlF^{2+}]}{[Al^{3+}]} + \frac{[AlL^{2+}]}{[Al^{3+}]}\right\}$$

$$= [Al^{3+}]\{1 + {}^cK_1[OH^-] + {}^cK_2[SO_4^{2-}] + {}^cK_3[F^-] + {}^cK_4[L^-]\}$$

$$(4.8)$$

The ratio of $[Al^{3+}]$ to Al_T, called the *distribution coefficient* for Al^{3+}, can be calculated with Eq. 4.8 if the concentrations of the *free-ion* species of the four complexing ligands are known:

$$\alpha_{Al} \equiv \frac{[Al^{3+}]}{Al_T} = \{1 + {}^cK_1[OH^-] + {}^cK_2[SO_4^{2-}] + {}^cK_3[F^-] + {}^cK_4[L^-]\}^{-1}$$

$$(4.9)$$

For OH^-, one can estimate the free-ion concentration from the pH value:

$$[OH^-] = \frac{{}^cK_w}{[H^+]} \approx \frac{10^{-14}}{10^{-pH}} = 10^{pH-14} \ mol \ dm^{-3} \qquad (4.10)$$

where ${}^cK_w$ is the ionization product of liquid water under the conditions that exist in the soil solution (hence the superscript c). For dilute solutions at 25°C and under 1 atm pressure, ${}^cK_w \approx 10^{-14} \ mol^{-2} \ dm^6$ and $[H^+] \approx 10^{-pH}$ numerically. Thus $[OH^-] \approx 10^{-9} \ mol \ dm^{-3}$ in the present example (pH 4.6).

For the other ligands in Eq. 4.9, the free-ion concentrations cannot be calculated directly. Given the large value of ${}^cK_4$ relative to ${}^cK_2$ and ${}^cK_3$, it is reasonable to expect that $[AlL^{2+}]$ will be nearly equal to Al_T and L_T in the present example. Thus in a first approximation, Eq. 4.7d can be used to estimate α_{Al}:

$$\frac{[AlL^{2+}]}{[Al^{3+}][L^-]} = \frac{\alpha_{AlL}}{\alpha_{Al}\alpha_L L_T} \approx \frac{1}{\alpha_{Al}^2 L_T} = {}^cK_4 = 10^{8.6} \ dm^3 \ mol^{-1} \qquad (4.11)$$

where

$$\alpha_{AlL} \equiv \frac{[AlL^{2+}]}{Al_T} \qquad \alpha_L \equiv \frac{[L^-]}{L_T} \qquad (4.12a)$$

are the distribution coefficients for AlL^{2+} and L^-, respectively, and L_T is the total "fulvic acid ligand" concentration. In Eq. 4.11, it has been assumed that $\alpha_{AlL} \approx 1$ and $\alpha_{Al} \approx \alpha_L$, with the result that:

$$\alpha_{Al}^2 \approx ({}^cK_4 L_T)^{-1} = 10^{-3.6}$$

and $\alpha_{Al} \approx 1.6 \times 10^{-2}$. Thus only about 2% of Al_T is in the form of Al^{3+}. This approximate result can be used to estimate the distribution coefficients for each inorganic complex (under the assumption that $\alpha_{AlL} \approx 1$):

$$\alpha_{AlOH} \equiv \frac{[AlOH^{2+}]}{Al_T} = \alpha_{Al} \frac{[AlOH^{2+}]}{[Al^{3+}]} = \alpha_{Al}{}^cK_1[OH^-] \approx 1.6 \times 10^{-2} \quad (4.12b)$$

$$\alpha_{AlSO4} \equiv \frac{[AlSO_4^+]}{Al_T} = \alpha_{Al} \frac{[AlSO_4^+]}{[Al^{3+}]} = \alpha_{Al}{}^cK_2[SO_4^{2-}] \approx 1.3 \times 10^{-3} \quad (4.12c)$$

$$\alpha_{AlF} \equiv \frac{[AlF^{2+}]}{Al_T} = \alpha_{Al} \frac{[AlF^{2+}]}{[Al^{3+}]} = \alpha_{Al}{}^cK_3[F^-] \approx 0.32 \quad (4.12d)$$

The assumption that $\alpha_{AlL} \approx 1$ is not consistent with the large value estimated for α_{AlF} in Eq. 4.12d, which implies $\alpha_{AlL} < 0.7$.

The estimates in Eq. 4.12 can be refined by considering the *ligand* speciation in more detail:

$$SO_{4T} = [SO_4^{2-}] + [AlSO_4^+] = [SO_4^{2-}]\{1 + {}^cK_2[Al^{3+}]\} \quad (4.13a)$$

$$F_T = [F^-] + [AlF^{2+}] = [F^-]\{1 + {}^cK_3[Al^{3+}]\} \quad (4.13b)$$

$$L_T = [L^-] + [AlL^{2+}] = [L^-]\{1 + {}^cK_4[Al^{3+}]\} \quad (4.13c)$$

where use has been made again of Eq. 4.7. Given $[Al^{3+}] = \alpha_{Al}Al_T \approx 1.6 \times 10^{-7}$ mol dm^{-3}, the ligand distribution coefficients are:

$$\alpha_{SO_4} \equiv \frac{[SO_4^{2-}]}{SO_{4T}} = \{1 + {}^cK_2[Al^{3+}]\}^{-1} \approx 1.0 \quad (4.14a)$$

$$\alpha_F \equiv \frac{[F^-]}{F_T} = \{1 + {}^cK_3[Al^{3+}]\}^{-1} \approx 0.38 \quad (4.14b)$$

$$\alpha_L \equiv \frac{[L^-]}{L_T} = \{1 + {}^cK_4[Al^{3+}]\}^{-1} \approx 0.015 \quad (4.14c)$$

The refined value of α_{AlF} that results from Eq. 4.14b is 0.38 times the value estimated in Eq. 4.12d, or 0.12. This approximate result and those in Eqs. 4.12b, 4.12c, 4.14a, and 4.14c are reasonably consistent with $\alpha_{AlL} \approx 0.9$. Thus about 90% of Al_T is organically complexed, and about 10% is complexed with inorganic ligands or in the free-ion form. This distribution of Al species is typical for acidic soil solutions containing dissolved organic matter at concentrations comparable to Al_T.

This example, despite the approximate nature of the calculations, illustrates all of the salient features of a more exact speciation calculation: *mass balance* (Eq. 4.6), *conditional stability constants* (Eq. 4.7), *distribution coefficients* (Eqs. 4.12 and 4.14), and the *iterative refinement* of the distribution coefficients through additional mass balance on the ligands (Eqs. 4.13 and 4.14). The approach can be applied to any soil solution for which the aqueous species and their conditional stability constants are known. Results of high precision are obtained if the iterative refinement is performed on a computer.

4.4 Predicting Speciation

The calculation of the distribution of chemical species in a soil solution can be done with a computer if three pieces of information are available: (1) the measured total concentrations of the metals and ligands, along with the pH values; (2) the conditional stability constants for all possible complexes of the metals and H^+ with the ligands; (3) expressions for the mass balance of each constituent in terms of chemical species (free ions and complexes). A flowchart outlining the method of calculation is shown in Fig. 4.1.

Total concentrations of the metals (M_T) and ligands (L_T), along with the pH value, are the basic input data for the calculation. They are presumed known for all important constituents of a soil solution. The speciation calculation then proceeds on the assumption that mass balance expressions like Eqs. 4.6 and 4.13 can be developed for each metal and ligand. The mass balance expressions are converted into a set of coupled algebraic equations with the free-ion concentrations as unknowns by substitution for the complex concentrations as illustrated in Eq. 4.8. This step requires access to a data base containing the values of all relevant conditional stability constants. In general, for the complex formation reaction:

$$v_c M^{m+}(aq) + \gamma H^+(aq) + v_a L^{1-}(aq) = M_{v_c} H_\gamma L_{v_a}(aq) \qquad (4.15)$$

the conditional stability constant is:

$$^c K_s = [M_{v_c} H_\gamma L_{v_a}]/[M]^{v_c}[H]^\gamma[L]^{v_a} \qquad (4.16)$$

where v_c, γ, and v_a are stoichiometric coefficients. Equation 4.16 can be rearranged to express $[M_{v_c} H_\gamma L_{v_a}]$ in terms of $^c K_s$ and the three free-ion concentrations. For example, the formation of $CaHCO_3^+$ can be expressed by the reaction:

$$Ca^{2+}(aq) + H^+(aq) + CO_3^{2-}(aq) = CaHCO_3^+(aq) \qquad (4.17a)$$

for which $^c K_s \approx 10^{11.2}$ dm^6 mol^{-2} at 25°C in a dilute solution. Thus numerically,

$$10^{11.2} = [CaHCO_3^+]/[Ca^{2+}][H^+][CO_3^{2-}] \qquad (4.17b)$$

and

$$[CaHCO_3^+] = 10^{11.2}[Ca^{2+}][H^+][CO_3^{2-}] \qquad (4.17c)$$

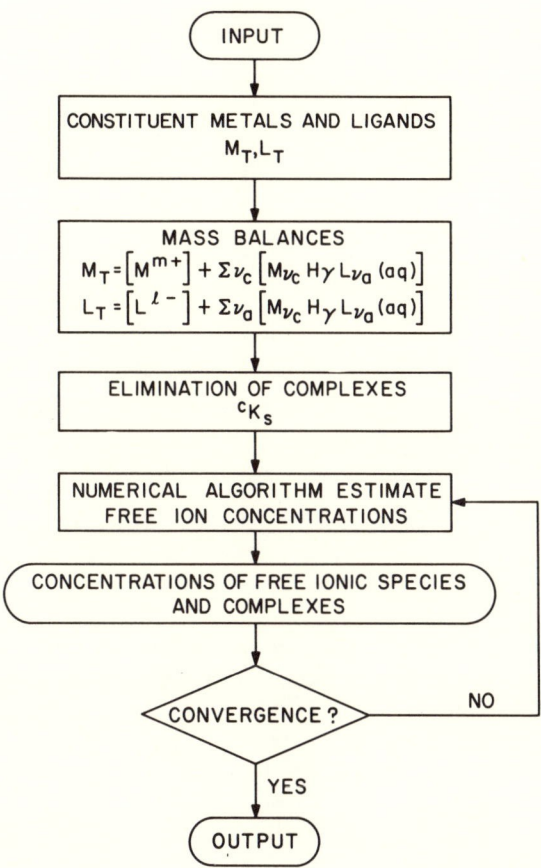

FIG. 4.1 Flow diagram for a speciation calculation based on mass balance and conditional stability constants for complexes.

The algebraic equations for the free-ion concentrations can be solved numerically by standard techniques based on estimated or "guessed" values, like the Newton–Raphson method. The resulting free-ion concentrations then are used to calculate the complex concentrations with expressions like Eq. 4.17c. The calculated species concentrations can be checked by introducing them into the mass balance equations to see if they sum numerically to the known total concentrations. If they do within some acceptable error (say, 0.01% difference from the input measured M_T or L_T), then the calculation is said to have "converged" and the speciation results may be printed. If convergence has not been achieved, then the numerical calculation is repeated using the speciation results to generate new input estimates for the free-ion concentrations in the numerical algorithm.

Table 4.4 shows a typical example of the results of a speciation calculation

TABLE 4.4 Composition and speciation of an acidic soil solution (pH 4)

Constituent	C_T (mmol m^{-3})	Percentage speciation
Ca	18	Ca^{2+} (99%), $CaSO_4^0$ (1%)
Mg	7	Mg^{2+} (100%)
K	37	K^+ (100%)
Na	3	Na^+ (100%)
Fe(II)	0.5	Fe^{2+} (99%), $FeSO_4^0$ (1%)
Mn(II)	2	Mn^{2+} (99%), $MnSO_4^0$ (1%)
Al	10	AlL^{2+} (94%), AlF^{2+} (6%)
CO_3	10	$H_2CO_3^*$ (100%)
SO_4	71	SO_4^{2-} (99%), HSO_4^- (1%)
Cl	13	Cl^- (100%)
F	2	F^- (57%), AlF^{2+}(38%), HF^0 (5%)
L[a]	10	AlL^{2+} (94%), HL^0 (6%)

[a]Organic ligands (like fulvic acid).

for an acidic soil solution. Measured concentrations of seven metals and five ligands were available and their speciation was calculated using a data base containing conditional stability constants for 92 complexes. The results confirm more precisely the approximate calculation of Al speciation presented in Section 4.3. These kinds of data may be useful in predicting the mobility or bioavailability of Al in the soil (see Chapter 13).

The *limitations* of speciation calculations based on the approach in Fig. 4.1 should not be forgotten:

1. Pertinent chemical equilibria like oxidation–reduction may have been ignored unintentionally in formulating mass balance, or important species in the soil solution may not have been considered.

2. Conditional stability constants for assumed species may be incorrect, or inadequate in some other way for soils.

3. Analytical methods for certain constituents in the soil solution may be inadequate to distinguish between various physical and chemical forms (e.g., dissolved vs. suspended, oxidized vs. reduced, monomeric vs. polymeric).

4. The rates of certain chemical reactions assumed to be at equilibrium on the basis of studies in simple systems may be so low in soil that equilibrium is not attained.

5. Temperature and pressure variations may need to be considered. Significant temperature and pressure gradients exist at times in nearly all natural soils. Adequate data on the temperature and pressure dependence of the relevant conditional stability constants often are not available. Usually the stability constant data will refer to 25°C and 1 atm pressure.

6. An equilibrium-based model may be a poor approximation for a particular soil solution because of the flows of matter and energy in natural soils. The appropriate time-invariant state might not be a state of equilibrium, but a steady state. With respect to this point, it is important to emphasize the essentially subjective, but critical, initial decision as to the "free-body cut" when applying the mass balance approach; that is, the choice of a *closed* model system whose behavior is to mimic the actual *open* system in nature.

4.5 The Thermodynamic Stability Constant

Speciation calculations are dependent on a high-quality data base of conditional stability constants. These parameters, as their name implies, vary with the composition and total electrolyte concentration of the soil solution. For example, in a very dilute solution, the conditional stability constant for the formation of $CaHCO_3^+$ (Eq. 4.17a) has the value 2×10^{11} dm^6 mol^{-2}. In a solution of 50 mol m^{-3} NaCl it is 0.4×10^{11} dm^6 mol^{-2}, and in 50 mol m^{-3} $CaCl_2$ it is 0.2×10^{11} dm^6 mol^{-2}. This variability requires the compilation of a different data base each time a speciation calculation is performed, which is not an efficient approach to the problem.

Instead, concepts in chemical thermodynamics (the branch of physical chemistry that investigates the macroscopic properties of chemical equilibria) may be called on to define a *thermodynamic stability constant*. This parameter is by definition independent of chemical composition at a chosen temperature and pressure, usually 25°C (298.15 K) and 1 atm. For the complex formation reaction in Eq. 4.15, the thermodynamic stability constant is defined by the equation:

$$K_s \equiv (M_{v_c}H_yL_{v_a})/(M)^{v_c}(H)^y(L)^{v_a} \qquad (4.18)$$

where boldface parentheses refer to the thermodynamic *activity* of the chemical species enclosed. Unlike ^cK_s in Eq. 4.16, K_s has a fixed value, regardless of the composition of the soil solution. To make this assertion a reality, the activity of a species is related to its concentration (in moles per cubic decimeter) through an *activity coefficient*:

$$(i) \equiv \gamma_i[i] \qquad (4.19)$$

where i is some chemical species like Ca^{2+} or $MnSO_4^0$ and γ_i is its activity coefficient. By convention, γ_i has the units dm^3 mol^{-1}, such that the activity has *no units* and the thermodynamic stability constant is *dimensionless*.

Conventions and laboratory methods have been developed to measure γ_i, (i), and K_s in electrolyte solutions. All species activity coefficients, for example, are required to approach the value 1.0 dm^3 mol^{-1} as a solution becomes infinitely dilute. Thus in the limit of infinite dilution, activities become equal *numerically* to concentrations and ^cK_s becomes equal *numerically* to K_s. With

Eqs. 4.16, 4.18, and 4.19, one can derive the relationship:

$$\log K_s = \log {}^c K_s + \log\{\gamma_{MHL}/\gamma_M^{v_c}\gamma_H^{y}\gamma_L^{v_a}\} \tag{4.20}$$

The second term on the right side must vanish in the limit of infinite dilution, so a graph of $\log {}^c K_s$ against a suitable concentration function must extrapolate to $\log K_s$ at zero concentration. Experiment and theory have shown that a useful concentration function for this purpose is the effective *ionic strength*, I:

$$I = \tfrac{1}{2}\Sigma_k Z_k^2[k] \tag{4.21}$$

where the sum is over all charged *species* (with valence Z_k) in a solution. The effective ionic strength is related closely to the electrolytic conductivity of a solution. Experimentation with soil solutions has indicated that the *Marion–Babcock equation*:

$$\log I = 1.159 + 1.009 \log \kappa \tag{4.22}$$

is accurate for ionic strengths up to about 0.3 mol dm^{-3}. In Eq. 4.22, I is in units of moles per cubic meter, and κ, the *electrolytic conductivity*, is in decisiemens per meter (dS m^{-1}). (For a discussion of the units used, see the Appendix.)

Experimental and theoretical studies of electrolyte solutions have led to semiempirical equations that relate species activity coefficients to the effective ionic strength. For charged species (free ions or complexes), there is the *Davies equation*:

$$\log \gamma_i = -0.512 Z_i^2 \left[\frac{\sqrt{I}}{1 + \sqrt{I}} - 0.3I \right] \tag{4.23}$$

where Z_i is the species valence. The accuracy of Eq. 4.23 for use in Eq. 4.18 can be tested by substituting it into Eq. 4.20:

$$\log K_s = \log {}^c K_s + 0.512 \left[\frac{\sqrt{I}}{1 + \sqrt{I}} - 0.3I \right] \Delta Z^2 \tag{4.24}$$

where

$$\Delta Z^2 \equiv v_c m^2 + \gamma + v_a \ell^2 - (v_c m + \gamma - v_a \ell)^2 \tag{4.25}$$

in terms of the valences of M, H, L, and $M_{v_c}H_y L_{v_a}$ in Eq. 4.15. According to the Davies equation, a graph of $\Delta \log K \equiv \log K_s - \log {}^c K_s$ against the parameter ΔZ^2 should be a straight line whose positive slope depends only on the value of I. Figure 4.2 shows a verification of this result at $I = 0.1$ mol dm^{-3} for 219 metal complexes for which $\Delta \log K$ has been measured and the corresponding ΔZ^2 calculated. The line through the data is Eq. 4.24 with $I = 0.1$ mol dm^{-3}.

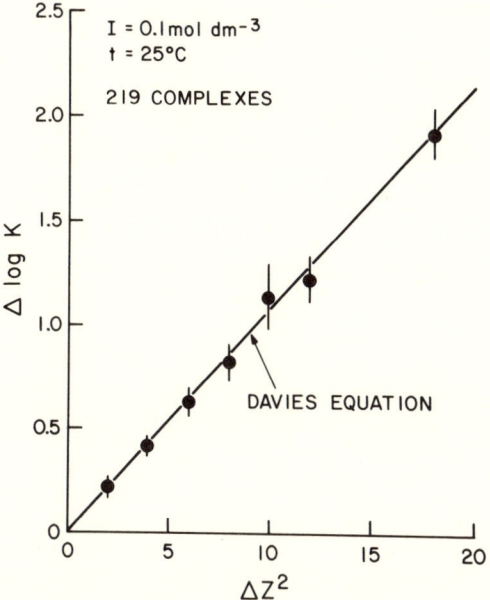

FIG. 4.2 A test of the Davies equation for metal–ligand complexes. [Reprinted with permission from G. Sposito, The future of an illusion: Ion activities in soil solutions, *Soil Sci. Soc. Am. J.* **48**:531–536 (1984).]

For uncharged monovalent metal–ligand complexes, uncharged proton–ligand complexes, and uncharged bivalent metal–ligand complexes, some semiempirical equations for $\log \gamma_i$ are:

$$\log \gamma_{ML} = \frac{-0.192I}{0.0164 + I} \qquad (M = Na^+, K^+, \text{etc.}) \qquad (4.26a)$$

$$\log \gamma_{HL} = 0.1I \qquad (4.26b)$$

$$\log \gamma_{ML} = -0.3I \quad (M = Ca^{2+}, Mg^{2+}, \text{etc.}) \qquad (4.26c)$$

for $I < 0.1$ mol dm^{-3}. These expressions conform to the theoretical requirement for neutral species, that $\log \gamma$ become proportional to I in the infinite-dilution limit.

With equations for γ_i like Eqs. 4.23 and 4.26, it is possible to calculate sets of conditional stability constants under varying composition from a single set of thermodynamic stability constants. For charged complexes, the necessary relationship is given in Eq. 4.24, whereas for uncharged complexes described with Eq. 4.26, one of the three expressions for $\log \gamma_i$ must be added to the right side of Eq. 4.24. For example, in the case of CaHCO$_3^+$ at $I = 0.05$ mol dm^{-3},

$K_s = 2 \times 10^{11}$ and

$$\log {}^c K_s = 11.3 - 0.512 \left[\frac{\sqrt{0.05}}{1 + \sqrt{0.05}} - 0.3(0.05) \right] + 8 = 11.3 - 0.69 = 10.6$$

According to Eq. 4.24 after rearrangement to calculate $\log {}^c K_s$. In the case of $MnSO_4^0$ at $I = 0.05$ mol dm^{-3}, Eq. 4.26c must be added to Eq. 4.24 and, with $K_s = 1.8 \times 10^2$,

$$\log {}^c K_s = 2.26 - 0.512[0.168] \times 8 + 0.3(0.05) = 2.26 - 0.69 + 0.15 = 1.72$$

In a speciation calculation following the flowchart in Fig. 4.1, a data base of K_s values would be used to create the required data base of ${}^c K_s$ values as illustrated earlier. An estimate of I (perhaps based on Eq. 4.22) would be needed to do this, and the ${}^c K_s$ data base would be refined in each iteration along with the species concentrations and the value of I in Eq. 4.21. Convergence of the calculation then would result in a mutually consistent set of species concentrations, ${}^c K_s$ values, and effective ionic strength.

The conceptual meaning of the activity of a chemical species stems from the formal similarity between Eqs. 4.16 and 4.18. The conditional stability constant is a convenient parameter with which to characterize equilibria, but it is composition-dependent in that it contains species concentrations only and does not correct for the electrostatic interactions among species that occur as their concentrations change. In the limit of infinite dilution these interactions must die out, and the extrapolated value of ${}^c K_s$ must represent the chemical equilibrium of an "ideal" solution wherein species interactions other than those involved to form a complex are unimportant. Thus the activity factors in Eq. 4.18 play the role of hypothetical concentrations of species in an "ideal" solution. But the real solution is not ideal as its concentration increases because species are brought closer together to interact more strongly. When this occurs, ${}^c K_s$ must begin to deviate from K_s. The activity coefficient then is introduced to "correct" the concentration factors in ${}^c K_s$ for nonideal species behavior and thereby restore the value of K_s via Eq. 4.20. This correction is expected to be larger for charged species than for neutral complexes (dipoles), and larger as the species valence increases. These trends are reflected in the model expressions in Eqs. 4.23 and 4.26.

FOR FURTHER READING

F. Adams, Soil solution, in E. W. Carson (ed.), *The Plant Root and Its Environment*, pp. 441–481. Univ. of Virginia Press, Charlottesville, VA, 1974. An excellent, thought-provoking discussion of the classic problem of obtaining and analyzing soil solutions.

T. M. Florence and G. E. Batley, Chemical speciation in natural waters, *Crit. Rev. Anal. Chem.* 9:219–296 (1980). A comprehensive survey of experimental and theoretical methodologies for speciation studies, with applications to more than a dozen chemical elements.

A. M. Elprince, *Chemistry of Soil Solutions*, van Nostrand Reinhold, New York, 1986. Parts IV and V of this useful anthology contain fundamental research papers on the modeling and sampling of soil solutions.

W. C. Gardiner, Jr., *Rates and Mechanisms of Chemical Reactions*, W. A. Benjamin, Menlo Park, CA, 1972. The first three chapters of this textbook provide a readable survey of rate laws and experimental methods in chemical kinetics.

A. L. Page, R. H. Miller, and D. R. Keeney, *Methods of Soil Analysis, Part 2, Chemical and Microbiological Properties*, American Society of Agronomy, Madison, WI, 1982. The first 34 chapters of this comprehensive treatise give reommendations for procedures to determine the chemical composition of soils, waters, and plants.

J. F. Pankow and J. J. Morgan, Kinetics for the aquatic environment, *Environ. Sci. Technol.* **15**:1155–1164 (1981). A useful introduction to the applications of chemical kinetics to complex formation and dissociation in natural waters.

L. D. Pettit and G. Brookes, Why stability constants? *Essays in Chemistry* **6**:1–33 (1977). A fine introduction to conditional and thermodynamic stability constants, with applications to metal–ligand complexes.

G. Sposito, The future of an illusion: Ion activities in soil solutions, *Soil Sci. Soc. Am. J.* **48**: 531–536 (1984). A brief survey of activities, activity coefficients, and speciation calculations in soil solutions.

PROBLEMS

The more difficult problems are indicated by an asterisk.

1. In the accompanying table are composition data for drainage water collected at the litter–soil interface and at a point 0.3 m below the interface in a soil supporting a deciduous forest. Discuss possible causes for the significant differences in pH and K and Ca concentrations between the two soil solutions. Calculate the total moles of cation and anion charge per cubic meter, as well as the net charge per cubic meter for each solution. Explain why the net charge in each case is not zero and why it is larger in absolute value for the litter solution than the subsoil solution.

					Constituent (mmol m^{-3})					
	Ca	Mg	Na	K	NH$_4$	NO$_3$	Cl	SO$_4$	PO$_4$	pH
Litter	50	37	11	63	5	1	36	62	1	4.86
Soil	23	33	19	39	8	2	40	50	1	5.98

2. Develop an appropriate kinetics expression like Eq. 4.3 for the formation of AlF^{2+} from Al^{3+} and F^-. The value of k_f for this reaction at 25°C is 110 dm^3 mol^{-1} s^{-1} at pH 3.9 and 726 dm^3 mol^{-1} s^{-1} at pH 4.9. Calculate the corresponding half-lives given $[Al^{3+}]_0 = [F^-]_0 = 10$ mmol m^{-3}.

3. The solvation water exchange reaction illustrated for Ca^{2+} in the last row in Table 4.3 has first-order rate constants as follows: Mg^{2+} (3×10^5 s^{-1}), Ca^{2+} (10^8 s^{-1}), Mn^{2+} (3×10^7 s^{-1}), Fe^{2+} (3×10^6 s^{-1}). Calculate the corresponding half-lives and compare the results with the ionic radii of the cations (Table 2.1). Explain any correlation you note in this comparison.

*4. The temperature dependence of rate constants often can be expressed mathematically by the *Arrhenius equation*:

$$\log k = A - B/RT$$

where A and B are constant parameters, R is the molar gas constant, and T is absolute temperature. For the AlF^{2+} formation reaction discussed in Problem 2, $B = 25$ kJ mol^{-1}. Calculate the value of k_f at 10°C (283.15 K) and pH 4, given the value at 25°C (298.15 K), then compute the corresponding half-life for $[Al]_0 = [F^-]_0 = 10$ mmol m^{-3}.

5. Develop an expression for α_{Ca} analogous to Eq. 4.9 based on the list of aqueous Ca species for alkaline soils in Table 4.1. What is the value of α_{Ca} at pH 8 if $[SO_4^{2-}] = [HCO_3^-] = 1$ mol m^{-3} and $\log {}^cK$ for the formation of HCO_3^- from H^+ and CO_3^{2-} is 10.3?

6. The mass balance of carbonate, ignoring complexes with metals, can be expressed:

$$CO_{3T} = [H_2CO_3^*] + [HCO_3^-] + [CO_3^{2-}]$$

Given the conditional stability constants

$${}^cK_1 = [H_2CO_3^*]/[H^+]^2[CO_3^{2-}] \approx 10^{16.7}$$

$${}^cK_2 = [HCO_3^-]/[H^+][CO_3^{2-}] \approx 10^{10.3}$$

derive expressions for the distribution coefficients for each of the three carbonate species. Use the approximation $[H^+] \approx 10^{-pH}$ to estimate the range of pH over which HCO_3^- is the dominant species.

7. Combine ${}^cK_1$ and ${}^cK_2$ in Problem 6 with K_H in Eq. 1.1 to derive the equation:

$${}^cK_1/{}^cK_2 K_H = P_{CO_2}/[H^+][HCO_3^-] \approx 10^{7.0} \text{ atm } dm^6 \text{ mol}^{-2}$$

where $CO_2(aq) \equiv H_2CO_3^*$ as in Eq. 2.9a. This equation shows that *the CO_2 pressure and $[HCO_3^-]$ serve to determine the pH*. Calculate the pH in equilibrium with $[HCO_3^-] = 1$ mmol m^{-3} and $P_{CO_2} = 10^{-3.5}$ or 10^{-2} atm (the range typical for soils).

*8. The *alkalinity* of a soil solution is defined by the equation:

$$Alk \equiv [HCO_3^-] + 2[CO_3^{2-}] + [OH^-] - [H^+]$$

Use the conditional stability constants in Problems 6 and 7 to calculate the alkalinity at pH 6 and $P_{CO_2} = 10^{-3}$ atm. What is the conceptual relationship between alkalinity and the acid-neutralizing capacity, defined in Eq. 3.7 for the case of dissolved humus?
(*Answer: Alk* $\approx [HCO_3^-] - [H^+] \approx 10^{-4.8} - 10^{-6} = 15$ mol m^{-3})

9. Plot a graph of log γ_i for univalent species against ionic strength in the range 0–1 mol dm^{-3} assuming that the Davies equation is accurate in this range.

10. Given that ΔZ^2 is usually positive, what general conclusion can be drawn from Eq. 4.24 about the effect of ionic strength on soluble-complex formation?

11. The electrolytic conductivity of a soil solution is 1.5 dS m^{-1}. Calculate the activities of Ca^{2+} and $CaSO_4^0$ in this solution if the respective concentrations of these species are 1 mol m^{-3} and 200 mmol m^{-3}.

12. Calculate the effect of increasing the electrolytic conductivity from 0.5 to 2.0 dS m^{-1} on the concentration of $Si(OH)_4^0$ (an uncharged proton–ligand complex) maintained at a constant activity of 10^{-4} in equilibrium with quartz (SiO_2).

13. The activity of Al^{3+}(aq) in equilibrium with gibbsite at pH 5 is $10^{-6.23}$. How does the concentration of Al^{3+}(aq) change if the ionic strength shifts from 5 mol m^{-3} to 30 mol m^{-3}?

14. Equation 4.26b is often applied to dissolved gases like CO_2(aq) as well as to neutral complexes. Given that the value of K_H in Table 1.6 actually represents $(CO_2$(aq))/P_{CO_2}, calculate the solubility of CO_2 in a soil solution whose ionic strength is 0.10 mol dm^{-3} and for which $P_{CO_2} = 10^{-3.5}$ atm.

*15. The value of log K_s for the formation of $AlSO_4^+$ from Al^{3+} and SO_4^{2-} is 3.21. Calculate log ^cK_s in a soil solution that has an electrolytic conductivity of 2.4 dS m^{-1}. Does increasing electrolytic conductivity enhance or diminish $AlSO_4^+$ formation?

5

Mineral Solubility

5.1 Dissolution Reactions

When water enters a dry soil, it begins at once to hydrate the surfaces of the solid phases present. On a microscopic scale, water molecules first invade small cracks or other structural imperfections on mineral surfaces. As they enter these disrupted sites, the water molecules are attracted to exposed ionic constituents of the mineral and begin to form solvation complexes with them. If the chemical bonds in the mineral have very little covalent character [e.g., NaCl (halite) or $CaSO_4 \cdot 2H_2O$ (gypsum)], the solvated ions will detach readily from the mineral structure and diffuse out into the soil solution. These ions then may form soluble complexes with other solutes, as described in Section 4.2.

Soil minerals like aluminosilicates and metal oxides have chemical bonds with short range covalent character (Section 2.1). Except at highly distorted structural faults on the surfaces of these minerals, hydrating water molecules are not likely to solvate and detach their Si, O, Al, or Fe ions into the aqueous phase. Exchangeable ions on the surfaces of these minerals (e.g., Na^+ or Mg^{2+} on a clay mineral or Cl^- on a metal hydroxide) can still solvate readily and diffuse away, but the framework ions cannot be dislodged so easily. For their removal it is necessary to create a strong perturbation of the bonds holding them in the mineral structure, and this can be accomplished only by a highly polarizing species like the proton or a solute molecule that forms inner-sphere complexes (Section 4.2).

Proton attack begins with adsorption by the anionic constituent of a mineral (e.g., OH in a metal hydroxide, CO_3 in a carbonate, or PO_4 in a

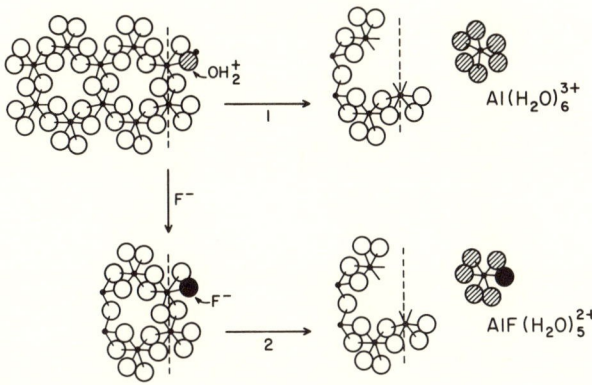

FIG. 5.1 Dissolution mechanisms for gibbsite: (1) Protonation of a surface OH and detachment of Al^{3+} as a solvation complex. (2) Ligand exchange of OH_2^+ for F^- and detachment of Al^{3+} as the AlF^{2+} complex.

phosphate). This relatively rapid reaction is followed by the slower process of polarizing the remaining metal–anion bonds at the site of proton adsorption and subsequent detachment of the metal–protonated anion complex into the aqueous phase. The two-step mechanism involved is illustrated schematically for the mineral gibbsite $[Al(OH)_3]$ in Fig. 5.1. A similar process is also shown in Fig. 5.1 for ligand attack. In this case, a strongly complexing ligand in the soil solution (e.g., fulvic acid, well-defined organic ligands like citrate, or inorganic anions like F^- and PO_4^{3-}) exchanges for a protonated OH bound to Al and the metal–ligand complex ultimately detaches into the aqueous phase.

For soil minerals whose ionic constituents are readily solvated and detached, or for the exchangeable ions on otherwise more covalently bonded minerals, the kinetics of dissolution can be described in terms of the film diffusion mechanism discussed in Section 3.3 and Special Topic 2. Thus the dissolution reactions of these minerals or exchangeable components can be termed *transport controlled*. For soil minerals like the clay minerals, metal hydrous oxides, and most carbonates, however, the rate of dissolution is *surface controlled* and is observed to follow zero-order kinetics, described mathematically in Table 4.2. If [A] is the aqueous-phase concentration of an ionic constituent of the mineral (e.g., Al^{3+}), then the rate of surface-controlled dissolution can be expressed by the equation:

$$\frac{d[A]}{dt} = k \tag{5.1}$$

where the parameter k is independent of [A], but is a function of temperature, pressure, the surface area of the mineral, $[H^+]$, and, if appropriate, the concentration of strongly-complexing ligand inducing dissolution via the second mechanism in Fig. 5.1. Typically the pH dependence of k is found to lead to an empirical expression of the form $k = k'[H^+]^n$, where k' ranges between 10^{-10} and 10^{-14} in magnitude, and n is a fractional exponent having values between 0 and 1. The precise values of k' and n depend on the nature of the mineral and the range of pH investigated. For example, $n \approx 1$ for albite ($NaAlSi_3O_8$) and gibbsite at pH <4, and $n \approx 0.5$ for dolomite [$CaMg(CO_3)_2$] at pH < 6. Near pH 7 it is observed often that n is very small.

The rate expression in Eq. 5.1 applies to a surface-controlled mineral dissolution reaction after any rapid ion exchange or solvation reactions have occurred, but well before equilibrium between the mineral and the soil solution is reached. The same consideration applies to transport-controlled dissolution reactions governed by an expression like Eq. s2.2 in Special Topic 2. As equilibrium approaches, the rate of dissolution becomes influenced by the stoichiometry of the dissolution reaction. Dissolution reactions for the minerals albite, anorthite, calcite, muscovite, vermiculite, and smectite were illustrated in Sections 1.5, 2.3, and 2.5, and in Special Topic 1. (Some of these reactions involved incongruent dissolution.) Two other examples are the dissolution reactions of gypsum and gibbsite:

$$CaSO_4 \cdot 2H_2O(s) = Ca^{2+}(aq) + 2H_2O(\ell) \tag{5.2}$$

$$Al(OH)_3(s) = Al^{3+}(aq) + 3OH^-(aq) \tag{5.3}$$

Following the chemical thermodynamic concepts introduced in Section 4.5, one can define a *dissolution equilibrium constant* for the reactions in Eqs. 5.2 and 5.3:

$$K_{dis} = (Ca^{2+})(SO_4^{2-})(H_2O)^2/(gypsum) \tag{5.4}$$

$$K_{dis} = (Al^{3+})(OH^-)^3/(gibbsite) \tag{5.5}$$

where the boldface parentheses indicate a thermodynamic activity. The solid-phase activities of gypsum and gibbsite that appear in Eqs. 5.4 and 5.5 are defined to have unit value if the minerals exist in pure, macrocrystalline form at $T = 298.15 \, K$ and 1 atm pressure. If, as often can be true in soils, the solid phases are "contaminated" with minor elements (Section 1.3) or are not well crystallized (Chapter 2), their activity will differ from 1.0.

The *solubility product constant* for gypsum or gibbsite is defined by the equation:

$$K_{so} \equiv K_{dis}(gypsum)/(H_2O)^2 = (Ca^{2+})(SO_4^{2-}) \tag{5.6}$$

$$K_{so} \equiv K_{dis}(gibbsite) = (Al^{3+})(OH^-)^3 \tag{5.7}$$

By convention, $K_{so} = K_{dis}$ numerically when the solid phase is pure and macrocrystalline (no structural imperfections), and the aqueous solution phase is infinitely dilute. In this case, the solid and water activities are both defined equal to 1.0. In the present example, $K_{so} = 2.4 \times 10^{-5}$ for gypsum, and $K_{so} = 1.29 \times 10^{-34}$ for gibbsite, according to published compilations of thermodynamic data like *Critical Stability Constants* (see the "For Further Reading" section at the end of this chapter). Usually K_{so} values for hydroxide solids are reported as $*K_{so}$, which is K_{dis} for the dissolution reaction obtained by replacing OH^- with H^+ via the formation reaction for liquid water. In the case of gibbsite, for example, one adds $3[OH^-(aq) + H^+(aq) = H_2O(\ell)]$ to Eq. 5.3 and replaces Eq. 5.7 with the definition:

$$*K_{so} \equiv *K_{dis}(\text{gibbsite})/(H_2O)^3 = (Al^{3+})/(H^+)^3 \tag{5.8}$$

Since the equilibrium constant for the water reaction is 10^{14}, $*K_{so} = 10^{42} \times 1.29 \times 10^{-34} = 1.29 \times 10^8$.

The right sides of Eqs. 5.6–5.8 contain the *ion activity product* corresponding to the solid phases dissolving. For the dissolution reaction of the solid M_aL_b,

$$M_aL_b(s) = M^{m+}(aq) + L^{\ell-}(aq) \tag{5.9}$$

the ion activity product (*IAP*) is defined by the equation:

$$IAP \equiv (M^{m+})(L^{\ell-}) \tag{5.10}$$

Evidently $IAP = (Ca^{2+})(SO_4^{2-})$ for gypsum and $IAP = (Al^{3+})(OH^-)^3$ or $(Al^{3+})(H^+)^{-3}$ for gibbsite.

The *IAP* can be calculated solely with data on the speciation of the soil solution according to the method discussed in Section 4.5. *Thus Eq. 5.10 can be applied regardless of whether the dissolution reaction in Eq. 5.9 is actually at*

FIG. 5.2 Time development of the relative saturation $[\Omega = (Al^{3+})(OH^-)^3/K_{so}]$ for gibbsite dissolution in Molokai clay. [Data from G. M. Marion, D. M. Hendricks, G. R. Dutt, and W. H. Fuller, Aluminum and silica solubility in soils, *Soil Sci.* **121**:76–85 (1976).]

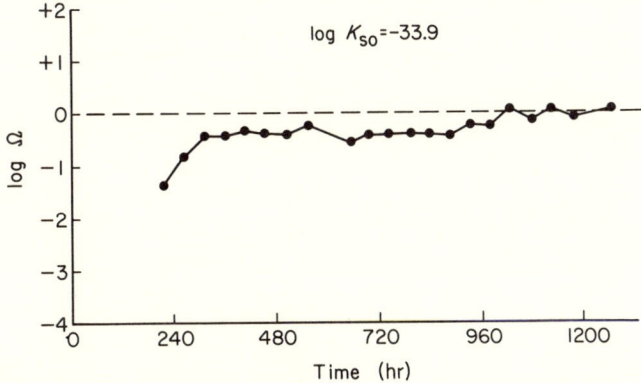

equilibrium. Used in this way, the *IAP* becomes a probe for determining whether dissolution equilibrium actually has been achieved. This kind of test is made by examining measured values of the *relative saturation*:

$$\Omega = IAP/K_{so} \qquad\qquad (5.11)$$

If $\Omega < 1$, the soil solution is "undersaturated"; if $\Omega > 1$, it is "supersaturated"; and when a dissolution reaction is at equilibrium, $\Omega = 1$. Near equilibrium, it is found typically that the rate of dissolution (or precipitation) can be expressed mathematically as some function of Ω. Figure 5.2 shows the approach of Ω from undersaturation to unit value in the soil solution of an Oxisol containing gibbsite. Equilibrium was reached after about 1000 h of reaction.

5.2 Activity-Ratio Diagrams

Investigations of the solubility of chemical elements in soils usually center on the question: Does a solid phase control the concentration of a particular element in the soil solution, and, if so, which solid phase is it? An answer to this question can be developed readily in terms of the dissolution equilibrium constant, K_{dis}, exemplified in Eqs. 5.4 and 5.5. The method of analyzing the problem leads to an *activity-ratio diagram*, which is constructed as follows.

1. Identify a set of solid phases that contain the chemical element of interest and may be controlling its solubility. Write a dissolution reaction for each solid, with the free ionic species of the element as one of the products. Be sure that the stochiometric coefficient of the free ion (metal or ligand) is 1.0.

2. Compile values of K_{dis} for the solid phases. Write an algebraic equation for each log K_{dis} in terms of log[activity] variables for the products and reactants in the corresponding dissolution reaction. Rearrange the equation to have log[(solid phase)/(free ion)] on the left side and all other log[activity] variables on the right side.

3. Choose an independent log[activity] variable against which log[(solid)/(free ion)] can be plotted for each solid phase. A typical example is pH $= -\log(H^+)$.

4. Select fixed values for all other log[activity] variables, corresponding to an assumed set of soil conditions. Use these values and that of log K_{dis} to develop a linear equation between log[(solid)/(free ion)] and the independent log[activity] variable for each solid phase. Plot all these equations on the same graph.

The construction of an activity-ratio diagram will be illustrated for the problem of calcium solubility in an arid-zone soil. The suite of solid phases to be examined includes anhydrite ($CaSO_4$), gypsum ($CaSO_4 \cdot 2H_2O$), and calcite

($CaCO_3$). Weathering reactions involving these minerals are in Eqs. 1.7, 2.8, 2.9b, and 5.2. The dissolution reaction for anhydrite is a special case of Eq. 5.9:

$$CaSO_4(s) = Ca^{2+}(aq) + SO_4^{2-}(aq) \tag{5.12}$$

with $\log K_{dis} = -4.38$ at 298.15 K (25°C) and 1 atm pressure. The dissolution reactions for gypsum and calcite are in Eqs. 5.2 and 2.9b, respectively, with $\log K_{dis} = -4.62$ (gypsum) and $+1.93$ (calcite). In the case of calcite, it is convenient to add to the reaction in Eq. 2.9b the carbonate species reactions (see Problem 6 in Chapter 4):

$$H^+(aq) + HCO_3^-(aq) = H_2CO_3^*(aq) \tag{5.13a}$$

$$H_2CO_3^* = CO_2(aq) + H_2O(\ell) \tag{5.13b}$$

for which $\log K$ equals 6.35 and 1.47, respectively. The resultant dissolution reaction is:

$$CaCO_3(s) + 2H^+(aq) = Ca^{2+}(aq) + CO_2(g) + H_2O(\ell) \tag{5.14}$$

with $\log K_{dis} = 1.93 + 6.35 + 1.47 = 9.75$.

For gypsum, the relation between K_{dis} and species activities is Eq. 5.4. The corresponding equation for $\log K_{dis}$ is:

$$\log K_{dis} = \log(Ca^{2+}) + \log(SO_4^{2-}) + 2\log(H_2O) - \log(gyp) \tag{5.15}$$

The activity ratio in this case is then:

$$\log[(gyp)/(Ca^{2+})] = -\log K_{dis} + \log(SO_4^{2-}) + 2\log(H_2O) \tag{5.16a}$$

In a similar way, one can derive expressions for the other two minerals:

$$\log[(anhy)/(Ca^{2+})] = -\log K_{dis} + \log(SO_4^{2-}) \tag{5.16b}$$

$$\log[(cal)/(Ca^{2+})] = -\log K_{dis} + 2\,pH + \log(CO_2) + \log(H_2O) \tag{5.16c}$$

The possible choices for the independent activity variable in this example are pH and the activity of SO_4^{2-}, $CO_2(g)$, or $H_2O(\ell)$. If pH is chosen, then fixed values of the other activity variables must be selected. The methods of chemical thermodynamics can be applied to show that the activity of $CO_2(g)$ is equal numerically to its pressure *in atmospheres* and that the activity of $H_2O(\ell)$ is equal to the relative humidity expressed as a decimal fraction [e.g., $(H_2O) = 0.6$ at 60% relative humidity]. In soils P_{CO_2} ranges between 3×10^{-4} and 10^{-2} atm (see Section 1.4 and Problem 7 in Chapter 4), whereas (H_2O) lies between 0.2 and 1.0, with the normal moisture range for plant growth corresponding to water activities >0.99. In arid-zone soils, $(SO_4^{2-}) \approx 3 \times 10^{-3}$ typically. This value along with $P_{CO_2} = 3 \times 10^{-4}$ and $(H_2O) = 1.0$ will be used in Eq. 5.16.

The "working form" of Eq. 5.16 for preparing an activity-ratio diagram is thus:

$$\log[(\text{gyp})/(Ca^{2+})] = 2.10 \tag{5.17a}$$

$$\log[(\text{anhy})/(Ca^{2+})] = 1.86 \tag{5.17b}$$

$$\log[(\text{cal})/(Ca^{2+})] = -13.27 + 2\ \text{pH} \tag{5.17c}$$

These linear relationships are graphed in Fig. 5.3 for pH values in the range 6.5–9.5, those typical of arid-zone soils. The interpretation of the three lines is based on a chemical thermodynamics principle, that *the solid phase that controls solubility is the one that produces the largest activity ratio for the free ionic species in solution.* In the present example, this governing concept means that the solubility-controlling, Ca-bearing mineral will be the one that has the *largest* value of $\log[(\text{solid})/(Ca^{2+})]$ at a given pH value under the conditions imposed on Eq. 5.16. Note that this criterion is equivalent to saying that (Ca^{2+}) will have the *smallest* value possible if the solid-phase activity has unit value. Thus the controlling Ca-bearing mineral should produce the smallest Ca^{2+} activity in the soil solution.

FIG. 5.3 Activity-ratio diagram for calcium solubility control in the calcite-gypsum-anhydrite system. Note that $(SO_4^{2-}) = 3 \times 10^{-3}$ for the solid lines and $(SO_4^{2-}) = 10^{-2}$ for the dashed lines.

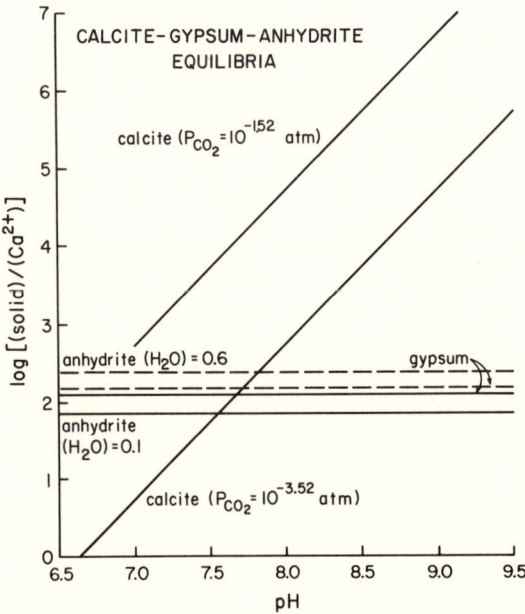

In respect to Eq. 5.17 and Fig. 5.3, the activity-ratio diagram indicates that, under the conditions imposed, gypsum would control Ca solubility for pH < 7.8, whereas calcite would control Ca solubility for pH > 7.8. The horizontal line for anhydrite always lies below that for gypsum, so only the hydrated calcium sulfate mineral is important when the water activity is near 1.0.

Soil solutions often equilibrate with CO_2 pressures well above that in the atmosphere because of biological activity (see Section 1.4). Therefore, it is of interest to examine the effect of increasing P_{CO_2} on Eq. 5.16c. If $P_{CO_2} = 3.2 \times 10^{-2}$ atm, about 100 times the atmospheric value, then Eq. 5.17c becomes

$$\log[(cal)/(Ca^{2+})] = -11.27 + 2 \text{ pH} \qquad (5.17d)$$

This equation is also plotted in Fig. 5.3. (There is no effect of P_{CO_2} on Eqs. 5.17a and 5.17b.) The result indicates that, at this elevated CO_2 pressure, calcite would control the solubility of Ca throughout the alkaline pH range.

A relative humidity of 60% for soil water is low but certainly possible in the surface horizon of an arid-zone profile. The effect of $(H_2O) = 0.6$ on Eqs. 5.17c and 5.17d is very small: the two corresponding lines in Fig. 5.3 drop vertically by only 0.2 log units. But the effect on gypsum and anhydrite is large if the sulfate ion activity is increased to 10^{-2} as a result of the lower water content accompanying the reduction in water activity. With $(H_2O) = 0.6$ and $(SO_4^{2-}) = 10^{-2}$ introduced, Eqs. 5.16a and 5.16b become

$$\log[(gyp)/(Ca^{2+})] = 2.18 \qquad (5.17e)$$

$$\log[(anhy)/(Ca^{2+})] = 2.38 \qquad (5.17f)$$

These two equations are plotted as dashed lines in Fig. 5.3. It is anhydrite now that would control Ca solubility up to pH 8 if $P_{CO_2} = 3 \times 10^{-4}$ atm. This result does not change if a larger sulfate ion activity is used in Eqs. 5.16a and 5.16b to represent a more concentrated soil solution. On the other hand, calcite will continue to control Ca solubility in the alkaline pH range if P_{CO_2} is increased to 3.2×10^{-2} atm.

This relatively simple example illustrates the usual development and interpretation of activity-ratio diagrams. It should be evident that "master variables" other than pH could be used (e.g., CO_2 pressure) and that the predictions made depend sensitively on the auxiliary conditions imposed, as well as the accuracy of the K_{dis} values selected.

5.3 Phosphate Fertilizer Reactions in Calcareous Soils

Suppose that a "neutral-reaction" phosphate fertilizer containing Ca-$HPO_4 \cdot 2H_2O$ (dicalcium phosphate dihydrate, or *brushite*) is applied to a calcareous soil. What solid phase is ultimately likely to control phosphate solubility in the soil? An answer to this important question has been found on the basis of experimental studies of the fate of phosphate fertilizers in soils. Depending on soil moisture conditions, there is a transformation of brushite to $CaHPO_4$ (dicalcium phosphate, or *monetite*), followed by a slow transformation to $Ca_8H_2(PO_4)_6 \cdot 5H_2O$ (*octacalcium phosphate*). Ultimately $Ca_{10}(OH)_2(PO_4)_6$ (*hydroxyapatite*) is expected, although octacalcium phosphate may persist for years if fertilizer is applied continually.

These phosphate transformations can be understood in terms of an activity-ratio diagram involving the four Ca phosphates and calcite. The relevant dissolution reactions are:

$$CaHPO_4 \cdot 2H_2O(s) = Ca^{2+}(aq) + HPO_4^{2-}(aq) + 2H_2O(\ell)$$

$$\log K_{dis} = -6.57 \qquad (5.18a)$$

$$CaHPO_4(s) = Ca^{2+}(aq) + HPO_4^{2-}(aq) \qquad \log K_{dis} = -6.90 \qquad (5.18b)$$

$$\tfrac{1}{6}Ca_8H_2(PO_4)_6 \cdot 5H_2O(s) + \tfrac{2}{3}H^+(aq) = \tfrac{4}{3}Ca^{2+}(aq) + HPO_4^{2-}(aq) + \tfrac{5}{6}H_2O(\ell)$$

$$\log K_{dis} = -3.28 \qquad (5.18c)$$

$$\tfrac{1}{6}Ca_{10}(OH)_2(PO_4)_6(s) + \tfrac{4}{3}H^+(aq) = \tfrac{5}{3}Ca^{2+}(aq) + HPO_4^{2-}(aq) + \tfrac{1}{3}H_2O(\ell)$$

$$\log K_{dis} = -2.38 \qquad (5.18d)$$

In this application, the free-ion activity of interest is (HPO_4^{2-}) and Eq. 5.18 have been arranged so that the stoichiometric coefficient of HPO_4^{2-} is 1.0, following the steps outlined in Section 5.2. The activity of Ca^{2+} in Eq. 5.18 is assumed to be controlled by calcite. Therefore, Eq. 5.14 with $\log K_{dis} = 9.75$ can be added to Eq. 5.18.

It will be assumed that the activities of $CaCO_3(s)$ and $H_2O(\ell)$ are 1.0. The HPO_4 activity ratios then can be expressed in logarithmic form showing a dependence on pH and P_{CO_2}. For brushite (DCPDH), the calculation runs as follows:

$$-6.57 = \log(Ca^{2+}) + \log(HPO_4^{2-}) - \log(DCPDH)$$

$$= 9.75 - 2\ pH - \log(CO_2) - \log[(DCPDH)/(HPO_4^{2-})]$$

and

$$\log[(DCPDH)/(HPO_4^{2-})] = 16.32 - \log P_{CO_2} - 2\ pH \qquad (5.19a)$$

Equation 5.16c [with $(CaCO_3) = (H_2O) = 1$] and $(CO_2) = P_{CO_2}$ were used to obtain Eq. 5.19a. In a similar fashion, one can derive expressions for the three other Ca phosphates:

$$\log[(DCP)/(HPO_4^{2-})] = 16.65 - \log P_{CO_2} - 2\ pH \qquad (5.19b)$$

$$\log[(OCP)/(HPO_4^{2-})] = 17.59 - \log P_{CO_2} - 2\ pH \qquad (5.19c)$$

$$\log[(HA)/(HPO_4^{2-})] = 21.24 - \log P_{CO_2} - 2\,pH \qquad (5.19d)$$

where DCP refers to monetite and obvious abbreviations have been used for the remaining two Ca phosphates.

Either pH or P_{CO_2} can be chosen as the "master variable" against which to plot $\log[(solid)/(HPO_4^{2-})]$ in an activity-ratio diagram based on Eq. 5.19. Regardless of which variable is chosen, it is apparent that Eq. 5.19 will plot as a series of four parallel lines, with that for HA at the top and that for DCPDH at the bottom of the graph. This relationship is illustrated in Fig. 5.4 for pH as the independent variable and $P_{CO_2} = 10^{-3.52}$ atm (appropriate for a surface soil). According to the criterion developed in Section 5.2 for interpreting activity-ratio diagrams, HA should control phosphate solubility at any pH in the alkaline range. This criterion, however, says nothing about the observed sequence of transformations leading from DCPDH to HA.

The parallel lines in Fig. 5.4 can be viewed as a sequence of HPO_4^{2-} activity "steps" in the sense that, at any fixed pH value, (HPO_4^{2-}) decreases as each line is traversed moving upward in the diagram. For example, at pH 7.5, (HPO_4^{2-}) equals successively $10^{-4.8}$, $10^{-5.2}$, $10^{-6.2}$, and $10^{-9.8}$ as the lines are crossed going from DCPDH to HA. This monotonic lowering of (HPO_4^{2-}) reflects the decreasing solubility of each phosphate solid and mimics the experimental sequence of solid-phase transformations described earlier. The formal description of this behavior is summarized in the *Gay-Lussac-Ostwald (GLO) Step Rule*, which can be stated as follows.

FIG. 5.4 Activity-ratio diagram for phosphate solubility control by Ca phosphates in a calcareous soil.

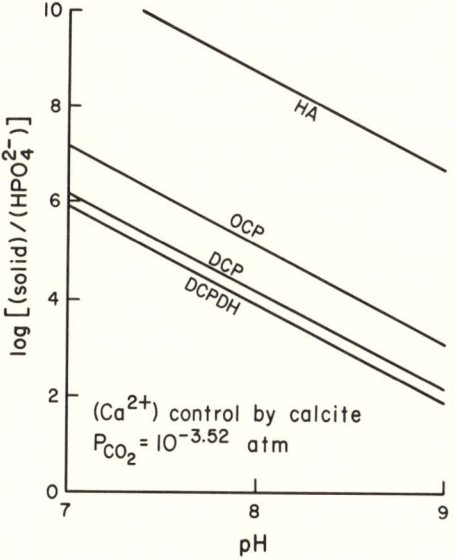

If the initial state of a soil is such that several solid phases can form potentially with a given ion, the solid phase that forms first will be the one for which the activity ratio is nearest above the initial value in the soil. Thereafter, the remaining accessible solid phases will form in order of increasing activity ratio, with the rate of formation of a solid phase in the sequence decreasing as its activity ratio increases. In an open system, any one of the solid phases may be maintained "indefinitely."

Applied to Fig. 5.4, the GLO Step Rule suggests that, if DCPDH is added to a calcareous soil, DCP (not HA) will form first by dissolution of DCPDH. Thereafter, DCP will dissolve and OCP will be formed, with this process occurring more slowly than the DCPDH $\rightarrow$ DCP transformation. Finally, OCP will slowly dissolve in favor of HA formation. This sequence of transformations is just what has been observed experimentally. In laboratory studies with Ca phosphate solutions maintained supersaturated with respect to OCP, but undersaturated with respect to DCPDH or DCP, OCP has been found to precipitate readily at a rate dependent on $\Omega = (Ca^{2+})^{4/3}(HPO_4^{2-})(H^+)^{-2/3}/K_{dis}$, the appropriate relative saturation variable. In field soils, continual fertilizer applications could maintain supersaturation with respect to OCP and thus stabilize this Ca phosphate in the soil for an indefinite period. The GLO Step Rule would predict this stability, as well as a very slow ultimate transformation of OCP to HA. This prediction, however, must be tempered by the fact that soluble phosphate or calcium complexes and plant uptake of phosphate could inhibit OCP formation, as could the precipitation of phosphate with other cations than Ca^{2+} in the soil solution.

5.4 Gypsum in Acid Soils

Gypsum ($CaSO_4 \cdot 2H_2O$) is applied to acid soils, like Ultisols and Oxisols, in order to enhance plant growth in them. The chemical effects of gypsum on these soils must derive ultimately from the dissolution reaction in Eq. 5.2, which releases calcium and sulfate ions into the soil solution. The calcium ions and, to a lesser extent, the sulfate ions enhance the plant nutrient status of the soil and replace nonessential, potentially hazardous ions, like Al^{3+} or OH^-, bound to soil particle surfaces. Typically, gypsum amendments do not change soil pH by > 0.3 log units, but they do result in much smaller concentrations of Al in the soil solution. This reduction in Al solubility can have a major positive impact on the fertility of acid soils.

One mechanism for the decrease in Al solubility following the application of gypsum is the precipitation of Al sulfate solid phases like jurbanite ($AlOHSO_4 \cdot 5H_2O$), basaluminite [$Al_4(OH)_{10}SO_4 \cdot 5H_2O$], or alunite [$KAl_3(OH)_6(SO_4)_2$], as described in Section 2.5. An activity-ratio diagram can help to indicate the soil chemical conditions under which these solid phases may form. The procedure for constructing this diagram follows the steps outlined in Section 5.2. The dissolution reactions for the three solid phases are:

$$AlOHSO_4 \cdot 5H_2O(s) + H^+(aq) = Al^{3+}(aq) + SO_4^{2-}(aq) + 6H_2O(\ell)$$

$$\log K_{dis} = -3.8 \qquad (5.20a)$$

$$\tfrac{1}{4}Al_4(OH)_{10}SO_4 \cdot 5H_2O(s) + \tfrac{5}{2}H^+(aq) = Al^{3+}(aq) + \tfrac{1}{4}SO_4^{2-}(aq) + \tfrac{15}{4}H_2O(\ell)$$

$$\log K_{dis} = 5.63 \qquad (5.20b)$$

$$\tfrac{1}{3}KAl_3(OH)_6(SO_4)_2(s) + 2H^+(aq)$$

$$= \tfrac{1}{3}K^+(aq) + Al^{3+}(aq) + \tfrac{2}{3}SO_4^{2-}(aq) + 2H_2O(\ell) \qquad \log K_{dis} = 0.2 \qquad (5.20c)$$

Since Al solubility is of interest, the expressions for $\log K_{dis}$ corresponding to the reactions in Eq. 5.20 are rearranged to calculate $\log[(solid)/(Al^{3+})]$:

$$\log[(jurbanite)/(Al^{3+})] = 3.8 + \log(SO_4^{2-}) + pH + 6\log(H_2O) \qquad (5.21a)$$

$$\log[(basaluminite)/(Al^{3+})] = -5.625 + \tfrac{1}{4}\log(SO_4^{2-}) + \tfrac{5}{2}pH + \tfrac{15}{4}\log(H_2O)$$

$$(5.21b)$$

$$\log[(alunite)/(Al^{3+})]$$

$$= -0.2 + \tfrac{1}{3}\log(SO_4^{2-}) + 2pH + \tfrac{1}{3}\log(K^+) + 2\log(H_2O) \qquad (5.21c)$$

There are four activity variables on the right side of Eq. 5.21 that can serve as the independent variable against which to plot the activity ratios. Since pH is not affected much by gypsum amendments, it and the activities of H_2O and K^+ will be given constant, representative values: $pH = 4.5$, $(H_2O) = 1.0$, and $(K^+) = 10^{-4}$. The log activity of SO_4^{2-} will be the "master variable" in this example.

With the fixed activities introduced, Eq. 5.21 become the linear relationships:

$$\log[(jurbanite)/(Al^{3+})] = 8.30 + \log(SO_4^{2-}) \qquad (5.22a)$$

$$\log[(basaluminite)/(Al^{3+})] = -5.62 + \tfrac{1}{4}\log(SO_4^{2-}) \qquad (5.22b)$$

$$\log[(alunite)/(Al^{3+})] = 7.47 + \tfrac{2}{3}\log(SO_4^{2-}) \qquad (5.22c)$$

These equations are plotted in Fig. 5.5. In order to interpret this activity-ratio diagram chemically, it is necessary to add some reference value of $\log[(solid)/(Al^{3+})]$ that represents the situation before gypsum application. Only then can it be ascertained whether the formation of an Al sulfate solid will lower the activity of Al^{3+}.

In acid soils, gibbsite $[Al(OH)_3]$ can be the solid phase that controls Al solubility. The dissolution reaction and corresponding expression for K_{dis} for

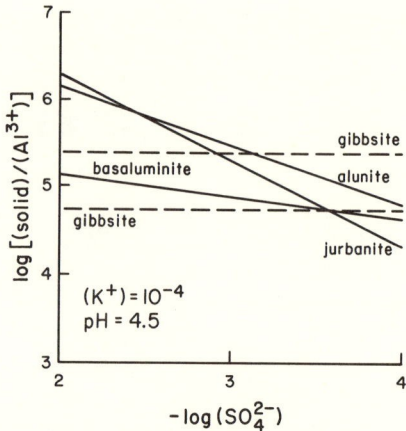

FIG. 5.5 Activity-ratio diagram for aluminum solubility control by Al sulfates in an acid soil. The lower dashed line represents "soil gibbsite."

gibbsite were given in Eqs. 5.3 and 5.5. Macrocrystalline gibbsite has a $*K_{so}$ value equal to 1.29×10^8 (see Eq. 5.8), but in acid soils this form of the mineral may not occur as often as some less perfectly crystalline precipitate does. Aged $Al(OH)_3$ in acidic soil environments has a $*K_{so}$ value equal to 5.89×10^8, whereas microcrystalline forms of the mineral have $*K_{so} = 2.24 \times 10^9$ and poorly crystallized surface precipitates may have $*K_{so} = 6 \times 10^{10}$. These larger $*K_{so}$ values imply that the activity of the less structurally perfect gibbsite is larger than that of macrocrystalline gibbsite, taken conventionally equal to 1.0. According to Eq. 5.8, the activity of aged "soil gibbsite" is related to that of macrocrystalline gibbsite through the equation:

$$(gibbsite)_{soil} = \frac{*K_{so}(soil)}{*K_{so}(lab)} \times (gibbsite)_{lab} = \frac{5.89 \times 10^8}{1.29 \times 10^8} = 4.6 \quad (5.23)$$

where "lab" refers to the macrocrystalline form that might be synthesized in the laboratory. This result is interpreted to mean that aged soil gibbsite has an activity nearly five times that of perfectly crystallized gibbsite. This larger activity is caused by smaller crystal size or structural imperfections that have caused the solid to become more soluble.

The activity-ratio equation that follows from Eq. 5.8 is:

$$\log[(gibbsite)/(Al^{3+})] = -\log *K_{so} + 3 \, pH + 3 \log(H_2O) \quad (5.24)$$

By convention, the solid-phase activity in Eq. 5.24 always refers to the pure macrocrystalline phase. Therefore, less than perfect crystallinity is represented by using $*K_{so}$ in the equation instead of $*K_{dis}$. Evidently, the first term on the right side of Eq. 5.24 can be either -8.11 or -8.77, corresponding to "lab" or "soil" gibbsite—both relatively well-crystallized minerals. These two choices are shown as dashed lines in Fig. 5.5. They are plots of Eq. 5.24 at pH = 4.5 and $H_2O = 1.0$; together they define the gibbsite "window" in the present application. Microcrystalline or poorly crystalline surface precipitates of $Al(OH)_3$ would produce lines below the lower dashed line in the figure (i.e., yet larger Al^{3+} activity).

Unamended acid soils have SO_4^{2-} activities near 10^{-4}. The activity-ratio diagram indicates that gibbsite should not convert to an Al sulfate solid under this condition (alunite is a possible exception, depending on the precise values of pH and K^+ activity). In gypsum-amended acid soils, $(SO_4^{2-}) \approx 10^{-2.5}$ is typical, and the diagram in Fig. 5.5 indicates that both jurbanite and alunite are more likely to control Al^{3+} activity than gibbsite; even basaluminite is more stable than "soil gibbsite." According to the GLO Step Rule, if gypsum were added to a soil with an Al^{3+} activity kept at $10^{-4.73}$ by gibbsite (lower dashed line), basaluminite should precipitate first, to be replaced eventually by alunite and jurbanite. If the initial activity were $10^{-5.39}$ (upper dashed line), only the latter two Al sulfates would be expected. Note that, for $(SO_4^{2-}) \approx 10^{-2.5}$, these two solid phases coexist and reduce (Al^{3+}) by about an order of magnitude. This kind of reduction in Al^{3+} activity has been observed in acid soils after gypsum application.

Figure 5.5 was constructed under the assumption that $(K^+) = 10^{-4}$. If alunite forms in a soil after gypsum application and coexists with gibbsite, however, the activity of $K^+(aq)$ may drop instead of remaining fixed. To see this point in detail, consider the expression for $\log(K^+)$ that develops after setting the right sides of Eqs. 5.21c and 5.24 equal [alunite–gibbsite coexistence at the same $Al^{3+}(aq)$ activity] with $*K_{so} = 1.29 \times 10^8$ for the gibbsite phase:

$$\log(K^+) = -23.73 + 3 \text{ pH} - 2 \log(SO_4^{2-}) + 3 \log(H_2O) \qquad (5.25)$$

Under the conditions pH = 4.5, $(SO_4^{2-}) = 10^{-2.5}$, and $(H_2O) = 1.0$, Eq. 5.25 predicts $(K^+) = 5.89 \times 10^{-6}$, which is considerably lower than $(K^+) = 10^{-4}$, a value representative of acid soil solutions. Thus the formation of alunite in equilibrium with gibbsite as a product of gypsum application could reduce the activity of K^+ without a reduction in Al^{3+} activity!

5.5 Clay Mineral Weathering

The Jackson–Sherman weathering scenario (Table 1.7) indicates that, as soil profiles are leached with fresh water, 2:1 layer-type clay minerals are replaced by 1:1 layer-type clay minerals, and ultimately by metal hydrous oxides. This sequence of clay mineral transformations can be represented by the successive dissolution reactions of smectite, kaolinite, and gibbsite (Tables 2.3 and 2.4).

The dissolution reactions of the smectite, beidellite (see Section 2.3), and kaolinite are exemplified by (with $H_4SiO_4^0$ representing silicic acid):

$$Mg_{0.328}[Si_{3.55}Al_{0.45}](Al_{1.41}Fe(III)_{0.385}Mg_{0.205})O_{10}(OH)_2(s)$$

$$+ 2.2H_2O(\ell) + 7.8H^+(aq) = 1.86Al^{3+}(aq) + 0.385Fe^{3+}(aq)$$

$$+0.533Mg^{2+}(aq) + 3.55H_4SiO_4^0(aq) \qquad \log K_{dis} = 9.6 \qquad (5.26)$$

$$Al_2Si_2O_5(OH)_4(s) + 6H^+(aq) = 2Al^{3+}(aq) + 2H_4SiO_4^0(aq) + H_2O(\ell)$$

$$\log K_{so} = 10.5 \qquad (5.27)$$

The solid-phase reactant in Eq. 5.26 is a beidellite with Mg^{2+} in the interlayer region (see also Eq. 2.7). The kaolinite in Eq. 5.27 has been assumed to be poorly crystallized—typical of intensive weathering conditions—with $K_{so} = 3.2 \times 10^{10}$. A well-crystallized kaolinite would dissolve with $K_{dis} = 1.6 \times 10^7$. The difference between these two dissolution equilibrium constants represents the kaolinite "window" in soils, as discussed in Section 5.4 for gibbsite.

Equations 5.26, 5.27, and 5.3 can be used to construct an activity-ratio diagram in respect to Al solubility as influenced by the leaching of silicic acid (see Section 1.5). The equations for $\log[(solid)/(Al^{3+})]$ are Eq. 5.24 (with $\log *K_{so} = 8.77$ for illustrative purposes) and the expressions:

$$\log[(kaolinite)/(Al^{3+})] = -5.25 + 3\ pH + \log(H_4SiO_4^0) + \tfrac{1}{2}\log(H_2O)$$

$$(5.28a)$$

$$\log[(Mg\text{-}beidellite)/(Al^{3+})]$$

$$= -5.16 + 0.207 \log(Fe^{3+}) + 0.287 \log(Mg^{2+})$$
$$+ 1.91 \log(H_4SiO_4^0) + 4.19\ pH - 1.18 \log(H_2O) \qquad (5.28b)$$

Note that Eq. 5.26 and its $\log K_{dis}$ values must be divided by 1.86, and that Eq. 5.27 and its $\log K_{so}$ value must be divided by 2, before Eq. 5.28 can be derived. If $(H_4SiO_4^0)$ is to be the "master variable," then pH, (H_2O), and the activities of Fe^{3+} and Mg^{2+} in the soil solution must be prescribed. Representative values are: pH = 5.6, $(H_2O) = 1.0$, $(Fe^{3+}) = 10^{-13}$, and $(Mg^{2+}) = 10^{-4}$. The resulting linear activity-ratio equations are then:

$$\log[(gibbsite)/(Al^{3+})] = 8.03 \qquad (5.29a)$$

$$\log[(kaolinite)/(Al^{3+})] = 11.55 + \log(H_4SiO_4^0) \qquad (5.29b)$$

$$\log[(Mg\text{-}beidellite)/(Al^{3+})] = 14.47 + 1.91 \log(H_4SiO_4^0) \qquad (5.29c)$$

The activity-ratio diagram resulting from Eq. 5.29 is shown in Fig. 5.6.

The effect of leaching is represented in the activity-ratio diagram by moving from left to right along the x axis. Amorphous silica supports a high $H_4SiO_4^0$

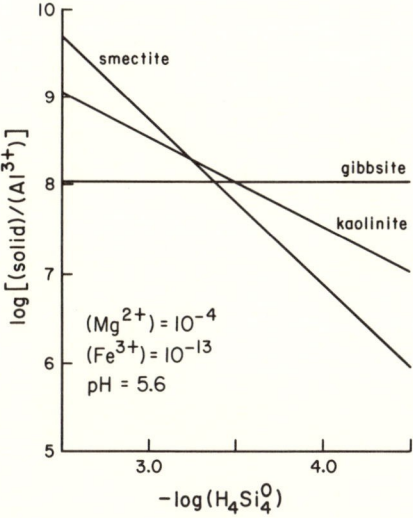

FIG. 5.6 Activity-ratio diagram for aluminum solubility control by clay minerals and gibbsite.

activity near $10^{-2.7}$. Under this condition, which might represent the intensive weathering of primary silicates in a soil, Fig. 5.6 indicates that smectite (Mg-beidellite) is the most stable solid phase with respect to solubility controls on Al. As leaching and loss of silica proceed, the silicic acid activity will drop (as the concentration of $H_4SiO_4^0$ drops), and when $(H_4SiO_4^0)$ approaches 10^{-4}—the silicic acid activity supported by quartz—gibbsite becomes the most stable Al-bearing solid phase. This progression agrees with field observations as summarized in the Jackson–Sherman weathering stages. Note that the rather narrow range of $(H_4SiO_4^0)$ over which kaolinite is stable derives in part from the choice of a poorly crystallized solid phase, represented by a large value of K_{so}. If a well-crystallized form of the mineral had been assumed instead, $\log K_{so}$ in Eq. 5.27 would have been 7.2, not 10.5, and the kaolinite line in Fig. 5.6 would have been 1.7 log units higher, with stability for $(H_4SiO_4^0)$ values between 2×10^{-4} and 8×10^{-4}. This kind of variability, together with the typical $(H_4SiO_4^0) \approx 8 \times 10^{-4}$ in slightly acidic soils, suggests that smectite and kaolinite will often coexist in active weathering environments.

FOR FURTHER READING

W. L. Lindsay, *Chemical Equilibria in Soils*, Wiley, New York, 1979. A thorough discussion of the preparation and use of activity-ratio diagrams to describe solid-phase solubility for 20 chemical elements of importance to soils.

A. E. Martell and R. M. Smith, *Critical Stability Constants*, Plenum Press, New York, 1974–1982. The five volumes of this exhaustive survey of chemical thermodynamic data provide critically selected stability constants for complexes and solubility product constants for macrocrystalline solids.

W. Stumm and J. J. Morgan, *Aquatic Chemistry*, Wiley, New York, 1981. Chapter 5 of this classic text should be read for discussions of all aspects of mineral solubility: kinetics of precipitation and dissolution, ion activity product, temperature and pressure effects on solubility, coprecipitation, activity-ratio diagrams, and solubility product constant data.

The following two articles give useful working discussions of the Gay-Lussac-Ostwald Step Rule and its applications.

B. S. Hemingway, Gibbs free energy of formation of Bayerite, Nordstrandite, $Al(OH)^{2+}$, and $Al(OH)_2^+$, aluminum mobility, and the formation of bauxites and laterites, *Adv. Phys. Geochem.* **2**:285–316 (1982).

G. Sposito, Chemical models of weathering in soils, in J. I. Drever (ed.), *The Chemistry of Weathering*, pp. 1–18. D. Reidel, Dordrecht, The Netherlands, 1985.

PROBLEMS

The more difficult problems are indicated by an asterisk.

1. The rate of dissolution of albite ($NaAlSi_3O_8$) for pH < 6 can be described with Eq. 5.1, where A is aluminum and $k = 10^{-9} (H^+)$ mol s^{-1}. Calculate the dissolution rates at pH 4.0 ("acid rain") and 5.6. Compare the expected residence times (see Problem 2 in Chapter 1) of albite in soil at the two pH values (i.e., calculate the ratio of residence times).

2. How many times larger is the rate of dissolution of dolomite [$CaMg(CO_3)_2$] at pH 5 than at pH 7?

*3. The rates of dissolution of pyroxenes (Table 2.2) for pH < 6 typically follow Eq. 5.1, with k proportional to the proton activity raised to a fractional power in the range 0.5–0.8 and A represented by Ca or Mg. This kinetics behavior has led to the conclusion that "decreases in pH of the soil solution are partially compensated by increases in base cation weathering, but not totally." Explain how this conclusion depends on the fractional power (as opposed to the power 1.0) dependence of the dissolution rate on (H^+).

   (*Hint:* Use Eq. 5.1 to express the logarithm of the dissolution rate.)

4. Near equilibrium, the rate of precipitation of calcite is proportional to $\Omega - 1$, where Ω is defined by Eq. 5.11. How many times larger is the rate of precipitation of calcite when the IAP is 10^{-7} than when it is 10^{-8}? (For calcite, $\log K_{dis} = -8.48$.)

*5. Typically the distribution coefficient for Ca^{2+} in the soil solutions of arid-zone soils is about 0.75 (see Section 4.3). Given this information and the data in Problem 6 of Chapter 4, calculate the IAP for calcite in a soil solution with a pH of 8, an electrolytic conductivity of 2.5 dS m^{-1}, an HCO_3^- concentration of 1 mol m^{-3}, and a total Ca concentration of 5 mol m^{-3}.

(*Answer:* $IAP = \gamma_{Ca^{2+}}[Ca^{2+}]\gamma_{CO_3^2}[CO_3^{2-}] = 4.4 \times 10^{-9}$)

6. The solubility of an element in the soil solution is usually equated experimentally with its total concentration. Given that the distribution coefficient of Ca^{2+} in a soil solution is 0.75 and that its activity coefficient is 0.5 L mol^{-1}, calculate the solubility of Ca in moles per cubic meter when controlled by calcite at pH 8 according to Eq. 5.17c. Repeat the calculation for gypsum control according to Eq. 5.17a.

7. As indicated in Section 1.3 (Table 1.5), cadmium may coprecipitate with calcite to form a solid solution of $CdCO_3$ (otavite) and $CaCO_3$. When this kind of mixed-solid formation occurs, the activity of $CdCO_3(s)$ is not 1.0, but instead is equal approximately to the molar ratio of Cd to Ca in the mixture. Given that $\log K_{dis} = -11.2$ for $CdCO_3(s)$, calculate the corresponding $\log K_{so}$ for a coprecipitate of otavite and calcite containing 5 mol % $CdCO_3$. Show that the activity of $Cd^{2+}(aq)$ produced in the soil solution by this mixed solid is 1/20 that which would be produced by pure otavite under the same conditions of temperature, pressure, and soil solution composition.

8. Construct an activity-ratio diagram for Ca solubility control by gypsum and calcite using $\log P_{CO_2}$ as the "master variable." The fixed conditions are: pH = 8, $(SO_4^{2-}) = 10^{-3}$, and $(H_2O) = 1.0$.

9. Construct an activity-ratio diagram analogous to Fig. 5.5, but with pH as the "master variable" and $(SO_4^{2-}) = 2.5 \times 10^{-3}$. Choose pH values in the range 3.5–5.5.

10. A soil is irrigated with wastewater containing 0.38 mol m^{-3} F, which produces a F$^-$ activity of 3×10^{-5} in the soil solution at pH 8. Given the dissolution reaction of fluorite (CaF_2),

$$CaF_2(s) = Ca^{2+}(aq) + F^-(aq) \qquad \log K_{dis} = -9.8$$

develop an activity-ratio diagram like Fig. 5.3 to determine whether fluorite would supersede gypsum or calcite as the solid phase controlling Ca solubility in the soil.

11. Combine Eqs. 5.16a and 5.16b to calculate the activity of water at which gypsum and anhydrite are in equilibrium (i.e., have the same Ca^{2+} activity). Use the result to verify the statement made following Eq. 5.17f, that anhydrite is favored over gypsum at $(H_2O) = 0.6$ regardless of the activity of $SO_4^{2-}(aq)$.

*12. The weathering of anorthite to form calcite and montmorillonite is described by the chemical reaction in Eq. 2.8. The rate of this reaction is thought to be limited by unfavorable kinetics of calcite precipitation, such that the activity of $Ca^{2+}(aq)$ remains larger than what K_{so} for calcite would predict for a given activity of $CO_3^{2-}(aq)$. This hypothesis assumes that the activity of $Ca^{2+}(aq)$ in equilibrium with anorthite is larger than that in the presence of calcite. Check the assumption by preparing an activity-ratio diagram for Ca solubility control by the two minerals following the method used to create Fig. 5.3. The dissolution reaction for anorthite is:

$$CaAl_2Si_2O_8(s) + 8H^+(aq) = Ca^{2+}(aq) + 2Al^{3+}(aq) + 2Si(OH)_4^0(aq)$$

$$\log K_{dis} = 23$$

Assume that Eq. 5.28b can be used to express the pH dependence of $\log(Al^{3+})$ for $(Si(OH)_4^0) = 10^{-4}$: $\log(Al^{3+}) = 16.6 - 4.2\,pH$.
(*Hint:* For anorthite, $\log[(solid)/(Ca^{2+})] = 2.2 - 0.4\,pH$.)

13. Suppose that Pb is carried into an acid soil as a solute dissolved in rainwater. Lead phosphates are often thought to be the solid phases controlling Pb solubility in acid soils, the two most important minerals being tertiary lead orthophosphate $[Pb_3(PO_4)_2]$ and chloropyromorphite $[Pb_5(PO_4)_3Cl]$. The dissolution reactions for these two solid phases are:

$$Pb(PO_4)_{\frac{2}{3}}(s) + \tfrac{4}{3}H^+(aq) = Pb^{2+}(aq) + \tfrac{2}{3}H_2PO_4^-(aq) \qquad \log K_{dis} = -1.80$$

$$Pb(PO_4)_{\frac{3}{5}}Cl_{\frac{1}{5}}(s) + \tfrac{6}{5}H^+(aq) = Pb^{2+}(aq) + \tfrac{3}{5}H_2PO_4^-(aq) + \tfrac{1}{5}Cl^-(aq)$$

$$\log K_{dis} = -5.06$$

Prepare an activity-ratio diagram for Pb solubility control by these two minerals. Use $(H_2PO_4^-) = 10^{-6}$ and $(Cl^-) = 10^{-3}$ as fixed conditions. Which solid phase is expected to control solubility? Does the conclusion change if $(Cl^-) = 10^{-5}$?

14. The dissolution reaction in Eq. 5.3 leads to $\log *K_{dis}$ values of 8.77 for "soil gibbsite," 9.35 for microcrystalline gibbsite, and 10.8 for amorphous $Al(OH)_3$. Prepare an activity-ratio diagram for Al solubility control by these three solid phases and use the GLO Step Rule to explain why amorphous $Al(OH)_3$ is likely to be the first solid phase of the three formed by precipitation of Al from mineral weathering in the pH range 4–7.

*15. The transformation of anorthite to montmorillonite and calcite (Eq. 2.8) is favored by $Si(OH)_4^0$ activities near 10^{-3} and pH values near 8.5. In calcareous soils, however, it is observed often that gibbsite forms instead of smectite when anorthite dissolves incongruently. Use Eqs. 5.24 and 5.28 to construct an activity-ratio diagram like Fig. 5.6 at pH 8, then apply the GLO Step Rule to explain why, when anorthite dissolves, gibbsite may form before smectite.

(*Hint:* Assume that weathering initially produces a large Al^{3+} activity in the soil solution.)

6

Electrochemical Phenomena

6.1 The Concept of pE

An *oxidation–reduction* (or *redox*) reaction is a chemical reaction in which electrons are transferred completely from one species to another. The chemical species that loses electrons in this charge transfer process is called *oxidized*, whereas the one receiving electrons is called *reduced*. For example, in the reaction involving iron species:

$$FeOOH(s) + 3H^+(aq) + e^-(aq) = Fe^{2+}(aq) + 2H_2O(\ell) \qquad (6.1)$$

the solid phase, goethite (Table 2.4 and Fig. 2.5), on the left side is the oxidized species, and $Fe^{2+}(aq)$ on the right side is the reduced species.

Equation 6.1 is a *reduction half-reaction*, in which an electron in aqueous solution, denoted $e^-(aq)$, serves as one of the reactants. This species, like the proton in aqueous solution, is understood in a formal sense to participate in charge transfer processes. The overall redox reaction in a soil always must be the combination of two reduction half-reactions, such that the species $e^-(aq)$ does not appear explicitly. Equation 6.1, for example, could be combined ("coupled") with the inverse of a half-reaction involving carbon species:

$$\tfrac{1}{2}CO_2(g) + \tfrac{1}{2}H^+(aq) + e^-(aq) = \tfrac{1}{2}CHO_2^-(aq) \qquad (6.2)$$

to cancel the aqueous electron and represent the reduction of iron via the oxidation of formate, CHO_2^-, in the soil solution:

$$FeOOH(s) + \tfrac{1}{2}CHO_2^-(aq) + \tfrac{5}{2}H^+(aq)$$
$$= Fe^{2+}(aq) + \tfrac{1}{2}CO_2(g) + 2H_2O(\ell) \qquad (6.3)$$

The aqueous electron is a very useful conceptual device for describing the redox status of soils, just as the aqueous proton is so useful for describing the acid–base status of soils. Soil acidity is expressed quantitatively by the negative common logarithm of the free-proton activity, the pH value. Similarly, soil oxidizability can be expressed by the negative common logarithm of the free-electron activity, *the pE value*:

$$pE \equiv -\log(e^-) \qquad (6.4)$$

Large values of pE favor the existence of electron-poor (i.e., oxidized) species, just as large values of pH favor the existence of proton-poor species (i.e., bases). Small values of pE favor electron-rich, or reduced, species, just as small values of pH favor proton-rich species, acids. Unlike pH, however, pE can take on negative values. This minor difference results from the separate conventions established for the interpretation of pH and pE measurements made with electrochemical cells (see Section 6.5).

The range of pE in soils is indicated by the shaded portion of the graph in Fig. 6.1. This graph, known as a *pE–pH diagram*, shows the domains of

FIG. 6.1 A pE–pH diagram showing the domain accessible to microorganisms (dashed perimeter) and that observed in soils (shaded area, with experimental data shown as points). [Based on data compiled by L. G. M. Baas Becking, I. R. Kaplan, and D. Moore, Limits of the natural environment in terms of pH and oxidation–reduction potential, *J. Geol.* **68**:243–284 (1960).]

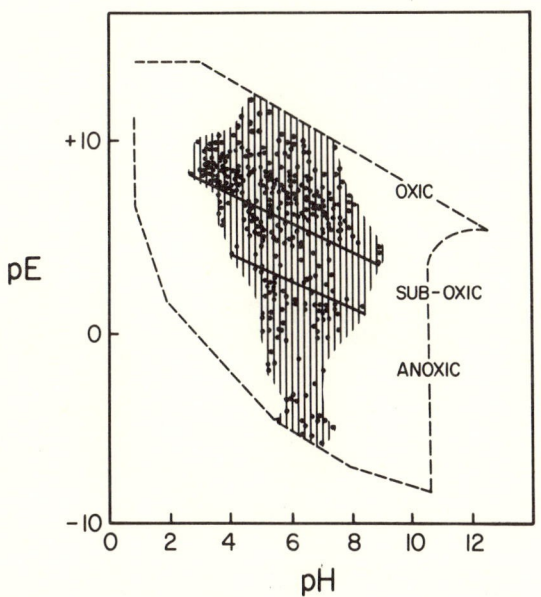

electron and proton activity that have been observed in soil environments worldwide. The largest pE value found is just below $+13.0$ and the smallest is near -6.0. This pE range can be divided into three parts that correspond to *oxic* soils (pE $> +7$ at pH 7), *suboxic* soils ($+2 < $ pE $< +7$ at pH 7), and *anoxic* soils (pE $< +2$ at pH 7). Suboxic soils differ from oxic soils in having pE values low enough to deplete $O_2(g)$, but not low enough to deplete sulfate ions (see Table 6.1).

The dashed perimeter of the irregular polygon in Fig. 6.1 encloses the pE–pH domain in which the microorganisms that catalyze redox reactions—principally bacteria—can flourish. These microorganisms increase the rate at which reactions like those in Eqs. 6.1 and 6.2 occur in soils. Figure 6.1 shows that the pE–pH conditions under which microorganisms occur include those observed in soils and, therefore, that *soil environments can support the microbial catalysis of redox reactions that may take place in them.* No "sterile" pE–pH domains are found in soils.

The most important chemical elements affected by soil redox reactions are C, N, O, S, Mn, and Fe. In contaminated soils, this list would grow to include As, Se, Cr, Hg, and Pb. If a soil is behaving effectively like a closed system (e.g., a flooded soil under stagnant ponding) and abundant sources of carbon and energy are available to support microbially mediated catalysis, there is a well-defined sequence of reduction of the inorganic elements among the principal six (see also Section 6.4). This sequence is illustrated in Table 6.1 by representative reduction half-reactions and the range of pE over which the reductions are *initiated* in neutral soils (pE_{init}). In each case, it may be imagined that the reduction half-reaction is coupled with the oxidation of organic

TABLE 6.1 The sequence of reduction reactions in neutral soils

Reduction half-reaction	Range of pE_{init}
$\frac{1}{4}O_2(g) + H^+(aq) + e^-(aq) = \frac{1}{2}H_2O(\ell)$	5.0–11.0
$\frac{1}{2}NO_2^-(aq) + H^+(aq) + e^-(aq) = \frac{1}{2}NO_2^-(aq) + \frac{1}{2}H_2O(\ell)$ $\frac{1}{5}NO_3^-(aq) + \frac{6}{5}H^+(aq) + e^-(aq) = \frac{1}{10}N_2(g) + \frac{3}{5}H_2O(\ell)$ $\frac{1}{8}NO_3^-(aq) + \frac{5}{4}H^+(aq) + e^-(aq) = \frac{1}{8}NH_4^+(aq) + \frac{3}{8}H_2O(\ell)$	3.4–8.5
$\frac{1}{2}MnO_2(s) + 2H^+(aq) + e^-(aq) = \frac{1}{2}Mn^{2+}(aq) + H_2O(\ell)$	3.4–6.8
$Fe(OH)_3(s) + 2H^+(aq) + e^-(aq) = Fe^{2+}(aq) + 3H_2O(\ell)$ $FeOOH(s) + 2H^+(aq) + e^-(aq) = Fe^{2+}(aq) + 2H_2O(\ell)$	1.7–5.0
$\frac{1}{8}SO_4^{2-}(aq) + \frac{9}{8}H^+(aq) + e^-(aq) = \frac{1}{8}HS^-(aq) + \frac{1}{2}H_2O(\ell)$ $\frac{1}{4}SO_4^{2-}(aq) + \frac{5}{4}H^+(aq) + e^-(aq) = \frac{1}{8}S_2O_3^-(aq) + \frac{5}{8}H_2O(\ell)$ $\frac{1}{8}SO_4^{2-}(aq) + \frac{5}{4}H^+(aq) + e^-(aq) = \frac{1}{8}H_2S(aq) + \frac{1}{2}H_2O(\ell)$	-2.5–0.0

matter, typified by the inverse of the reduction of CO_2 to glucose:

$$\tfrac{1}{4}CO_2(g) + H^+(aq) + e^-(aq) = \tfrac{1}{24}C_6H_{12}O_6(aq) + \tfrac{1}{4}H_2O(\ell) \qquad (6.5)$$

or the reduction of CO_2 to formate (Eq. 6.2).

As the pE of the soil solution drops below $+11.0$, enough electrons become available to reduce $O_2(g)$ to $H_2O(\ell)$, as indicated in Table 6.1. Below pE 5, oxygen is not stable in neutral soils. Above pE 5, it is consumed in the respiration processes of aerobic microorganisms. As the pE decreases below 8, electrons become available to reduce NO_3^-. This reduction is catalyzed by nitrate respiration (i.e., NO_3^- serving as a biochemical electron acceptor like O_2) involving bacteria that ultimately excrete NO_2^-, N_2, N_2O, or NH_4^+. *Denitrification* is the special case of nitrate respiration in which $N_2(g)$ and other nitrogenous gases are produced.

As the soil pE value drops into the range 7 to 5, electrons become plentiful enough to support the reduction of Fe and Mn in solid phases. Iron reduction does not occur until O_2 and NO_3^- are depleted, but Mn reduction can be initiated in the presence of nitrate. *Thus Fe and Mn reduction are characteristic of a suboxic soil environment.* As the pE value decreases below $+2$, a soil becomes anoxic and, when pE < 0, electrons are available for sulfate reduction catalyzed by a variety of anaerobic bacteria. Typical products in aqueous solution are H_2S, bisulfide (HS^-), or thiosulfate ($S_2O_3^-$) ions, as indicated in Table 6.1.

The chemical reaction sequence for the reduction of O, N, Mn, Fe, and S induced by changes in pE is also a microbial ecology sequence for the biological catalysts that mediate the reactions. Aerobic microorganisms that utilize O_2 to oxidize organic matter do not function below pE 5. Denitrifying bacteria thrive in the pE range between $+10$ and 0, for the most part. Sulfate-reducing bacteria do not grow at pE values above $+2$. These examples show that pE–pH diagrams portray domains of stability for both chemical and microbial species.

6.2 Redox Reactions

Redox reactions that go to equilibrium can be described in terms of equilibrium constants analogously to the approach used in Chapter 5 for dissolution reactions. The only new feature is the need to account for electron transfer. This is done by associating oxidation numbers with oxidized or reduced species and by careful balancing of the overall redox reaction in terms of reduction half-reactions, as explained in Special Topic 3, at the end of this chapter. A list of important reduction half-reactions and their equilibrium constants is provided in Table 6.2. These equilibrium constants have the same formal meaning as those discussed in Chapters 4 and 5, even though the

TABLE 6.2 Some important reduction half-reactions

Reduction half-reaction	$\log K_{298}$
$\frac{1}{4}O_2(g) + H^+(aq) + e^-(aq) = \frac{1}{2}H_2O(\ell)$	20.8
$H^+(aq) + e^-(aq) = \frac{1}{2}H_2(g)$	0.0
$\frac{1}{2}NO_3^-(aq) + H^+(aq) + e^-(aq) = \frac{1}{2}NO_2^-(aq) + \frac{1}{2}H_2O(\ell)$	14.1
$\frac{1}{4}NO_3^-(aq) + \frac{5}{4}H^+(aq) + e^-(aq) = \frac{1}{8}N_2O(aq) + \frac{5}{8}H_2O(\ell)$	18.9
$\frac{1}{5}NO_3^-(aq) + \frac{6}{5}H^+(aq) + e^-(aq) = \frac{1}{10}NO_2(aq) + \frac{3}{5}H_2O(\ell)$	21.1
$\frac{1}{8}NO_3^-(aq) + \frac{5}{4}H^+(aq) + e^-(aq) = \frac{1}{8}NH_4^+(aq) + \frac{3}{8}H_2O(\ell)$	14.9
$\frac{1}{2}MnO_2(s) + 2H^+(aq) + e^-(aq) = \frac{1}{2}Mn^{2+}(aq) + H_2O(\ell)$	20.7
$\frac{1}{2}MnO_2(s) + \frac{1}{2}HCO_3^-(aq) + \frac{3}{2}H^+(aq) + e^-(aq) = \frac{1}{2}MnCO_3(s) + H_2O(\ell)$	20.2
$Fe(OH)_3(s) + 3H^+(aq) + e^-(aq) = Fe^{2+}(aq) + 3H_2O(\ell)$	16.4
$FeOOH(s) + 3H^+(aq) + e^-(aq) = Fe^{2+}(aq) + 2H_2O(\ell)$	11.3
$\frac{1}{2}Fe_3O_4(s) + 4H^+(aq) + e^-(aq) = \frac{3}{2}Fe^{2+}(aq) + 2H_2O(\ell)$	14.9
$\frac{1}{2}Fe_2O_3(s) + 3H^+(aq) + e^-(aq) = Fe^{2+}(aq) + \frac{3}{2}H_2O(\ell)$	11.1
$\frac{1}{4}SO_4^{2-}(aq) + \frac{5}{4}H^+(aq) + e^-(aq) = \frac{1}{8}S_2O_3^{2-}(aq) + \frac{5}{8}H_2O(\ell)$	4.9
$\frac{1}{8}SO_4^{2-}(aq) + \frac{9}{8}H^+(aq) + e^-(aq) = \frac{1}{8}HS^-(aq) + \frac{1}{2}H_2O(\ell)$	4.3
$\frac{1}{8}SO_4^{2-}(aq) + \frac{5}{4}H^+(aq) + e^-(aq) = \frac{1}{8}H_2S(aq) + \frac{1}{2}H_2O(\ell)$	5.1
$\frac{1}{2}CO_2(g) + \frac{1}{2}H^+(aq) + e^-(aq) = \frac{1}{2}CHO_2^-(aq)$	−3.8
$\frac{1}{4}CO_2(g) + \frac{7}{8}H^+(aq) + e^-(aq) = \frac{1}{8}C_2H_3O_2^-(aq) + \frac{1}{4}H_2O(\ell)$	1.2
$\frac{1}{4}CO_2(g) + \frac{1}{12}NH_4^+(aq) + \frac{11}{12}H^+(aq) + e^-(aq) = \frac{1}{12}C_3H_4O_2NH_3(aq) + \frac{1}{3}H_2O(\ell)$	0.8
$\frac{1}{4}CO_2(g) + H^+(aq) + e^-(aq) = \frac{1}{24}C_6H_{12}O_6(aq) + \frac{1}{4}H_2O(\ell)$	−0.2
$\frac{1}{8}CO_2(g) + H^+(aq) + e^-(aq) = \frac{1}{8}CH_4(g) + \frac{1}{4}H_2O(\ell)$	2.9

reactions to which they refer contain the aqueous electron. The reason for this is the convention in which the reduction of the proton (the second reaction listed in Table 6.2) is *defined* to have $\log K = 0$. Thus every half-reaction in Table 6.2 may be combined with the inverse of the proton reduction reaction to cancel $e^-(aq)$ while leaving $\log K$ for the half-reaction completely unchanged numerically. In this sense, each half-reaction in Table 6.2 is a formal way of expressing an overall redox reaction which always includes the oxidation of $H_2(g)$.

The $\log K$ data in Table 6.2 can be combined in the usual way to calculate $\log K$ for an overall redox reaction. Consider, for example, the combination of Eqs. 6.1 qnd 6.2 to produce Eq. 6.3. According to Table 6.2, the *reduction* of goethite has $\log K = 11.3$, and the *oxidation* of formate has $\log K = 3.8$. It follows that the reduction of goethite by formate oxidation has $\log K = 11.3 + 3.8 = 15.1$. This equilibrium constant can be expressed in terms of activities related to Eq. 6.3:

$$K = \frac{(Fe^{2+})(CO_2)^{\frac{1}{4}}(H_2O)^2}{(FeOOH)(H^+)^{\frac{5}{2}}(CHO_2^-)^{\frac{1}{2}}} = 10^{15.1} \tag{6.6}$$

If the activities of goethite and water are set equal to 1.0, and the usual expressions for the activities of $CO_2(g)$ and $H^+(aq)$ are used (Section 5.2), then

Eq. 6.6 can be written in the form:

$$(Fe^{2+})P_{CO_2}^{1/2}10^{5/2pH}/(CHO_2^-)^{\frac{1}{4}} = 10^{15.1} \tag{6.7}$$

At pH 6 and under a CO_2 pressure of $10^{-3.52}$ atm, this equation reduces to the simpler expression:

$$\frac{(Fe^{2+})}{(CHO_2^-)^{\frac{1}{4}}} = 10^{1.86} \tag{6.8}$$

Equation 6.8 leads one to conclude that the equilibrium state for the redox reaction in Eq. 6.3 requires the activity of Fe^{2+} in the soil solution to be about 70 times the square root of the activity of formate in the soil solution. For example, if $(Fe^{2+}) = 10^{-7}$, then Eq. 6.8 predicts that $(CHO_2^-) = 2 \times 10^{-18}$. This result shows that formate would be depleted entirely from the soil solution after goethite reduction and dissolution.

The reduction half-reactions in Table 6.2 can also be used individually to predict the ranges of pE and pH over which one redox species or another predominates. Nearly all half-reactions are special cases of the general reaction:

$$mA_{ox} + nH^+(aq) + e^-(aq) = pA_{red} + qH_2O(\ell) \tag{6.9}$$

where A is some chemical species in any phase [e.g., $CO_2(g)$ or $Fe^{2+}(aq)$] and "ox" or "red" designates oxidized or reduced species, respectively. The equilibrium constant for this general half-reaction in Eq. 6.9 is:

$$K = \frac{(A_{red})^p(H_2O)^q}{(A_{ox})^m(H^+)^n(e^-)} \tag{6.10}$$

This equation can be rearranged to provide an expression for calculating pE or pH in terms of other log activity variables. Consider, for example, the reduction of sulfate to bisulfide:

$$\tfrac{1}{8}SO_4^{2-}(aq) + \tfrac{9}{8}H^+(aq) + e^-(aq) = \tfrac{1}{8}HS^-(aq) + \tfrac{1}{2}H_2O(\ell) \tag{6.11}$$

for which $\log K = 4.3$. Under the assumption that $(H_2O) = 1.0$, Eq. 6.10 becomes in this case:

$$K = \frac{(HS^-)^{\frac{1}{8}}}{(SO_4^{2-})^{\frac{1}{8}}(H^+)^{\frac{9}{8}}(e^-)} \tag{6.12}$$

In logarithmic form, Eq. 6.12 can be rearranged to give an expression for $pE = -\log(e^-)$:

$$pE = 4.3 - \tfrac{9}{8}pH + \tfrac{1}{8}\log\left[\frac{(SO_4^{2-})}{(HS^-)}\right] \tag{6.13}$$

Suppose a soil solution was at pH 7 and contained sulfate and bisulfide ions at equal activities. The corresponding pE would be $4.3 - 9(\frac{7}{8}) + 0 = -3.6$, according to Eq. 6.13. Alternatively, one could use $pE_{init} = -2.5$ for sulfate reduction (Table 6.1) to calculate the sulfate–bisulfide activity ratio when reduction begins at pH 7:

$$\log\left[\frac{(SO_4^{2-})}{(HS^-)}\right] = 8(pE - 4.3 + \tfrac{9}{8}pH) = 8(-2.5 - 4.3 + 7.9) = 8.8 \tag{6.14}$$

If $(SO_4^{2-}) = 10^{-3}$ when sulfate reduction begins, then the activity of bisulfide ions would be about 10^{-12}. The calculation illustrated in Eq. 6.14 can be done for any pE and pH values of interest.

If the pE value is known, Eq. 6.12 can be solved in logarithmic form to calculate pH instead. For the case of sulfate reduction, Eq. 6.13 can be rearranged to derive the expression:

$$pH = 3.8 - \tfrac{8}{9}pE - \tfrac{1}{9}\log\left[\frac{(SO_4^{2-})}{(HS^-)}\right] \tag{6.15}$$

If $pE = -2$, the pH value at which sulfate and bisulfide ion activities are the same is equal to $3.8 + 1.8 - 0 = 5.6$. Note that increasing the bisulfide ion activity relative to that of sulfate at fixed pE would increase the pH, according to Eq. 6.15. This trend is an example of a general feature of reduction half-reactions represented by Eq. 6.9. *The formation of reduced species almost always results in proton consumption.* Thus each reduction half-reaction in Table 6.2 represents a mechanism by which free protons can be removed from the soil solution. Reduction is therefore an important way in which excess soil acidity can be regulated. Conversely, oxidation can create free protons and increase soil acidity (see Section 11.4).

It is very important to understand that the data in Table 6.2 indicate that certain redox reactions *can* occur in soils, but not that they *will* occur: a chemical reaction that is favored by a large value of $\log K$ is not necessarily favored kinetically. This fact is especially applicable to redox reactions because they are often extremely slow, and because reduction and oxidation half-reactions often do not couple well to each other. For example, the coupling of the half-reaction for $O_2(g)$ reduction with that for a glucose oxidation leads to a $\log K$ value of 21.0 for the overall redox reaction:

$$\tfrac{1}{4}O_2(g) + \tfrac{1}{24}C_6H_{12}O_6(aq) = \tfrac{1}{4}CO_2(aq) + \tfrac{1}{4}H_2O(\ell) \tag{6.16}$$

For a soil solution that is in equilibrium with the earth's atmosphere $(P_{O_2} = 0.21 \text{ atm})$, the above value of $\log K$ predicts complete oxidation of carbon from C(0) to C(IV) at any pH value. But this prediction is contradicted by the observed persistence of dissolved organic matter in soil solutions under surface terrestrial conditions.

The lack of effective coupling and the slowness of redox reactions mean that catalysis is required if equilibrium is to occur. In soil solutions, as mentioned in Section 6.1, the catalysis of redox reactions is mediated by microbial organisms. In the presence of the appropriate microbial species, a reduction or oxidation half-reaction can proceed quickly enough in a soil to produce activity values of the reactants and products that agree with equilibrium predictions. Of course, this possibility is dependent entirely on the growth and ecological behavior of the soil microbial population and the degree to which the products of biochemical reactions can diffuse in the soil solution. In some cases, redox reactions will be controlled by the highly variable dynamics of an open biological system, with the result that redox speciation at best will correspond to local conditions of partial equilibrium. In other cases, including the important one of the flooded soil, redox reactions will be controlled by the behavior of a closed chemical system that is catalyzed effectively by bacteria and for which an equilibrium description is especially apt (Section 6.4). Regardless of which of these two extremes is the more appropriate to characterize the redox reactions in a soil, the role of organisms deals only with the kinetics aspect of redox. *Soil organisms affect the rate of a redox reaction, not its equilibrium constant.* If a redox reaction is not favored by a positive log K, microbial intervention cannot change that fact!

6.3 pE–pH Diagrams

The pE–pH diagram in Fig. 6.1 showed the domains of aqueous electron and proton activity that were accessible in soils and utilized by different classes of microorganisms. This kind of diagram can also be used to indicate the relative predominance of different redox species composing the same chemical element. In this application, the pE–pH diagram is divided into geometric regions whose interiors are domains of stability of either an aqueous species or a solid phase, and whose boundary lines are generated by transforming Eq. 6.10 (or another suitable expression for an equilibrium constant) into pE–pH relationships like Eq. 6.13. By comparing a pE–pH diagram for a chemical element (e.g., Mn or S) with Fig. 6.1, one can predict the redox species expected at equilibrium under oxic, suboxic, or anoxic conditions in a soil at a given pH value.

An example of a pE–pH diagram for manganese is presented in Fig. 6.2. This diagram was constructed as follows. First, a suite of redox species was chosen: $MnO_2(s)$, a synthetic compound analogous to the naturally occurring Mn(IV) solid phase, birnessite (Tables 1.3 and 2.4); $MnCO_3(s)$, which occurs in alkaline soils; and $Mn^{2+}(aq)$. Chemical reactions that relate these species are

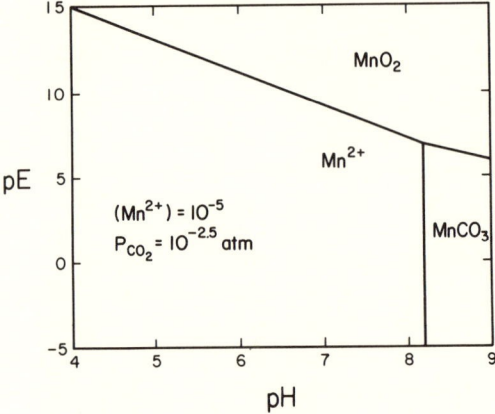

FIG. 6.2 A pE–pH diagram for manganese.

listed in Table 6.2 along with their equilibrium constants. The equilibrium constant that relates $MnO_2(s)$ to $Mn^{2+}(aq)$ can be expressed in the logarithmic form:

$$20.7 = \tfrac{1}{2}\log(Mn^{2+}) + \log(H_2O) + 2\,pH + pE - \tfrac{1}{2}\log(MnO_2) \quad (6.17)$$

It is customary to set solid-phase and liquid water activities equal to 1.0, whereupon Eq. 6.17 reduces to the pE–pH relationship:

$$pE = 20.7 - \tfrac{1}{2}\log(Mn^{2+}) - 2\,pH \qquad (6.18)$$

Equation 6.18 is not strictly a relation between pE and pH until the activity of $Mn^{2+}(aq)$ has been specified. A value that is relevant to soil solutions is 10^{-5}. Then Eq. 6.18 becomes:

$$pE = 23.2 - 2\,pH \qquad (6.19)$$

This equation is plotted as a line on the upper left in Fig. 6.2. It defines the boundary between MnO_2 and Mn^{2+} in respect to the predominance of one species or another. Above the line, Mn(IV) in the form of MnO_2 predominates; below the line, Mn(II) in the form of $Mn^{2+}(aq)$ (or soluble complexes of Mn) predominates.

The reaction connecting MnO_2 to $MnCO_3$ in Table 6.2 is expressed more conveniently if Eqs. 5.13a and 5.13b are combined with it to produce the equation:

$$\tfrac{1}{2}MnO_2(s) + \tfrac{1}{2}CO_2(g) + H^+(aq) + e^-(aq) = \tfrac{1}{2}MnCO_3(s) + H_2O(\ell) \qquad (6.20)$$

for which $\log K = 20.2 - \tfrac{1}{2}(6.35 + 1.47) = 16.3$ (see Section 5.2). The pE–pH relationship analogous to Eq. 6.18 is then:

$$pE = 16.3 + \tfrac{1}{2} \log P_{CO_2} - pH \tag{6.21}$$

The activity of $CO_2(g)$ has been set equal numerically to the CO_2 pressure *in atmospheres* (see Section 5.2). A relevant value of P_{CO_2} for soils is $3 \times 10^{-3}\,atm = 10^{-2.5}\,atm$. Then Eq. 6.21 becomes:

$$pE = 15.0 - pH \tag{6.22}$$

This equation defines the boundary between MnO_2 and $MnCO_3$. It is plotted on the upper right in Fig. 6.2. Note that this line extends to the left, and the line representing Eq. 6.19 extends to the right, until they intersect at $pE = 6.8$, $pH = 8.2$ in the pE–pH diagram. The reduced solid species, $MnCO_3$, lies below the line represented by Eq. 6.22.

Finally, Eq. 6.20 can be combined with the MnO_2–Mn^{2+} reaction in Table 6.2 to create the dissolution reaction for $MnCO_3$:

$$MnCO_3(s) + 2H^+(aq) = Mn^{2+}(aq) + CO_2(aq) + H_2O(\ell) \tag{6.23}$$

for which $\log K = 2(20.7 - 16.3) = 8.8$. This reaction is like that for calcite dissolution in Eq. 5.14. The corresponding pE–pH relation is:

$$0 = 8.8 - \log(Mn^{2+}) - \log P_{CO_2} - 2\,pH = 16.3 - 2\,pH \tag{6.24}$$

The pE variable does not appear because no change in the oxidation number of Mn occurs in Eq. 6.23. Eq. 6.24 plots as a vertical line at pH 8.2 in Fig. 6.2. This line intersects the point $pE = 6.8$, $pH = 8.2$ on the boundary lines representing Eqs. 6.19 and 6.22.

Comparison of Figs. 6.1 and 6.2 reveals that Mn(II) species are expected to predominate in all but the most oxic soil conditions ($pE + pH > 15$). In calcareous soils, $MnCO_3$ precipitates if $pE < 7$ (i.e., in suboxic and anoxic soils). Note that increasing the $Mn^{2+}(aq)$ activity or the CO_2 pressure will shift the vertical line in Fig. 6.2 to the left. An increase of one log unit in either variable would move the line to pH 7.8, as can be deduced from Eq. 6.24.

The construction of pE–pH diagrams can be a useful guide to the prediction of redox speciation in soils if effective microbial catalysis of the component half-reactions is expected. The example in Fig. 6.2 illustrates the general procedure for developing pE–pH diagrams:

1. Establish a set of redox species and obtain values of $\log K$ for all of the reactions between the species.

2. Unless other information is available, set the activities of liquid water and all solid phases equal to 1.0. Set all gas-phase pressures at values appropriate to soil conditions.

3. Develop each expression for $\log K$ into a pE–pH relation. In one relation involving an aqueous species and a solid phase wherein a change in

oxidation number is involved, choose a value for the activity of the aqueous species.

4. In each relation involving two aqueous species, set the activities of the two species equal.

6.4 Flooded Soils

A soil inundated by water becomes nearly a closed system insofar as redox reactions are concerned. This condition leads to the depletion of oxygen and the buildup of carbon dioxide from microbial respiration processes. Oxygen depletion must regulate the electron activity in the soil solution according to the first reaction in Table 6.1, just as carbon dioxide evolution will regulate the proton activity in the soil solution according to the reactions in Eq. 5.13. If labile organic matter is available as a source of electrons in the soil, the increase in electron activity brought on by oxygen depletion can continue, with effects on N(V), Mn(IV), Fe(III), and S(VI) redox species.

Figure 6.3 shows the typical sequence of reduction processes in flooded soil as observed in a silty clay amended with rice straw and incubated in suspension without oxygen supply for 1 week. During the first day, the content of $O_2(g)$

FIG. 6.3 Relative changes in O_2, NO_3^-, Mn(IV), and Mn(III) solid phases, and Fe(II) content of a soil with time elapsed after flooding. Changes in pE are also shown and labeled on the right side. [Data from F. T. Turner and W. H. Patrick, Chemical changes in waterlogged soils as a result of oxygen depletion, *Trans. IX Congress, Int. Soil Sci. Soc.* (Adelaide, Australia) **4**:53–65 (1968).]

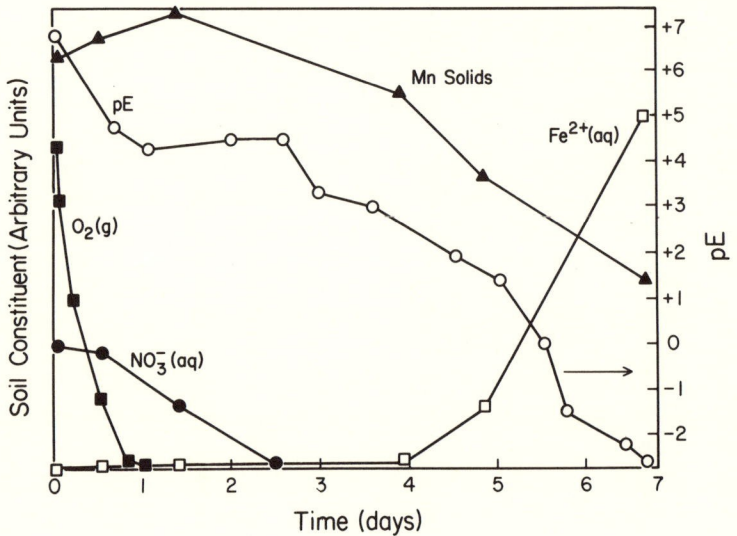

dropped to a negligible value, and the pE decreased to about 3.5 (at pH 7). According to the first equation in Table 6.2, the $pE-P_{O_2}$ relation

$$pE = 20.8 + \tfrac{1}{4} \log P_{O_2} - pH \tag{6.25}$$

at equilibrium, and $pE = 3.5$ at pH 7 leads to $P_{O_2} = 10^{-40}$ atm, which is quite negligible.

The depletion of nitrate in Fig. 6.3 begins shortly before oxygen disappears. The principal reactions involved in nitrate reduction are listed in Table 6.2. That involving NO_3^- and NO_2^- as redox species can be considered briefly to illustrate the effect of pE on aqueous nitrogen speciation. Following the approach taken in Section 4.3, one can express the total N concentration in the soil solution as the sum of nitrate and nitrite concentrations:

$$N_T = [NO_3^-] + [NO_2^-] \tag{6.26}$$

This representation is accurate for pE values greater than about 6. Equation 6.26 can be transformed like Eq. 4.8 into an expression for $[NO_3^-]$ alone:

$$N_T = [NO_3^-]\left\{1 + \frac{[NO_2^-]}{[NO_3^-]}\right\} = [NO_3^-]\{1 + 10^{28.2-2pH-2pE}\} \tag{6.27}$$

where

$$K = \frac{(NO_2^-)^{\frac{1}{2}}(H_2O)^{\frac{1}{2}}}{(NO_3^-)^{\frac{1}{2}}(H^+)(e^-)} = 10^{14.1} \tag{6.28}$$

has been used to estimate the concentration ratio:

$$\frac{[NO_2^-]}{[NO_3^-]} = \frac{(NO_2^-)}{(NO_3^-)} = 10^{28.2}(H^+)^2(e^-)^2 = 10^{28.2-2pH-2pE} \tag{6.29}$$

The definition of the distribution coefficient (see Eq. 4.9),

$$\alpha_{NO_3} \equiv \frac{[NO_3^-]}{N_T} \tag{6.30a}$$

then permits Eq. 6.27 to be written in the form:

$$\alpha_{NO_3} = \{1 + 10^{28.2-2pH-2pE-1}\}^{-1} \tag{6.31a}$$

In a similar fashion to Eq. 4.12,

$$\alpha_{NO_2} \equiv \frac{[NO_2^-]}{N_T} \tag{6.30b}$$

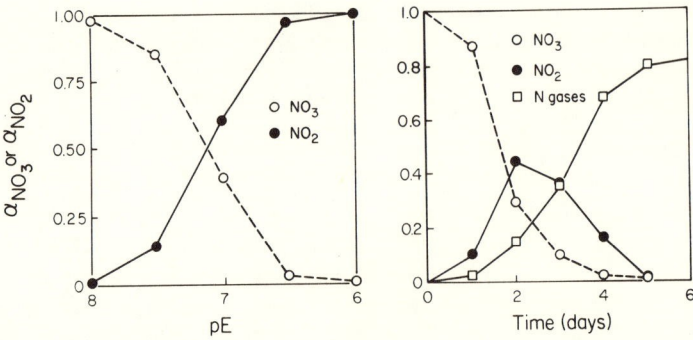

FIG. 6.4 Calculated distribution coefficients for NO_3^-(aq) and NO_2^-(aq) as a function of pE (left). Observed distribution coefficients for nitrate, nitrite, and gaseous N compounds as a function of time after waterlogging a soil (right). [Data from W. H. Patrick and I. C. Mahapatra, Transformation and availability to rice of nitrogen and phosphorus in waterlogged soil S, *Adv. Agron.* **20**:323–359 (1968).]

results in the equation:

$$\alpha_{NO_2} = \alpha_{NO_3} 10^{28.2 - 2pH - 2pE} \qquad (6.31b)$$

Equations 6.31a and b are plotted as functions of pE at pH 7 in Fig. 6.4, where an observed sequence of nitrogen transformations in a soil after flooding also is shown for comparison. Qualitative agreement between the two graphs is apparent. The observed decrease in α_{NO_2} after its peak at 2 days could also be represented mathematically by including, say, N_2O(aq) as a third species in Eq. 6.27 and recalculating α_{NO_2} in terms of NO_2^- production by NO_3^- reduction and NO_2^- loss by denitrification.

Figure 6.3 shows a continuing decrease in Mn(IV) [and Mn(III)] solid phases after the pE value drops below 4. A pronounced increase in Fe^{2+} in the soil solution is evident only after pE decreases below 3. These trends are consistent with the reduction sequence in Table 6.1 and with the log K data in Table 6.2, when expressed as in Eq. 6.18 and in the pE–pH relation for $Fe(OH)_3$(s):

$$pE = 16.4 - \log(Fe^{2+}) - 3\,pH \qquad (6.32)$$

To provide a Mn^{2+}(aq) activity of 10^{-5} at pH 7 requires a pE value of 9.3 according to Eq. 6.18, whereas the same conditions for Fe^{2+}(aq) require pE = 0.4 according to Eq. 6.32.

In the case of Mn and Fe, decreasing pE results in solid-phase dissolution because the stable forms of Mn(IV) and Fe(III) are solid species (see Fig. 6.2 for Mn). Besides the increase in Mn and Fe solubility expected from this effect of lowered pE, one usually observes a marked increase in the soil solution concentrations of metals like Cu, Zn, or Cd, and of ligands like $H_2PO_4^-$ or $HMoO_4^-$, accompanying Mn and Fe reduction. The principal cause of this secondary phenomenon is the desorption of metals and ligands that occurs when the adsorbents to which they are bound become unstable and dissolve. Typically, the metals released in this fashion, including Mn and Fe, are soon readsorbed by solids that are stable at low pE (e.g., clay minerals or soil organic matter) and become exchangeable surface species. These surface speciation changes have an obvious influence on the bioavailability of the chemical elements involved, particularly phosphorus (see Section 13.4).

As pE becomes negative in a flooded soil, the reduction of sulfur can take place. According to Eq. 6.13, sulfate reduction at pH 7 becomes significant at $pE = -3.6$. In practice, sulfide concentrations are detectable in soil solutions at pH 7 when $pE = -2.5$, which corresponds to $(HS^-)/(SO_4^{2-}) \approx 10^{-9}$ according to Eq. 6.14. If metals like Mn, Fe, Cu, or Zn are present in the soil solution at high enough concentrations, they can react with bisulfide to form metal sulfides that are quite insoluble ($K_{so} < 10^{-20}$). Thus anoxic soil conditions can diminish significantly the solubility of micronutrient metals.

6.5 Measuring pE and pH

Most reduction half-reactions are special cases of Eq. 6.9 with an equilibrium constant given by Eq. 6.10. The pE relationship that follows from this latter equation is:

$$pE = \log K + \log \frac{(A_{ox})^m}{(A_{red})^p} - n\,pH \tag{6.33}$$

if $(H_2O) = 1.0$. When a certain reduction process is known to be at equilibrium in a soil, Eq. 6.33 can be used to calculate pE with measured values of pH and the activities of the redox species involved. This kind of calculation was done in connection with Eqs. 6.13, 6.25, and 6.32. As another illustration, consider a calcareous soil in which both MnO_2 and $MnCO_3$ are present in equilibrium with $P_{CO_2} = 10^{-3}$ atm at pH 8.5. Then Eq. 6.21 can be applied to estimate $pE = 6.3$ in the soil, which is therefore suboxic (see Fig. 6.1). Uncertainty associated with this estimate derives from errors in the values of log K, pH, and the redox species activities, as well as inaccuracy in the assumption that equilibrium exists. Often it is found that the application of Eq. 6.33 to a pair of redox half-reactions believed to couple well in a soil produces very different estimates of pE. No chemical significance can be attributed to the estimated pE in these cases.

If the aqueous redox species involved in a half-reaction assumed to regulate pE are electroactive, then it may be possible to measure pE with an electrode assembly. The *electrode potential* corresponding to a pE value is *defined* by the equation:

$$E_H \equiv \frac{RT \ln 10}{F} pE = 0.05916 \; pE \quad (T = 298 \; K) \quad (6.34)$$

where R is the molar gas constant, T is absolute temperature, F is the Faraday constant, and the electrode potential E_H is in volts (see the Appendix). Equation 6.34 is a useful definition only if an electrochemical cell can be devised for measuring E_H unambiguously. One example of a cell used often in soil redox investigations is that containing a clean platinum electrode and a saturated calomel electrode. The Pt electrode responds to the electron transfer associated with the oxidation half-reaction whose associated pE value is to be measured, whereas the calomel electrode undergoes the reduction half-reaction:

$$\tfrac{1}{2}Hg_2Cl_2(s) + e^-(aq) = Cl^-(aq) + Hg(\ell) \quad (6.35)$$

for which $\log K = 4.53$ at 298.15 K. The methods of chemical thermodynamics can be used to show that the electrical potential difference produced across the two electrodes in the cell is:

$$E = B + E_J - E_H \quad (6.36)$$

where B is a parameter that depends on $\log K$ for the reaction in Eq. 6.35 and the activity of Cl^- in the electrode filling solution (usually KCl). The parameter E_J is the *liquid junction potential* created by the variation in chemical composition that occurs in passing from the aqueous solution in the cell to a saturated KCl solution ("salt bridge") that acts as a barrier to Cl^- transfer from inside the calomel electrode into the cell solution. For a given calomel electrode, B is a fixed parameter. The value of E_J, however, depends on the precise nature of the steady-state charge transfer across the liquid junction between the cell solution and the KCl salt bridge. The salt bridge is required to prevent $Cl^-(aq)$ in Eq. 6.35 from contributing charge to the Pt electrode, which is supposed to respond only to the redox half-reaction in the cell. Thus E_J is a "necessary evil" in the typical electrochemical cell used to measure E_H. The value of $B + E_J$ can be determined, for example, by calibrating the cell in a solution of ferrous and ferric ammonium sulfate, each at a concentration of 100 mol m^{-3} in 1.0 mol L^{-1} H$_2$SO$_4$ ("redox buffer"). The accepted value of E_H for this solution is 0.430 V at 298 K. Equation 6.36 and the measured E for the solution can then be used to calculate $B + E_J$ for the cell.

As an example of the use of Eq. 6.36, suppose that by calibration one finds $B + E_J = 0.25$ V at 298 K and that the Pt–calomel electrode cell is used to measure an E_H governed by the half-reaction relating $MnO_2(s)$ to $Mn^{2+}(aq)$ in Table 6.2. If the measured cell potential difference is 0.510 V, then $E_H = 0.26$ V according to Eq. 6.36, and pE = 4.4 according to Eq. 6.34. The corresponding activity of $Mn^{2+}(aq)$ is 1.6×10^{-5} at pH 7, according to Eq. 6.18.

Though straightforward, the measurement of pE in soil solutions via an electrode potential is subject to great uncertainty. The principal difficulties are that (1) the Pt electrode responds to more than one redox half-reaction, (2) the concentrations of redox species are too small to provide electron transfers detectable by the Pt electrode, (3) the redox species are not electroactive (e.g., N, S, and C species), (4) the Pt electrode becomes contaminated by oxide or other impurity coatings, and (5) the value of E_J differs greatly from that in the redox buffer used to calibrate the cell. The net result of these difficulties is the consensus that E_H *measurements give only a qualitative indication of pE in most soil solutions.* They are useful to classify soils as oxic, suboxic, or anoxic, but little beyond that.

Just as pE is analogous to pH, an electrochemical cell analogous to the Pt–calomel electrode cell for measuring electron activity can be developed to measure proton activity in soil solutions. In this cell, a glass membrane electrode replaces the Pt electrode. The membrane serves as an adsorbing surface for protons in the soil solution. The electrode potential it creates comes from differences, across the membrane, of proton activity in the soil solution and in the inner solution of the electrode, as modified by the slow diffusion of charged species through the membrane. As before, the salt bridge is required to prevent the glass electrode from responding to charge transfers produced by the reaction in Eq. 6.35. The electrical potential differences for the cell comprising the glass–calomel electrode cell is:

$$E = A + E_J - \frac{RT}{F} \ln(H^+) = A + E_J + 0.05916 \, pH \tag{6.37}$$

at 298 K, where A is a parameter that depends on log K for the reaction in Eq. 6.35, as well as on the activities of Cl^- and H^+ in the inner solutions of the electrodes. In the case of proton activity, E is calibrated directly in terms of the assigned pH values of buffer solutions. Therefore, by convention, only relative values of pH are measured in electrochemical cells:

$$pH(soil \, solution) = pH(buffer) + \frac{E(buffer) - E(soil \, solution)}{0.05916} \tag{6.38}$$

at 298 K, where the E values are in volts. [This convention differs very much from that used for the electron activity, which assigns pE = 0 to the reduction of the proton (second reaction in Table 6.2) at pH = 1 and $P_{H_2} = 1$ atm, according to Eq. 6.33.]

The principal difficulty in applying Eqs. 6.37 and 6.38 to soil solutions is uncertainty as to the magnitude of E_J. Since soil solution compositions differ greatly from those of pH buffers, the ions diffusing across the salt bridge in the former differ from those in the latter, and the corresponding liquid junction potentials can be quite different as well. It is virtually impossible to know precisely how large this difference in E_J will be, since no unambiguous method exists for measuring or calculating E_J independently. If the difference in E_J implicit in Eq. 6.38 is large and unknown, the "pH value" of the soil solution measured electrochemically would be of *no chemical significance*. This conclusion is even stronger if one attempts to apply Eq. 6.38 (or Eq. 6.36) to soil pastes or suspensions, since then E_J is certainly very different in the soil system from what it is in a standard buffer solution.

FOR FURTHER READING

L. G. M. Baas Becking, I. R. Kaplan, and D. Moore, Limits of the natural environment in terms of pH and oxidation–reduction potentials, *J. Geol.* **68**:243–284 (1960). The classic study of pE and pH in natural waters.

R. G. Bates, The modern meaning of pH, *Crit. Rev. Anal. Chem.* **10**:247–278 (1981). An authoritative review of the concept and measurement of pH, including discussions of liquid junction potentials and pH standards.

R. H. Dowdy et al., *Chemistry in the Soil Environment*, ASA Spec. Publ. No. 40, American Society of Agronomy, Madison, WI, 1981. Chapters 5 and 6 of this edited collection of presented papers, written by R. J. Bartlett and by R. W. Blanchar and C. E. Marshall, respectively, give useful discussions of the application of pE to soils.

D. L. Rowell, Oxidation and reduction, in D. J. Greenland and M. H. B. Hayes (eds.), *The Chemistry of Soil Processes*, pp. 401–461. Wiley, Chichester, U.K., 1981. This chapter presents a comprehensive discussion of redox phenomena in soils. It is an essential follow-up for the introductory discussion in the present chapter.

W. Stumm and J. J. Morgan, *Aquatic Chemistry*, Wiley, New York, 1981. Chapter 7 of this classic text gives an encyclopedic discussion of redox phenomena in natural water systems, including the concept and measurement of pE and the role of microbial catalysis in redox kinetics.

T. R. Yu, *Physical Chemistry of Paddy Soils*, Springer-Verlag, Berlin, 1985. This book provides an excellent summary of redox processes in flooded soils, with emphasis on many useful data obtained under laboratory and field conditions.

PROBLEMS

The more difficult problems are indicated by an asterisk.

1. Given the following two reactions, show that $Fe^{3+}(aq)$ and $S^{2-}(aq)$ are unstable equilibrium species in soil solutions. How might ferric iron ions and sulfide ions be stabilized in a soil solution?

$$Fe^{3+}(aq) + e^-(aq) = Fe^{2+}(aq) \qquad \log K = 13.0$$

$$S^{2-}(aq) + H^+(aq) = HS^-(aq) \qquad \log K = 13.92$$

2. Calculate the CO_2 pressure required to form the amino acid alanine ($C_3H_4O_2NH_3$) at a concentration of 1 mmol m^{-3} at pH 5 and pE $= -4$ in the presence of 3 mmol m^{-3} NH_4^+.

3. Use the data in Table 1.6 to calculate $\log K$ values for the NO_3^-/N_2O and SO_4^{2-}/H_2S half-reactions in Table 6.2 with gas-phase species instead of $N_2O(aq)$ or $H_2S(aq)$.

4. Develop an explanation for the fact that the lines separating the suboxic region from the oxic and anoxic regions are not horizontal (i.e., parallel to the pH axis). Consider the reduction of $MnO_2(s)$ to $Mn^{2+}(aq)$ (Eq. 6.18) and show how pE depends on pH if the activity of $Mn^{2+}(aq)$ is to remain fixed. The suboxic region boundary lines are drawn to maintain *fixed* redox species activities.

*5. A soil containing gypsum ($CaSO_4 \cdot 2H_2O$) and siderite ($FeCO_3$) has a CO_2 pressure of 10^{-3} atm and $(Ca^{2+}) = 10^{-3.65}$ in the soil solution. Calculate the pE value at which FeS ($K_{so} = 3.7 \times 10^{-19}$) will precipitate if pH = 8.2.

    (*Hint:* Use the reactions in Problems 1 and 9, along with Eq. 6.13 and the K_{so} value for gypsum (Section 5.2).)

6. In alkaline, suboxic soils, the important aqueous species of selenium are SeO_4^{2-} and SeO_3^{2-}. Given the half-reaction:

$$\tfrac{1}{2}SeO_4^{2-}(aq) + H^+(aq) + e^-(aq)$$
$$= \tfrac{1}{2}SeO_3^{2-}(aq) + \tfrac{1}{2}H_2O(\ell) \qquad \log K = 14.5$$

    determine whether SeO_3^{2-} can be oxidized to SeO_4^{2-} (the more toxic, mobile species) through the reduction of $Mn(IV)$.

7. Bacteria of the genus *Nitrobacter* catalyze the oxidation of nitrite to nitrate using $O_2(g)$ as an electron acceptor in respiration. Write a balanced overall redox reaction for this process and calculate $\log K$. What will be the concentration ratio of NO_3^- to NO_2^- when pE = 7 at pH 6?

8. Develop a balanced reaction for sulfate reduction to bisulfide via glucose oxidation to bicarbonate. Use the data in Table 6.2 and Problem 7 of Chapter 4 to calculate $\log K$ for this reaction.

9. Because of favorable kinetics, $Fe(OH)_3(s)$ is usually regarded as the principal Fe(III) redox species in flooded soils within a few months after inundation. Given the information on $Fe(OH)_3$ in Table 6.2 and the reaction:

$$Fe^{2+}(aq) + CO_2(g) + H_2O(\ell)$$
$$= FeCO_3(s) + 2H^+(aq) \qquad \log K = -7.5$$

    construct a pE–pH diagram for the species $Fe^{2+}(aq)$, $FeCO_3(s)$ and $Fe(OH)_3(s)$. Take $(Fe^{2+}) = 10^{-7}$ and $P_{CO_2} = 10^{-2}$ atm.

*10. How would the pE–pH diagram called for in Problem 9 differ if the reaction:

$$\tfrac{1}{2}Fe(OH)_2 \cdot (Fe(OH)_3)_2(s) + 4H^+(aq) + e^-(aq)$$
$$= \tfrac{3}{2} Fe^{2+}(aq) + 4H_2O(\ell) \qquad \log K = 21.2$$

were included? The species $Fe_3(OH)_8$ on the left in the preceding reaction is called ferrosoferric hydroxide.

11. Given the appropriate reactions in Table 6.2, prepare a pE–pH diagram for sulfur based on the aqueous species SO_4^{2-}, H_2S, and HS^-.

12. Calculate ranges of "initiating" electrode potentials at 298 K that correspond to the pE_{init} values in Table 6.1. Compare your result for sulfate reduction with the statement that "little or no sulfide is produced at $E_H > -150\,mV$."

13. A Pt–calomel electrode cell is calibrated to have $B + E_J$ in Eq. 6.36 equal to 0.16 V at 298 K. If this cell is used to measure E_H for a soil solution at pH 7 containing sulfate and bisulfide ions at equal activity, what electric potential difference will it develop?

14. By convention, the value of $A + E_J$ in Eq. 6.37 for a hydrogen electrode–calomel electrode cell immersed in a mixture of 10 mol m^{-3} HCl–90 mol m^{-3} KCl is 0.3358 V. The electric potential difference of the cell for this solution is 0.4583 V. Calculate the pH value of the solution.

*15. Liquid junction potential differences of 10 mV are not unusual between soil solutions and pH buffers. Calculate the error (in pH units) that would attend pH measurements made in the presence of this junction potential difference. A general rule of thumb is that pH measurements in natural systems can be no more accurate than $\pm 0.05\,pH$ units. To what magnitude of liquid junction potential difference (in mV) does this correspond?
(*Hint:* Rewrite Eq. 6.38 in the form $\Delta pH = \Delta E_J/59.16$.)

Special Topic 3

Balancing Redox Reactions

Redox species differ from other chemical species in that their status as "oxidized" or "reduced" molecular entities must be noted along with their other chemical properties. The redox status of the atoms in a redox species is quantified through the concept of *oxidation number*, the hypothetical valence denoted by a positive or negative roman numeral and assigned to an atom according to the following rules.

1. For a monoatomic species, the oxidation number equals the valence.
2. For a molecule, the sum of oxidation numbers of the constituent atoms equals the net charge on the molecule expressed in units of protonic charge.

3. For a chemical bond in a molecule, the shareable, bonding electrons are assigned entirely to the more electronegative atom participating in the bond. If no difference in electronegativity exists, each atom receives half the bonding electrons.

These rules can be illustrated by working out the oxidation numbers for the atoms in the redox species, $FeOOH$, CHO_2^-, N_2, SO_4^{2-}, and $C_6H_{12}O_6$. In $FeOOH$, oxygen is more electronegative than Fe or H and is conventionally designated $O(-II)$. Thus oxygen has oxidation number "minus two." The hydrogen atom in OH is designated $H(I)$ (oxidation number "plus one"). By Rule 2, the iron atom is designated $Fe(III)$, since $FeOOH$ has zero net charge and $3 + 2(-2) + 1 = 0$. (Recall that this notation was used already in Chapter 2 to distinguish ferric from ferrous iron in soil minerals.) A similar computation is done for CHO_2^-, in which oxygen and hydrogen are designated as above and carbon must be $C(II)$, since $2 + 2(-2) + 1 = -1$ = the net number of protonic charges on the formate anion.

In the case of N_2, there is no difference in electronegativity between the two identical atoms in the molecule, and, by Rule 3 just given, neither can be assigned all of the bonding electrons. Since the molecule is neutral, Rule 2 then leads to the designation $N(O)$ for each constituent nitrogen atom. For sulfate, oxygen is again $O(-II)$ and sulfur must be $S(VI)$, according to Rule 2. Finally, in glucose, carbon must be $C(O)$ because the designations $O(-II)$ and $H(I)$ lead by themselves to a neutral $C_6H_{12}O_6$ molecule.

Redox reactions must obey the same mass and charge balance laws as were described for other chemical reactions in Special Topic 1, at the end of Chapter 1. The only new feature is the need to account for changes in oxidation number when charge balance is imposed. Consider, for example, the weathering of olivine as represented by the reaction in Eq. 2.2. The essential characteristic of this reaction in the present context is the oxidation of $Fe(II)$ to $Fe(III)$. The reduced species is olivine, $Mg_{1.63}Fe_{0.37}SiO_4$, and the oxidized species is goethite. Equation 2.2 also displays $O_2(g)$ as a reactant. This species must have been reduced to water in order that the electrons released by Fe oxidation will be absorbed in the weathering process. Thus the redox aspect of Eq. 2.2 is captured by considering how to balance the postulated reaction:

$$Mg_{1.63}Fe_{0.37}SiO_4(s) + O_2(g) \rightarrow FeOOH(s) + H_2O(\ell) \qquad (s3.1)$$

The schematic reaction in Eq. s3.1 can be balanced by first dividing it into reduction and oxidation half-reactions. The Fe half-reaction is the inverse of the reduction half-reaction:

$$FeOOH(s) + e^-(aq) \rightarrow Mg_{1.63}Fe_{0.37}SiO_4(s) \qquad (s3.2)$$

which is analogous to Eq. 6.1. Mass balance for Fe is obtained by giving olivine the stoichiometric coefficient $1/0.37 = 2.7$, after which mass balance can be imposed on Mg and Si:

$$FeOOH(s) + 4.4Mg^{2+}(aq) + 2.7Si(OH)_4^0(aq) + e^-(aq) \rightarrow$$

$$2.7Mg_{1.63}Fe_{0.37}SiO_4(s) \qquad (s3.3)$$

Mass balance for oxygen can be achieved by adding two water molecules to the right side of Eq. s3.3. Proton balance then would require $1 + 2.7(4) - 4 = 7.8H^+(aq)$ to be added also to the right side:

$$FeOOH(s) + 4.4Mg^{2+}(aq) + 2.7Si(OH)_4^0(aq) + e^-(aq)$$

$$= 2.7Mg_{1.63}Fe_{0.37}SiO_4(s) + 2H_2O(\ell) + 7.8H^+(aq) \qquad (s3.4)$$

This reaction can be shown quickly to meet the requirement of overall charge balance. Note that $e^-(aq)$ is essential for this.

To develop a redox reaction without the aqueous electron, one need only add to the inverse of Eq. s3.4 the reduction half-reaction for $O_2(g)$ in Table 6.1:

$$2.7Mg_{1.63}Fe_{0.37}SiO_4(s) + 0.25O_2(g) + 8.8H^+(aq) + 1.5H_2O(\ell)$$

$$= FeOOH(s) + 4.4Mg^{2+}(aq) + 2.7Si(OH)_4^0(aq) \qquad (s3.5)$$

where $0.5H_2O(\ell)$ has been canceled from both sides of the result. Equation s3.5 is a balanced redox reaction showing the weathering of olivine in an oxic environment to form goethite, aqueous magnesium ions, and silicic acid. The procedure by which it was developed can be described as follows:

1. Identify the two pairs of oxidized and reduced species participating in the overall redox reaction.
2. For each redox pair, develop a balanced reduction half-reaction in which 1 mol of aqueous electrons is transferred.
3. Combine the two half-reactions developed in Step 2 to cancel the aqueous electron and produce the required reactant and product redox species in the overall reaction.

7

Soil Particle Surfaces

7.1 Surface Functional Groups

In Table 3.5, the most important functional groups in soil humus are listed. Examples include the carboxyl group, $-CO_2H$, and the phenolic hydroxyl group (aromatic ring OH), both of which can dissociate a proton and become negatively charged in the soil solution. Of the variety of functional groups present in the organic compounds that polymerize to form soil humus (Sections 3.1 and 3.2), it can be expected that some ultimately will come to reside on the interface between solid soil organic matter and the aqueous phase in soil. These molecular units that protrude from the solid surface into the soil solution are *surface functional groups*. In the case of soil organic matter, the surface functional groups are necessarily organic molecular units. But in general they can be bound to either organic or inorganic solids, and they can have any molecular structural arrangement that is possible for them were they bound to small molecules instead of polymeric materials like soil humus or clay minerals. Unlike the situation for small molecules, however, functional groups on surfaces cannot be diluted infinitely, even in aqueous suspension. Unless the substrate to which they are bound decomposes, surface functional groups remain separated by more-or-less fixed distances, regardless of how dilute a suspension of the substrate may be. Thus the groups remain closely associated and can influence one another in nearly all circumstances.

Examples of organic surface functional groups can be imagined in Fig. 3.1 by considering the carbonyl ($C=O$) at the top of the figure to be at an interface, or from Fig. 3.2 by imagining the cellulose OH to protrude from a surface. Because of the variety of possible functional group compositions (Table 3.5), a broad spectrum of surface functional group reactivity is likely. Superimposed on this intrinsic variability is that created by the wide range of stereochemical and charge distribution characteristics possible in a heterogeneous organic matrix (Sections 3.2 and 3.3). For this reason, it is entirely conceivable that no particular organic surface functional group (e.g., carboxyl) possesses well-defined quantitative chemical properties, like the proton dissociation equilibrium constant, but instead can be characterized only by ranges of values for these properties. This "smearing-out" of their chemical behavior is another important feature that distinguishes organic surface functional groups from functional groups bound to small organic molecules.

Surface functional groups on *inorganic* solids also are common in soils. In Fig. 5.1, for example, a water molecule bearing positive charge is shown bound to an Al^{3+} ion at the periphery of the mineral gibbsite (upper left). This combination of metal cation and water molecule at an interface is a *Lewis acid site*, with the metal cation identified as the Lewis acid. ("Lewis acid" is the name given to metal cations and protons when their reactions are considered from the perspective of the electron orbitals in ions.) Lewis acid sites can exist also on the surface of goethite (Fig. 2.5) if peripheral Fe^{3+} ions are bound to water molecules there. Thus any metal hydrous oxide, as well as the edge surfaces of clay minerals like kaolinite (Fig. 7.1), can expose Lewis acid sites to the soil solution. These surface functional groups are very reactive, since the positively charged water molecule is quite unstable and is exchanged readily for an organic or inorganic anion in the soil solution, which then can form a more stable bond with the metal cation. This *ligand exchange reaction* is described by Eqs. 3.15 and 3.16 for the example of carboxylate and a metal–hydroxide Lewis acid site.

The inorganic surface functional group of greatest abundance and reactivity in soil clays is the hydroxyl group exposed on the outer periphery of a mineral. This kind of OH group is found on metal oxides, oxyhydroxides, and hydroxides (Fig. 2.5), on clay minerals (Fig. 7.1), and on amorphous silicate minerals like allophane. Usually, more than one kind of mineral surface OH

FIG. 7.1 Surface functional groups on layer silicates: the siloxane cavity (left), the Lewis acid site [Al(III)·H_2O], aluminol, and silanol groups (right).

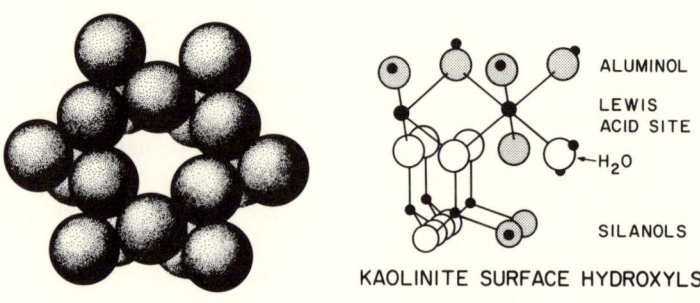

ALUMINOL

LEWIS
ACID SITE

←H_2O

SILANOLS

KAOLINITE SURFACE HYDROXYLS

group can be distinguished on the basis of stereochemistry, and these different kinds of surface OH group, in turn, will have properties (e.g., their reactivity with protons) that set them apart from OH groups inside the bulk mineral structure.

The general characteristics of the inorganic surface hydroxyl group can be illustrated with the example of goethite, whose molecular structure is presented in Figs. 2.5 and 7.2. The surface of goethite consists primarily of exposed planes of differing orientation. The surface OH groups on these planes are denoted A, B, and C (Fig. 7.2). The type A hydroxyl group is a former oxygen ion that is coordinated to one Fe^{3+} cation in the bulk mineral structure and has become protonated upon exposure to an aqueous solution as a surface group. The type C hydroxyl group is formed in the same way, except that it is coordinated to two Fe^{3+} cations. The type B hydroxyl group is the same as hydroxyl of the bulk structure coordinated to three Fe^{3+} cations (Fig. 2.5), but is exposed on a surface. These three surface OH groups exhibit different reactivities: only the type A hydroxyls are found to protonate or dissociate protons and to complex metal cations.

Clay minerals also expose singly coordinated OH groups on the edge surfaces created when crystallites are broken apart. These edge–surface hydroxyls are illustrated for kaolinite on the right side of Fig. 7.1. At the edge of the octahedral sheet, OH groups are singly coordinated to Al^{3+} cations and at the edge of the tetrahedral sheet they are singly coordinated to Si^{4+} cations. Because of the greater valence of silicon, these latter OH groups tend only to dissociate protons, as opposed to the OH groups coordinated to Al^{3+}, which bind as well as dissociate protons. The two types of edge–surface hydroxyl group, which are distinguished by the names *aluminol* and *silanol*, also differ in their reactivity toward oxyanions such as COO^- and HPO_4^{2-}. The SiOH groups do not undergo the two-step ligand exchange reaction in Eqs. 3.15 and 3.16. Reactions with metal cations, however, are possible for either kind of OH group by exchange with a proton.

The plane of oxygen atoms on the surface of a 2:1 layer silicate (Fig. 2.4) is called a *siloxane surface*. This plane is characterized by a distorted hexagonal symmetry among its constituent oxygen atoms (Section 2.3). The functional group associated with the siloxane surface is the roughly hexagonal cavity formed by six corner-sharing silica tetrahedra, shown on the left in Fig. 7.1. This cavity has a diameter of about 0.26 nm and is bordered by six sets of electron orbitals emanating from the surrounding ring of oxygen atoms.

The reactivity of the siloxane cavity depends on the nature of the electronic charge distribution in the layer silicate structure. If there are no neighboring isomorphic cation substitutions to create local deficits of positive charge in the underlying layer, the siloxane cavity will function as a very mild electron donor that can complex only neutral, dipolar molecules such as water molecules. The complexes formed are not very stable, an example being the easily reversed entrapment of a water molecule having one of its hydroxyl groups directed into

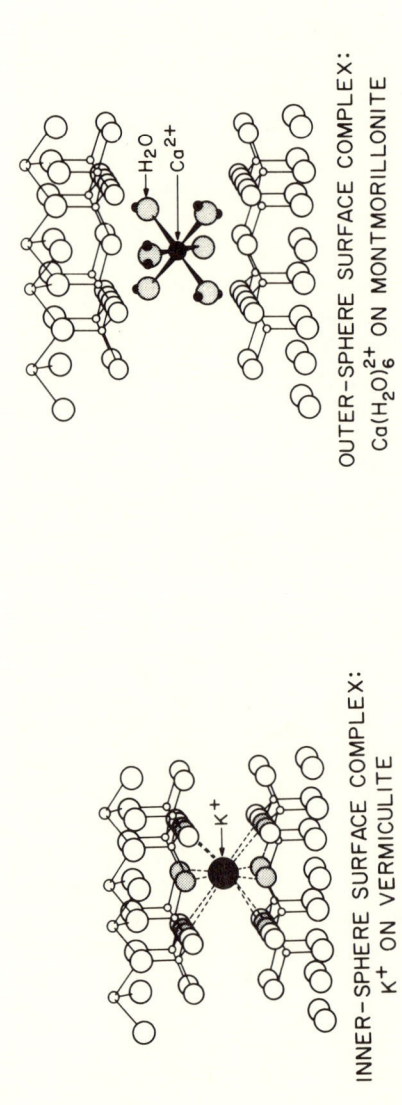

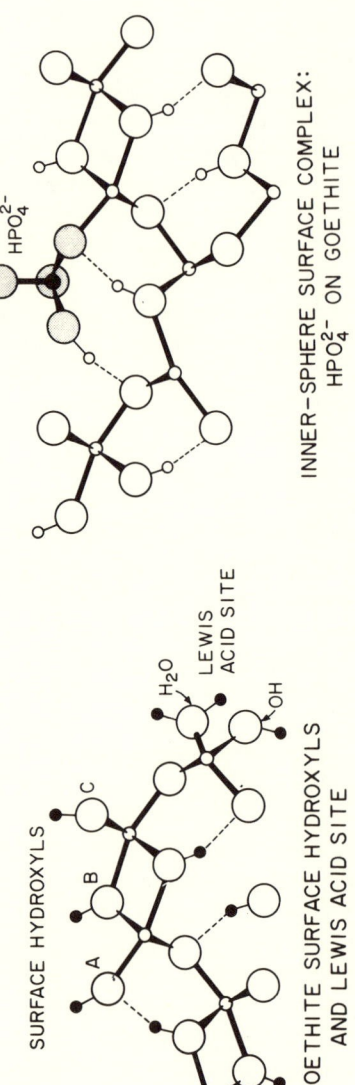

OUTER-SPHERE SURFACE COMPLEX:
Ca(H₂O)²⁺₆ ON MONTMORILLONITE

INNER-SPHERE SURFACE COMPLEX:
K⁺ ON VERMICULITE

INNER-SPHERE SURFACE COMPLEX:
HPO₄²⁻ ON GOETHITE

GOETHITE SURFACE HYDROXYLS
AND LEWIS ACID SITE

FIG. 7.2 Examples of surface complexes. (After G. Sposito, *The Surface Chemistry of Soils*, Oxford Univ. Press, New York, 1984.)

a cavity perpendicularly to the siloxane surface. If isomorphic substitution of Al^{3+} by Fe^{2+} or Mg^{2+} occurs in the octahedral sheet, the resulting excess negative charge on a nearby siloxane cavity makes it possible to form reasonably strong complexes with cations as well as dispolar molecules. If isomorphic substitution of Si^{4+} by Al^{3+} occurs in the tetrahedral sheet, the excess negative charge is located much nearer to the surface oxygen atoms, and much stronger complexes with cations and molecules become possible because of this localization of charge.

The complexes formed between surface functional groups and constituents of the soil solution can be classified analogously to the complexes that form entirely among aqueous species (Section 4.2). If a surface functional group reacts with an ion or a molecule dissolved in the soil solution to form a stable molecular unit, a *surface complex* is said to exist and the formation reaction is termed *surface complexation*. Two broad categories of surface complex are distinguished on structural grounds. If no water molecule is interposed between the surface functional group and the ion or molecule it binds, the complex is *inner-sphere*. If at least one water molecule is interposed between the functional group and the bound ion or molecule, the complex is *outer-sphere*. As a general rule, outer-sphere surface complexes involve electrostatic bonding mechanisms and therefore are less stable than inner-sphere surface complexes, which necessarily involve either ionic or covalent bonding, or some combination of the two. These concepts are parallel with those developed in Section 4.2 for strictly aqueous species.

Examples of surface complexes are shown schematically in Fig. 7.2. An inner-sphere surface complex involving K^+ on vermiculite appears in "exploded view" at the upper left in the figure. This surface complex requires the coordination of a potassium ion with 12 oxygen atoms bordering two opposing siloxane cavities. The layer charge in soil vermiculite is large enough (Table 2.3) that each siloxane cavity in a basal plane of the mineral can complex a K^+ cation. Moreover, the ionic radius of K^+ (Table 2.1) is almost precisely equal to that of a cavity. This combination of charge distribution and stereochemical factors gives K–vermiculute surface complexes great stability in soils and is the molecular basis for potassium fixation (Section 13.2). An outer-sphere surface complex with a Ca^{2+} cation is illustrated on the upper right side in Fig. 7.2. In this example, the two-layer hydrate of Ca-montmorillonite, two opposing siloxane cavities complex a Ca^{2+} cation solvated by six water molecules in octahedral coordination.

As mentioned earlier, the type A hydroxyl group in goethite can be protonated to form a Lewis acid site. It can then be exchanged as in Eq. 3.16 to allow the formation of an inner-sphere surface complex with the oxyanion, HPO_4^{2-}. This surface complex is illustrated on the lower right in Fig. 7.2. It consists of a HPO_4^{2-} bound through its oxygen ions to a pair of adjacent Fe^{3+} cations ("binuclear surface complex"). The configuration of the *o*-phosphate unit is especially compatible with the grooved structure of the goethite surface, thus providing stereochemical enhancement of the stability of the inner-sphere complex. Inner-sphere complexes can also form through the ligand exchange of other oxyanions with protonated OH groups on goethite.

These conceptual remarks suggest that the surface chemistry of a soil will be determined in large measure by the nature and reactivity of its surface functional groups. But the surface reactions of soils are also conditioned by pedochemical weathering (Section 1.5). The dissolution and precipitation of minerals (Chapters 2 and 5), and the oxidation and synthesis of organic matter (Chapters 3 and 6) will necessarily alter the categories and reactivities of surface functional groups. The weathering of smectite to kaolinite, and ultimately thereafter, to gibbsite and colloidal silica (Table 1.7), certainly will transform the surface chemistry of the clay fraction of a soil. In this context, *the sequential progress of pedochemical weathering can be regarded as an evolution of surface functional groups from preeminence of the siloxane cavity to pre-eminence of the inorganic hydroxyl group.* This surface chemical interpretation of soil formation has broad implications in soil fertility (see Chapters 11–13).

7.2 Adsorption

Adsorption is the net accumulation of matter at the interface between a solid phase and an aqueous solution phase. It differs from precipitation (Chapter 5) because it does not include the development of a three-dimensional molecular structure, even if such a structure grows on a surface ("surface precipitate"). The matter that accumulates in two-dimensional molecular arrangements at an interface is the *adsorbate*. The solid surface on which it accumulates is the *adsorbent*. A molecule or an ion in the soil solution that potentially can be adsorbed is termed an *adsorptive*.

Adsorption on soil particle surfaces can take place via the three mechanisms illustrated in Fig. 7.3 for a monovalent cation (e.g., K^+) on the siloxane surface of a 2:1 layer silicate like montmorillonite. The inner-sphere surface complex shown involves the siloxane cavity, as described in Section 7.1. The outer-sphere surface complex shown includes the cation solvation shell and is similar to that depicted in Fig. 7.2. If a solvated ion does not form a complex with a charged surface functional group, but instead neutralizes surface charge only in a delocalized sense, it is said to be adsorbed in the *diffuse-ion swarm*, also shown in Fig. 7.3. This last adsorption mechanism involves ions that are fully dissociated from surface functional groups and are, accordingly, free to move about nearby in the soil solution. The diffuse-ion swarm and the outer-sphere surface complex mechanisms of adsorption involve almost exclusively electrostatic bonding, whereas inner-sphere complex mechanisms are likely to involve ionic as well as covalent bonding. Since covalent bonding depends significantly on the particular electron configurations of both the surface group and the complexed ion, it is appropriate to consider inner-sphere surface

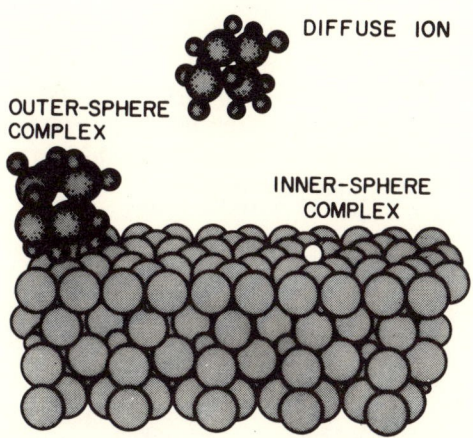

FIG. 7.3 The three mechanisms of cation adsorption on a siloxane surface (e.g., montmorillonite).

complexation as the molecular basis of the term, *specific adsorption.* Correspondingly, diffuse-ion association and outer-sphere surface complexation are the molecular basis for the term, *nonspecific adsorption.* The "nonspecificity" implied by this definition refers to the weak dependence on the electron configuration of the surface group and adsorbed ion to be expected for the interaction of solvated species.

Readily exchangeable ions in soil are those that can be replaced easily be leaching with an electrolyte solution of prescribed composition, concentration, and pH value. Despite the empirical nature of this concept, there is a consensus that ions adsorbed specifically (like K^+ and HPO_4^{2-} in Fig. 7.2) are not to be considered "readily exchangeable." Thus experimental methods to determine readily exchangeable adsorbed ions must avoid extracting specifically adsorbed ions. From this point of view, *only fully solvated ions adsorbed on soils are readily exchangeable ions,* with the molecular definition of "readily exchangeable" based on the diffuse-ion swarm and outer-sphere complex mechanisms of adsorption. (Ion exchange is described in more general terms in Chapter 9 and Section 3.3.)

Some of the mechanisms through which dissolved soil organic matter binds to soil minerals (Table 3.6) can be restated in terms of surface complexation ideas and the adsorption mechanisms described earlier. The cation exchange mechanism (Eq. 3.13) evidently would involve surface complexation of organic B^+ ions, whereas the anion exchange mechanism would involve the surface complexation of COO^-. Water bridging (Eq. 3.14) involves solvated surface species, but there is no exchange reaction: the organic molecular unit B merely forms an outer-sphere surface complex with an adsorbed cation M^{m+}. Cation bridging occurs if B forms an inner-sphere surface complex directly with adsorbed M^{m+}. Ligand exchange (Eq. 3.16), as discussed in Section 7.1, involves the coordination of an organic anion to the metal cation in a Lewis acid site after displacement of an OH or H_2O bound previously to the cation. These brief remarks are intended to show how the surface complex idea can be useful to unify diverse mechanisms of surface reactions in soils. Section 3.5 can be reread with profit from this perspective.

7.3 Surface Charge

Solid-particle surfaces in soils develop electrical charge in two principal ways: either from isomorphic substitutions among ions of differing valence in soil minerals, or from the reactions of surface functional groups with ions in the soil solution. The electrical charge developed by these two mechanisms is expressed conventionally in moles of charge per kilogram (mol$_c$ kg^{-1}, see the Appendix). Four different types of surface charge contribute to the *net total particle charge* in soils, denoted σ_p. This important parameter can be positive, zero, or negative, depending on soil chemical conditions.

The *permanent structural charge* of a soil, denoted σ_0, is the moles of charge per kilogram created by isomorphic substitutions in soil minerals (Sections 1.3 and 2.3). These substitutions occur in both primary and secondary soil minerals, but they produce significant surface charge only in the 2:1 layer silicates. The contribution to σ_0 from isomorphic substitutions in hydrous oxides and 1:1 layer silicates (like kaolinite) is typically less than about 0.02 mol$_c$ kg^{-1}. On the other hand, the clay mineral groups illite, vermiculite, and smectite can each contribute permanent structural charge up to 100 times larger than this value. Estimates of these contributions can be made directly from Table 2.3. For example, in the smectite group, the (negative) layer charge produced by isomorphic substitutions ranges from 0.5 to 1.2. The corresponding value of σ_0 is calculated with the equation (see Problem 6 in Chapter 2):

$$\sigma_0 = -(x/M_r) \cdot 10^3 \qquad (7.1)$$

where x is the layer charge, and M_r is relative molecular mass (see the Appendix). The value of M_r is computed with the chemical formula and the relative molecular masses of each element that appears in the formula. In the case of smectite, from Table 2.3:

$$M_r = 8(28.09) + 3.2(27) + 0.2(55.9) + 0.6(24.3) + 24(16) + 4(1) = 725$$
$$\text{Si} \qquad \text{Al} \qquad \text{Fe} \qquad \text{Mg} \qquad \text{O} \qquad \text{H}$$

Therefore, according to Eq. 7.1 and the range of x in Table 2.3, σ_0 ranges between -0.7 and -1.7 mol$_c$ kg^{-1} for smectites. In a similar way, σ_0 is found to range from -1.9 to -2.8 mol$_c$ kg^{-1} for illites and from -1.6 to -2.5 mol$_c$ kg^{-1} for vermiculites. In a soil, these contributions to permanent structural charge could be multiplied by the respective fraction of the total solid mass attributed to each kind of clay mineral, then summed to give an overall value of σ_0.

The *net proton charge* of a soil, denoted σ_H, is the difference between the moles of protons and the moles of hydroxide ions *complexed* by surface functional groups:

$$\sigma_H = q_H - q_{OH} \tag{7.2}$$

where q_i is moles of ion i ($i = H^+$ or OH^-) complexed by surface groups. Diffuse-swarm protons are *not* included in σ_H.

The surface complexation of an hydroxide ion is chemically the same as the dissociation of a proton. For example, if SH represents a mole of surface functional groups bearing a proton (e.g., carboxyl groups with $S \equiv COO$), then OH^- complexation results in the same reaction product as proton dissociation if solvation water is ignored:

$$SH(s) + OH^-(aq) = SHOH^-(s) = S^-(s) \tag{7.3a}$$

Similarly, if SM^+ represents a mole of surface functional groups that have complexed the bivalent metal cation M^{2+}, then the complexation of OH^- via hydrolysis of the adsorbed metal is like proton dissociation:

$$SM^+(s) + H_2O(\ell) = SMOH^0(s) + H^+(aq) \tag{7.3b}$$

The most important surface functional groups in soils that complex protons are on soil humus, hydrous oxides, and $1:1$ aluminosilicates (e.g., kaolinite). (The siloxane cavity forms only very weak proton complexes.) Values of σ_H on these adsorbents can be measured as a function of pH in titration experiments like that described for soil humus in Section 3.3. The formation function, calculated by putting the titration data into Eq. 3.5, is in fact the same as σ_H in Eq. 7.2. When this method is applied to a soil, however, the difficult problem arises of avoiding proton consumption or production in nonsurface reactions, like mineral dissolution. These unwanted side reactions will contribute unknown errors to the practice of setting the difference between protons added and free protons measured equal to protons complexed, as in Eq. 3.5. There is also the unknown error in pH measurement contributed by liquid junction potential differences between the titrated soil system and buffer solutions, discussed in Section 6.5. These difficulties have not been resolved by any method that is applicable to all soils. Nonetheless, titration methods are used widely and reported measurements of σ_H range from -9 to $+1$ mol$_c$ kg^{-1} for soil humus and from -0.7 to $+0.4$ mol$_c$ kg^{-1} for soil minerals bearing surface OH groups. The very large negative values of σ_H observed for soil humus indicate that organic matter is the most important source of free protons in soil, just as it is also the most important source of free electrons (Section 6.1).

The algebraic sum of σ_0 and σ_H is the *intrinsic surface charge* of a soil. This terminology is used to emphasize that σ_0 and σ_H arise principally from structural components of the adsorbents in soils, that is, ionic constituents of soil minerals and surface OH groups on inorganic and organic solids. The relative contribution of σ_0 and σ_H to the intrinsic surface charge in a soil will depend on the extent of weathering of its minerals and on its organic matter content. In respect to soil mineralogy, since the dominant source of permanent structural surface charge is the siloxane cavity, it is evident that σ_0 will be important in soils at the early and intermediate stages of weathering (Table 1.7). For this reason, these soils often are termed *permanent-charge soils*. On the other hand, soils at the advanced stage of weathering are enriched in minerals bearing reactive OH groups (Section 7.1 and Table 1.7) and therefore are termed *variable-charge soils*. In these soils, σ_H will dominate σ_0, and changes in soil pH will influence strongly the development of surface charge.

Besides the intrinsic surface charge, soil particles bear the *inner-sphere complex charge*, σ_{IS}, and the *outer-sphere complex charge*, σ_{OS}. Contributing to σ_{IS} is the net total charge of the ions, other than H^+ or OH^-, which are bound into inner-sphere surface complexes. For example, the inner-sphere surface complexes illustrated in Fig. 7.2 would contribute $+1$ mol_c (K^+) and -2 mol_c (HPO_4^{2-}) to σ_{IS}. Similarly, σ_{OS} is the net total charge of the ions, other than H^+ or OH^-, which are bound into outer-sphere surface complexes. The outer-sphere complex with Ca^{2+} in Fig. 7.2, for example, would contribute $+2$ mol_c to σ_{OS}. Unlike the intrinsic surface charge, σ_{IS} and σ_{OS} arise solely from constituents of the soil solution that are adsorbed by particle surfaces. They can be measured by suitable ion-selective electrode techniques, comparable to those used to determine σ_H, or by ion displacement methods like those used to determine ion exchange capacities (see Sections 3.3, 8.1, and 10.1 for additional discussion).

With its four components now defined, the net total particle charge in soil can be represented mathematically by the equation:

$$\sigma_P = \sigma_0 + \sigma_H + \sigma_{IS} + \sigma_{OS} \tag{7.4}$$

Note that σ_0 is virtually always negative in sign, whereas σ_H, σ_{IS}, and σ_{OS}—like σ_P itself—can be either positive, zero, or negative, depending on the soil solution composition. Regardless of its value, σ_P represents the net surface charge on the solid particles in a soil. Each of its components is, in turn, a resultant of the contributions from a variety of adsorbents or adsorbates, both inorganic and organic.

Although soil particles may bear electrical charge, soils themselves are always electrically neutral. Thus σ_P in Eq. 7.4 must be balanced, when it is nonzero, by another kind of surface charge. This balancing charge arises from the ions in the soil solution that are not bound into surface complexes but still are adsorbed by soil particles, that is, *the diffuse-ion swarm* (Fig. 7.3). These ions move about freely in the soil solution while remaining near enough to solid surfaces to create the effective surface charge σ_D that balances σ_P. On the molecular scale, this effective surface charge can be apportioned to each diffuse-swarm ion according to the equation:

$$\sigma_{Di} = \frac{Z_i}{m_s} \int_V [c_i(\mathbf{x}) - c_{0i}]dV \tag{7.5}$$

where Z_i is the valence of the ion, $c_i(\mathbf{x})$ is its concentration at point $\mathbf{x}$ in the soil solution, and c_{0i} is its concentration in the soil solution far enough from any particle surface to avoid adsorption in the diffuse-ion swarm. The integral in Eq. 7.5 is over the entire volume V of soil solution contacting the mass m_s of soil particles. Thus Eq. 7.5 represents the *excess* charge of ion i in the soil solution: if $c_i(\mathbf{x}) = c_{0i}$ uniformly, there would be no contribution of ion i to σ_D. Note that Eq. 7.5 applies to all ions in the soil solution, including H^+ and OH^-, and that σ_D is the sum of all σ_{Di}. This sum is required to balance σ_P to maintain the soil electrically neutral:

$$\sigma_P + \sigma_D = 0 \tag{7.6}$$

Equation 7.6 is a statement of the *balance of surface charge* for a soil.

7.4 Points of Zero Charge

Points of zero charge are *pH values* at which one or more of the surface charge components in Eq. 7.6 vanishes. The three most important points of zero charge are summarized in Table 7.1. The most general *point of zero charge* (PZC) is the pH value at which the net total particle charge vanishes: $\sigma_P = 0$. At this pH value, there is no net charge contributed by the ions adsorbed in the diffuse swarm, according to Eq. 7.6. This condition can be ascertained experimentally by establishing the pH value at which soil particles do not move in an applied electric field ("electrophoretic mobility measurement") or at which settling occurs in a suspension of soil particles ("flocculation measurement"). As is explained in detail in Section 10.5, the PZC signals the absence of freely moving adsorbed ions and the enhancement of interparticle forces that produce coagulation effects. Thus the PZC plays an important role in soil aggregate formation and in the retention of adsorbed ions against

TABLE 7.1 Some points of zero charge

Symbol	Name	Defining condition
PZC	Point of zero charge	$\sigma_P = 0$
PZNPC	Point of zero net proton charge	$\sigma_H = 0$
PZNC	Point of zero net charge	$\sigma_{IS} + \sigma_{OS} + \sigma_D = 0$

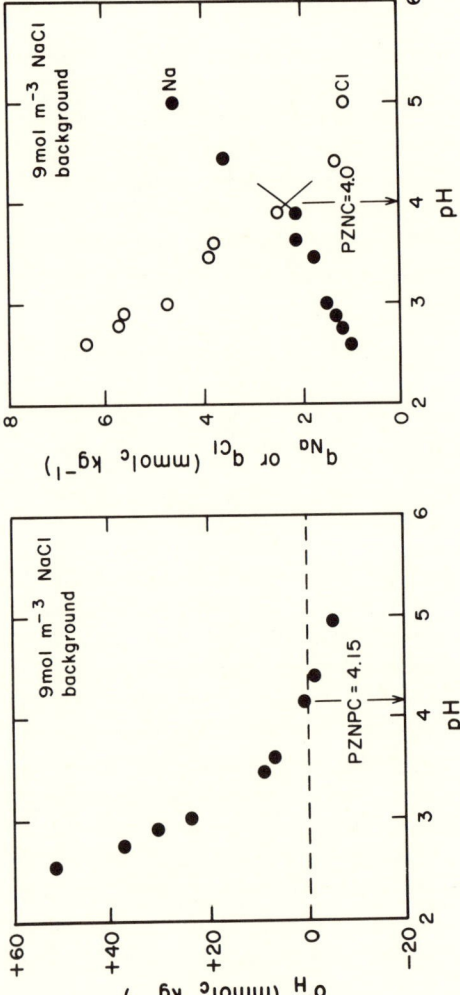

FIG. 7.4 The point of zero net proton charge (PZNPC) and the point of zero net charge (PZNC) for an Oxisol. [Based on data in L. Charlet, Adsorption of some macronutrient ions on an Oxisol. An application of the Triple Layer model, Ph.D. dissertation, Univ. of California, Riverside, 1986.]

leaching losses, particularly in variable-charge soils. It is observed often that these soils tend to weather to pH values that are close to the PZC.

The *point of zero net proton charge* (PZNPC) is the pH value at which σ_H vanishes. Figure 7.4 illustrates the measurement of the PZNPC for an Oxisol. Note also the general property of σ_H, that it *decreases* as the pH value *increases* (i.e., $\Delta\sigma_H/\Delta pH$ is always negative). This trend exists regardless of the composition or ionic strength of the soil solution and independently of the nature of the soil particles (whether inorganic or organic, etc.).

Analogously to the PZNPC, one can define the *point of zero net charge* (PZNC) as the pH value at which the net adsorbed ion charge, other than that represented by σ_H, vanishes. If q_+ and q_- represent the moles of adsorbed cation and anion charge, respectively, then $q_+ = q_-$ at the PZNC. This point of zero charge is also illustrated in Fig. 7.4. In this example, the soil, an Oxisol, was saturated with Na^+ and Cl^-; then the values of q_{Na} and q_{Cl} were measured as a function of varying pH while the ionic strength was maintained constant. The PZNC is the pH value when $q_{Na} = q_{Cl}$. It is common practice to utilize "index ions" like these in the measurement of the PZNC. Evidently, the value of the PZNC will depend on the choice of index ions, although experience shows that this dependence is very small if the ions chosen are adsorbed only nonspecifically ("indifferent electrolyte"). Ions such as Li^+, Na^+, Cl^-, ClO_4^-, and NO_3^- are examples. Note that, according to the concepts developed in Section 7.2,

$$q_+ - q_- \equiv \sigma_{IS} + \sigma_{OS} + \sigma_D = 0 \qquad (pH = PZNC) \qquad (7.7)$$

at the PZNC (although σ_{IS} will not contribute if the soil is saturated with an indifferent electrolyte). Thus mobile adsorbed ions exist at the PZNC, whereas they do not at the PZC. This fact may have important implications for the retention of nutrient ions in soils. Typical values of PZNC for soil minerals are listed in Table 7.2.

General relationships among the component surface charge densities and points of zero charge can be derived from Eq. 7.6 using the definitions in Table

TABLE 7.2 Representative values of PZNC for soil minerals[a]

Soil mineral	PZNC	Soil mineral	PZNC
Quartz and silica	2.0–3.0	Goethite	7.0–8.0
Birnessite	1.5–2.5	Hematite	8.0–8.5
Kaolinite	4.0–5.0	Gibbsite	8.0–9.0

[a]Chemical formulas of the minerals are in Table 1.3.

7.1. For example, if pH = PZNPC, then

$$\sigma_0 = -(q_+ - q_-) \qquad (\text{pH} = \text{PZNPC}) \tag{7.8}$$

according to Eq. 7.4 and the definition of "adsorbed ion" given in Section 7.2. Equation 7.8 states that adsorbed ions, other than H^+ or OH^- contributing to σ_H, balance the permanent structural charge when pH = PZNPC. Thus σ_0 can be measured by adjusting the pH to this point of zero charge and determining the net adsorbed ion charge. This method can be used without pH adjustment if $|\sigma_H| \ll |\sigma_0|$, that is, in permanent-charge soils.

As another example, consider Eqs. 7.6 and 7.7 at the PZC. At this pH value, $\sigma_D = 0$. The PZNC will occur at the same pH value if and only if $\sigma_{IS} + \sigma_{OS}$ also vanishes, according to Eq. 7.7. Thus *PZC = PZNC only when the net surface complex charge vanishes at the PZNC.* If the cations and anions in surface complexes balance in charge at the PZNC, there will be no need for a diffuse-ion swarm contribution and the PZC will have been achieved as well. Note that, if an indifferent electrolyte is used in the determination of the PZNC, then it is necessary to partition the adsorbed ions into outer-sphere complexes and diffuse-swarm categories in order to establish whether PZC = PZNC.

Finally, the condition that pH = PZC applied to Eqs. 7.5 and 7.6 can be expressed in the equation:

$$\sigma_H = -(\sigma_0 + \sigma_{IS} + \sigma_{OS}) \qquad (\text{pH} = \text{PZC}) \tag{7.9}$$

Since σ_H varies inversely with pH, the pH value at which Eq. 7.9 holds must increase as $\sigma_{IS} + \sigma_{OS}$ increases or decrease as $\sigma_{IS} + \sigma_{OS}$ decreases. Thus if $\sigma_{IS} + \sigma_{OS}$ increases, say through the adsorption of additional K^+ ions in inner-sphere surface complexes (Fig. 7.2), then σ_H at the PZC will decrease according to Eq. 7.9 and the PZC itself will increase, since $\Delta\sigma_H/\Delta\text{pH}$ is always negative. The general rule is:

The formation of surface complexes will shift the PZC in the same direction as the change in net surface complex charge.

Additional adsorption of cations in surface complexes will thus raise the PZC, whereas additional adsorption of anions in surface complexes will lower it. Note that *these shifts in the PZC do not require specific adsorption, but only surface complexation.* Thus a shift in the PZC on additional adsorption of an ion *cannot* be interpreted as evidence for specific adsorption. This kind of shift is evidence only that surface complexes have formed in obedience to the law of charge balance.

Frequently, σ_H is measured by titration in the presence of an indifferent electrolyte at several ionic strengths. The resulting σ_H–pH curves are plotted as on the left in Fig. 7.4, and the pH value at which the curves intersect (if they intersect at a *single* point) is determined. This pH value is the *point of zero salt effect* (PZSE). Unlike the points of zero charge in Table 7.1, however, the PZSE is strictly a kind of invariance point for σ_H, not a pH value at which surface charge necessarily vanishes. The relationship between PZSE and the points of zero charge is therefore indirect and is not interpretable without either additional model assumptions about soil particle surfaces or additional experiments. Its use in understanding surface charge in soils is thereby limited.

7.5 Adsorbed Water

Soil particle surfaces develop electrical charge when they are in contact with a liquid phase. A clear understanding of the effect of this charge on ion adsorption requires not only information about the structural nature of surface functional groups, but also a comprehension of the influence the liquid phase may have on them. In a first approximation, one can assume that these effects are the same as observed for functional groups on small molecules in bulk aqueous solution. This assumption should be correct for dilute soil suspensions. But it may be quite wrong for soil particles enveloped by thin films of water, because the solid surfaces could perturb the water molecules in the films enough to alter their molecular configuration from what exists in the bulk liquid phase. The altered water structure, in turn, could exhibit different properties from the bulk liquid and therefore could affect surface functional groups in a different manner. The interfacial region in which water molecules take on a structure that differs from that in bulk water (or in an aqueous electrolyte solution) defines the region of *adsorbed water* on a soil particle surface.

The molecular structure of water adsorbed by soil minerals bearing siloxane surfaces has been investigated extensively with a variety of spectroscopic techniques. Some of the structural concepts that have emerged from these investigations are illustrated in Figs. 7.5–7.7. In the case of kaolinite group minerals, a monolayer of water molecules is weakly adsorbed to form the hexagonal-network configuration shown in Fig. 7.5. The stability of the network comes from hydrogen bonding, both among the water molecules and between them and the atoms in the basal planes that bound the interlayer region. Because of the constraints imposed by the requirement of fitting the network to the mineral surface atom arrangement, the hydrogen bonds within the monolayer must be about 0.3 nm long, and those between the water molecules and the mineral surface must deviate considerably from linearity. These structural characteristics are similar to those in bulk liquid water. Thus the water monolayer on kaolinite is a strained, hydrogen-bonded network whose structure reflects a compromise between the relative disorder of liquid water and the rigid configuration demanded by a strict fit with the atoms in the basal planes of the clay mineral. The monolayer is liquidlike in not exhibiting a rigid orientation of its constituent molecules and in comprising relatively weak, distorted hydrogen bonds. However, the clay mineral surface is able to "slow

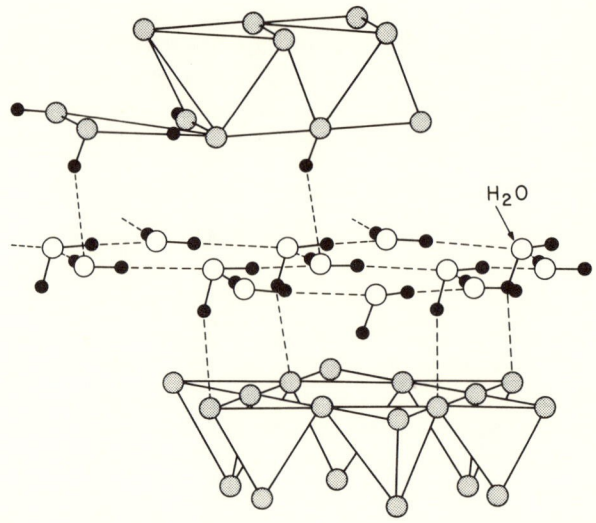

FIG. 7.5 The molecular structure of adsorbed water in the interlayers of halloysite (see Table 2.3). (Reprinted with permission from G. Sposito, *The Surface Chemistry of Soils*, Oxford Univ. Press, New York, 1984.)

down" the molecular motions in the monolayer because of hydrogen bonding.

The configuration of water molecules between the siloxane surfaces of vermiculite depends sensitively on the fact that a cation is adsorbed on these surfaces to balance negative charge produced by isomorphic substitution of Al^{3+} for Si^{4+} in the tetrahedral sheet. Because of this substitution pattern, the siloxane cavities bear a relatively localized charge that can interact strongly with both cations and water molecules, as mentioned in Section 7.1. The overall effect of charge localization is the close proximity of an adsorbed cation to a tetrahedral site containing Al^{3+} and the formation of relatively strong hydrogen bonds between interlayer water molecules and surface oxygen atoms. Aside from these special effects of the vermiculite surface, however, the organization of adsorbed water molecules in the interlayer region conforms to the behavior observed in relatively concentrated aqueous solutions. In particular, water molecules are coordinated in a single solvation shell about monovalent cations and in two shells about bivalent cations (Fig. 7.6) in fully hydrated vermiculite. The solvation water molecules near monovalent cations are relatively mobile, whereas those in the first solvation shells about bivalent cations are bound rigidly and move with the cation as a unit. Thus hydrated vermiculites contain *interlayer ionic solutions*, with both the cations and their solvating water molecules influenced by the coulomb fields from tetrahedral sites containing Al^{3+}.

As with the water molecules on the surface of vermiculite, the behavior of water on smectite surfaces depends sensitively on the type of adsorbed cation

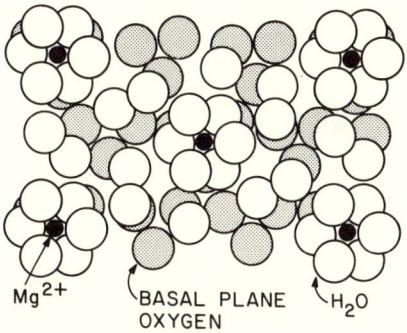

FIG. 7.6 The molecular structure of adsorbed water in the interlayers of Mg-saturated vermiculite. (Reprinted with permission from G. Sposito, *The Surface Chemistry of Soils*, Oxford Univ. Press, New York, 1984.)

and the location of isomorphic cation substitutions in the layer structure. In many respects, the configuration of water molecules hydrating smectites is quite parallel to that on vermiculite. The structure of adsorbed water in the one-layer hydrate of monovalent cation-saturated smectite is illustrated in Fig. 7.7 for the case of montmorillonite. This structure is a strained, icelike configuration of water molecules around the adsorbed M^+ cations, with weak hydrogen bonds formed both intermolecularly and with the siloxane cavities in the montmorillonite surface. The nearest-neighbor distance between water molecules is 0.32 nm, and five water molecules are assigned to each exchangeable cation in the completed monolayer. Since some of the water molecules have a hydroxyl group proton entrapped by a siloxane cavity, the network of hydrogen bonds among the water molecules is broken in places, thereby permitting a variety of rotational motions that are not possible in ice. These properties also suggest that the adsorbed water is organized like a two-dimensional aqueous electrolyte solution, with hydrogen bonding affected by the charge distribution on the siloxane surface.

The definition of *adsorbed water* requires an arrangement of water molecules on a surface that differs significantly from that in an appropriate reference aqueous phase. For water on the surfaces of kaolinite group minerals, the reference phase is bulk liquid water, whereas for water on vermiculite and smectite surfaces, the reference phase is an aqueous electrolyte solution because of the presence of adsorbed cations on the 2:1 layer silicates. On this basis, the consensus is that adsorbed water exists at least within the first nanometer out from a siloxane surface. Experiments have shown that this water has a much lower dielectric constant than in bulk water (between 2 and 50 in adsorbed water vs. 80 in bulk water). This lower dielectric constant implies that adsorbed water molecules are less free to rotate than are the molecules in bulk liquid water, evidently because of orientations among the

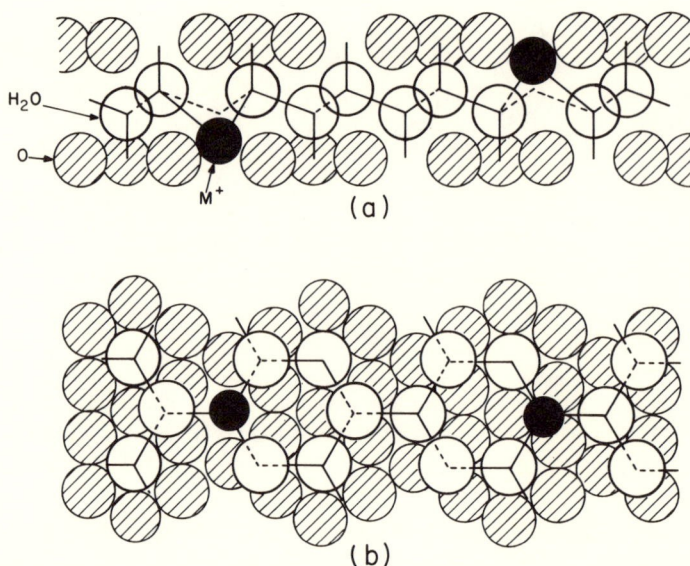

FIG. 7.7 The molecular structure of adsorbed water in the interlayers of K-saturated montmoril-lonite. (a) View in cross section (basal-plane oxygen atoms shown as shaded circles). (b) View in the basal plane (water molecules nearest the upper basal plane indicated by dashed lines). (Reprinted with permission from G. Sposito, *The Surface Chemistry of Soils*, Oxford Univ. Press, New York, 1984.)

water molecules imposed by the strong coulomb fields of adsorbed cations and the requirements of hydrogen bonding. The most significant chemical effect of this loss of rotational mobility and consequent lower dielectric constant is the enhancement of surface complex formation. It is well known that, with a lower dielectric constant in aqueous systems, ion association into complexes is favored and the development of a diffuse-ion swarm near a colloidal particle is reduced.

Another important chemical property of adsorbed water on vermiculite and smectite surfaces is its enhanced acidity. This property arises principally from the acidity of the solvated adsorbed cations, as illustrated by the reaction:

$$M(H_2O)^{m+} = MOH^{(m-1)+} + H^+ \tag{7.10}$$

The equilibrium constant for this hydrolysis reaction has been measured for a variety of metal cations in aqueous solutions and is known to correlate positively with their *ionic potential* (the ratio of valence to ionic radius). As the ionic potential increases, the intensity of the positive coulomb field of the cation increases and repulsion of a solvating water proton becomes more likely. As the number of water molecules on a siloxane surface is reduced by drying to the point where only primary solvation shells can form, the coulomb field of an adsorbed cation can produce surface proton dissociation as in Eq. 7.10. Experiments on smectites and vermiculites have indicated that, in adsorbed water, the degree of proton dissociation is about two orders of magnitude larger than in bulk water. This excess of free protons in adsorbed water can have an important effect on surface protonation reactions (Section 3.5) and other aspects of soil acidity (Chapter 11).

FOR FURTHER READING

G. W. Brindley and G. Brown, *Crystal Structures of Clay Minerals and Their X-ray Identification*, Mineralogical Society, London, 1980. Chapter 3 of this standard reference, coauthored by D. M. C. MacEwan and M. J. Wilson, gives an excellent discussion of adsorbed water and surface complexes with clay minerals.

D. J. Greenland (ed.), *Characterization of Soils*, Clarendon Press, Oxford, 1981. Chapter 4 of this fine book on the soils of the humid tropics, written by A. S. R. Juo, presents a concise but thoughtful review of the chemistry of variable-charge soils.

D. J. Greenland and M. H. B. Hayes (eds.), *The Chemistry of Soil Constituents*, Wiley, New York, 1978. Chapter 4 of this fine treatise on soil chemistry, coauthored by D. J. Greenland and C. J. B. Mott, provides a detailed review of the structures of soil mineral surfaces and the origins of the electric charge they bear.

M. M. Mortland and K. V. Raman, Surface acidity of smectites in relation to hydration, exchangeable cation, and structure, *Clays and Clay Minerals* **16**:393–398 (1968). This article describes the classic protonation experiments that helped to establish the existence of enhanced acidity in adsorbed water on smectites.

G. Sposito, *The Surface Chemistry of Soils*, Oxford Univ. Press, New York, 1984. The first three chapters of this monograph discuss surface functional groups, adsorbed water, and surface charge in detail.

B. K. G. Theng, *Soils with Variable Charge*, New Zealand Society of Soil Science, Lower Hutt, New Zealand, 1980. Part III of this useful edited volume contains four chapters on the chemistry of variable-charge soils: adsorption, ion exchange, fertility aspects, and organic matter interactions.

PROBLEMS

The more difficult problems are indicated by an asterisk.

1. Ammonium ions can form inner-sphere surface complexes with 2:1 layer-type clay minerals like that shown in Fig. 7.2 for K^+. Given this fact, discuss the advantages and disadvantages of using ammonium acetate solution to extract the readily exchangeable cations in a soil.

2. Verify the calculation of the range of σ_0 for vermiculite given in Section 7.3.

*3. Through a titration experiment, it is found that the net proton surface charge density of a soil humus sample can be described mathematically by the equation:

$$\sigma_H = \frac{B_1 K_1 10^{-pH}}{1 + K_1 10^{-pH}} + \frac{B_2 K_2 10^{-pH}}{1 + K_2 10^{-pH}} - C$$

where $C = 8.0$ mol$_c$ kg^{-1}, $B_1 = 6$ $B_1 = 6$ mol$_c$ kg^{-1}, $B_2 = 2$ mol$_c$ kg^{-1}, $K_1 = 10^{4.7}$, and $K_2 = 10^{9.2}$. Calculate the PZNPC for the humus sample. (*Hint:* For pH < 4.0, the second term on the right side equals 2 mol$_c$ kg^{-1} and the first term equals $B_1[1 - (10^{pH}/K_1)]$ to a good approximation. Then one can assume that pH $=$ PZNPC for all practical purposes when the factor in square brackets equals 0.95.)

*4. Given the equation for σ_H in Problem 3, calculate the value of q_+ (adsorbed cation charge) at pH 7 on the soil humus sample. What would be the contribution of this component to the adsorbed cation charge in a soil containing 20 g humus per kilogram of soil? (*Answer:* $q_+ = 6$ mol$_c$ kg^{-1} in the humus and 12 cmol$_c$ kg^{-1} in the soil at pH 7).

5. The data in the accompanying table give the pH dependence of σ_H for an Oxisol suspension in 30 mol m^{-3} NaCl. Plot these data as in Fig. 7.4, and determine the PZNPC.

σ_H (mmol$_c$ kg^{-1})	pH	σ_H (mmol$_c$ kg^{-1})	pH
58	2.6	12	3.6
42	2.8	1.2	4.0
34	2.9	-4.1	4.4
26	3.1	-7.5	4.6
19	3.3	-11	4.8

6. The data here show the pH dependence of q_K and q_{NO_3} (in mmol kg^{-1}) for an Oxisol suspended in 10 mol m^{-3} KNO$_3$. Determine the PZNC for the soil.

q_K	q_{NO_3}	pH	q_K	q_{NO_3}	pH
0.7	6.6	2.5	4.0	3.5	3.9
1.3	5.1	2.8	4.2	3.0	4.1
2.5	4.7	3.2	4.9	1.5	4.6
2.3	4.3	3.6	6.3	1.2	4.9
2.6	4.0	3.7	9.5	0.6	5.3

7. Oxisols tend to have essentially no permanent structural surface charge. Use Eq. 7.5 to show that PZNPC $=$ PZNC for these soils. This result applies also to the variable-charge minerals in Table 7.2.

8. The value of q_{Cs} on gibbsite suspended in CsCl solution was found to increase linearly from 0 to 20 mmol$_c$ kg^{-1} as the pH increased from 7.7 to 9.0. The value of q_{Cl} decreased linearly from 13 to 0 mmol$_c$ kg^{-1} as the pH increased from 4 to 9. Estimate the PZNC and the PZNPC for this mineral.

9. A permanent-charge soil typically will have $<20\%$ of its intrinsic surface charge contributed by σ_H. Explain, using Eqs. 7.6 and 7.8, why the measurement of the PZNPC is likely to be very difficult for these soils.

10. The PZC of a variable-charge soil is 5.0. After a phosphate fertilizer is applied to the soil, it is observed that the soil retains more adsorbed cations at pH 5 than before. Explain in terms of surface charge balance why this effect occurs.

11. Potassium fertilizer is added to a vermiculitic soil, and it is observed that the retention of NO_3^- by the soil at a given pH increases. Explain why this should occur on the basis of surface charge balance concepts.

12. Discuss the mineralogical basis for the statement: "In variable-charge soils, the greater the degree of desilication (silica removal), the higher will be the PZNC." What are the principal implications of this fact for K^+ and NO_3^- adsorption by these soils? What is its relation to the Jackson–Sherman weathering stages (Table 1.7)?

*13. The PZSE of a soil can be defined mathematically by the equation:

$$(\partial \sigma_H / \partial I)_{pH=PZSE} = 0$$

where I is the ionic strength of an electrolyte solution in which the soil is suspended. Show that PZSE = PZNC if $\partial(q_+ - q_-)\partial I$ is zero at the PZNC.

(*Hint:* Calculate the derivative of both sides of Eq. 7.6 with respect to I.)

14. Given the value $\sigma_0 = -0.91$ mol$_c$ kg^{-1} and the structure of the two-layer Ca-montmorillonite hydrate in Fig. 7.2, calculate the water content of the clay mineral in grams H_2O per kilogram of clay.

15. The data in the accompanying table give the yield of NH_4^+ found from the reaction: $NH_3 + H^+ = NH_4^+$ on the surface of Ca- or Mg-montmorillonite equilibrated with water at a given thermodynamic activity (relative humidity/100). Explain these results in terms of the ionic potentials of Ca^{2+} and Mg^{2+}.

Adsorbed cation	(H_2O)	NH_4^+ formed (mol kg^{-1})
Mg^{2+}	0.20	1.01
Mg^{2+}	0.98	0.74
Ca^{2+}	0.20	0.80
Ca^{2+}	0.98	0.16

8

Soil Adsorption Phenomena

8.1 Measuring Adsorption

Adsorption as defined in Section 7.2 is studied experimentally in soils by means of two basic laboratory operations: (1) reaction of the soil with a solution of known composition at fixed temperature and applied pressure for a prescribed period of time and (2) chemical analysis of the reacted soil, the soil solution, or both, to determine their compositions. The reaction in step 1 can take place either with the solution mixed uniformly with the soil particles ("batch process") or with the solution in uniform motion relative to a column or pad of soil particles ("flow-through process"). The reaction time should be long enough to permit a detectable accumulation of the adsorbate, but short enough to avoid unwanted side reactions, such as redox, precipitation, or dissolution reactions. In batch processes, the chemical analysis in step 2 is usually carried out after the isolation of the soil from the reacted solution by centrifugal or gravitational force (Section 4.1). Some of the reactant solution will always be entrained with the soil in this kind of separation. In flow-through processes, the composition of the effluent solution is analyzed to determine the changes caused by adsorption.

The moles of chemical species i adsorbed per kilogram of dry soil contacting an aqueous solution is calculated with the equation:

$$q_i = n_i - M_w m_i \tag{8.1}$$

where n_i is the total moles of species i per kilogram of dry soil in the soil–entrained solution slurry (batch process), or in the wet soil column or pad (flow-through process); M_w is the gravimetric water content of the slurry or soil column (kilograms water per kilogram of dry soil); and m_i is the molality

148

(moles per kilogram of water) of species i in the supernatant solution (batch process) or in the effluent solution (flow-through process). For a discussion of the units of n_i, M_w, and m_i, see the Appendix. Equation 8.1 represents the *surface excess*, q_i, of a chemical species. Even though no direct reference to an adsorbing surface appears in the equation, a simple calculation shows that q_i is identically zero if species i is water. Therefore, strictly speaking, q_i refers to an interface at which there is no net adsorption of water.

The value of q_i in Eq. 8.1 can be positive, zero, or negative. Consider, for example, a permanent-charge soil that has been reacted in a batch process with a $CaCl_2$ solution. After the reaction, the soil and bulk aqueous solution are separated by centrifugation. The soil slurry is found to contain 0.053 mol Ca kg^{-1} and to have a gravimetric water content of 0.45 kg kg^{-1}. The supernatant solution contains Ca at a molality of 0.01 mol kg^{-1}. According to Eq. 8.1,

$$q_{Ca} = 0.053 - (0.45)(0.01) = +0.049 \text{ mol kg}^{-1}$$

is the *positive* surface excess of Ca in the soil. By the same token, suppose that the molality of Cl in the supernatant solution is 0.02 mol kg^{-1} and that the soil slurry contains 0.0028 mol Cl kg^{-1}. Then,

$$q_{Cl} = 0.0028 - (0.45)(0.02) = -0.006 \text{ mol kg}^{-1}$$

is the *negative* surface excess of Cl in the soil. In both examples, q_i is the *excess* moles of species i (per kilogram of dry soil), relative to an aqueous solution containing M_w kilograms of water and species i at the molality m_i. This net excess is attributed to the presence of adsorbing soil particles.

If the initial molality of species i in the reactant aqueous solution is m_i^0 and the total mass of water in this solution that is mixed with 1 kg of dry soil, in either a batch or a flow-through process, is M_{Tw}, then the condition of mass balance for species i can be expressed:

$$\underset{\text{(moles added initially)}}{m_i^0 M_{Tw}} \quad = \quad \underset{\text{(moles in slurry)}}{n_i} \quad + \quad \underset{\text{(moles in supernatant solution)}}{m_i(M_{Tw} - M_w)} \qquad (8.2)$$

Equations 8.1 and 8.2 can be combined to yield:

$$q_i = \Delta m_i M_{Tw} \qquad (8.3)$$

where $\Delta m_i \equiv m_i^0 - m_i$ is the change in molality attributed to adsorption. Equation 8.3 is applied frequently to calculate the surface excess as the product of the change in adsorptive concentration times the mass of water *added* per unit mass of dry soil. Note that the right side of Eq. 8.3 refers only to the aqueous solution phase. In practice, the difference between molality and concentration in moles per liter (cubic decimeter) can often be neglected in applying the equation. (See the Appendix for a discussion of molality and concentration units.)

Implicit in the use of macroscopic expressions like Eq. 8.1 or 8.3 is the assumption that the soil–aqueous solution reaction *is* an adsorption process, as defined in Section 7.2. This assumption in general cannot be validated by investigating q_i alone. Ideally, there should be experimental spectroscopic evidence for surface species (Fig. 7.3) in addition to surface excess measurements. This kind of evidence is not easy to obtain for soils, but often the results found in studies with reference soil minerals (Fig. 7.2) can be extrapolated to apply to soils.

8.2 Adsorption Kinetics and Equilibria

Experiments with both cation and anion adsorptives have shown that adsorption reactions in soils are typically rapid, operating on time scales of minutes or hours, but that they can exhibit long-time "tails" that extend over days or even weeks. Readily exchangeable ions (Section 7.2) adsorb and desorb very rapidly, with a rate governed by a film diffusion mechanism (Section 3.3 and Special Topic 2). Specifically adsorbed ions show much more complicated behavior, in that they often adsorb by multiple mechanisms that differ from those involved in their desorption, and their rates of adsorption or desorption are described by more than one equation during the course of either process. It is these ions whose adsorption reactions have the long-time "tails."

The adsorption kinetics for cations or anions are usually *assumed* to be represented mathematically by the difference of two terms, as in Eq. 4.2:

$$\frac{d\Delta c_i}{dt} = R_f - R_b \tag{8.4}$$

where $\Delta c_i \equiv c_i^0 - c_i$ is the change in adsorptive i concentration (molality or moles per liter) attributed to adsorption. In Eq. 8.4, Eq. 8.3 has been used to replace q_i in the time derivative, with the constant parameter M_{Tw} absorbed into the forward and backward rate functions, R_f and R_b. A consensus does not exist as to which rate laws should be applied to model R_f and R_b; many different empirical formulations appear in the soil chemistry literature. One popular choice has been the first-order kinetics law (Table 4.2), for which

$$R_f = k_f c_i \qquad R_b = k_b' q_i = k_b \Delta c_i \tag{8.5}$$

where k_f and k_b are rate constants (Eq. 4.3). As indicated in Table 4.2, a graph of $\ln c_i$ against time can be used to calculate either k_f or k_b if data are taken under conditions such that either R_b or R_f is negligible. First-order rate laws do not reflect a unique mechanism of adsorption of desorption. They are empirical mathematical models whose molecular significance must be established by independent experiments on the detailed nature of the surface reactions they describe.

An illustration of the required interplay between kinetics data and molecular data is provided by the reaction of phosphate ions with calcite ($CaCO_3$). The increase in Δc_{PO_4} (i.e., the loss of phosphate from aqueous solution) in the presence of calcite is pronounced on a time scale of tens of minutes, and is enhanced by increasing temperature or pH. Thereafter, on a time scale of hours or days, the value of Δc_{PO_4} increases gradually, then rises sharply again. This behavior is interpreted mechanistically as the adsorption of phosphate at selective sites on calcite, followed by the nucleation of a calcium phosphate solid on the surface. The gradual increase in Δc_{PO_4} persists longer if the initial phosphate concentration is low, and it represents the rearrangement of adsorbed phosphate clusters into Ca phosphate nuclei. Rapid three-dimensional growth of calcium phosphate crystals then follows. Scanning electron micrographs of calcite taken during the rearrangement period show hemispheric growths of phosphate (identified by microprobe chemical analysis) at edge sites and dislocations on the calcite crystal surface. The adsorption kinetics are found to be second-order (Table 4.2), whereas the rearrangement kinetics are first-order. Thus in this example, first-order kinetics did not reflect an adsorption process, but instead were caused by a molecular rearrangement process on a surface. The second-order kinetics are thought to reflect a dependence of the adsorption rate both directly on the solution concentration of phosphate and indirectly through a dependence on the surface excess.

A graph of q_i against m_i or c_i at fixed temperature and applied pressure is an *adsorption isotherm*. Adsorption isotherms are convenient for representing the effects of adsorptive concentration on the surface excess, especially if other variables such as pH and ionic strength are controlled along with temperature and pressure. Figure 8.1 shows four categories of adsorption isotherm observed commonly in studies with soils.

The *S-curve* isotherm is characterized by an initially small slope that increases with adsorptive concentration. This behavior suggests that the affinity of the soil particles for the adsorbate is less than that of the aqueous solution for the adsorptive. In the example of copper adsorption given in Fig. 8.1, the S-curve is thought to result from competition for Cu^{2+} ions between soluble organic matter and the soil particles. Once the concentration of Cu(II) exceeds the complexing capacity of the organic ligands, the soil particle surface gains in the competition and begins to adsorb copper ions significantly. Thereafter, the isotherm takes on its characteristic S shape. In some instances, especially when organic compounds are adsorbed, the S-curve isotherm is the result of cooperative interactions among the adsorbed molecules. These interactions (e.g., surface polymerization or stereochemical interactions) cause the adsorbate to become stabilized on a solid surface, and thus they produce an enhanced affinity of the surface for the adsorbate as its surface excess increases.

The *L-curve* isotherm is characterized by an initial slope that does not increase with the concentration of adsorptive in the soil solution. This type of

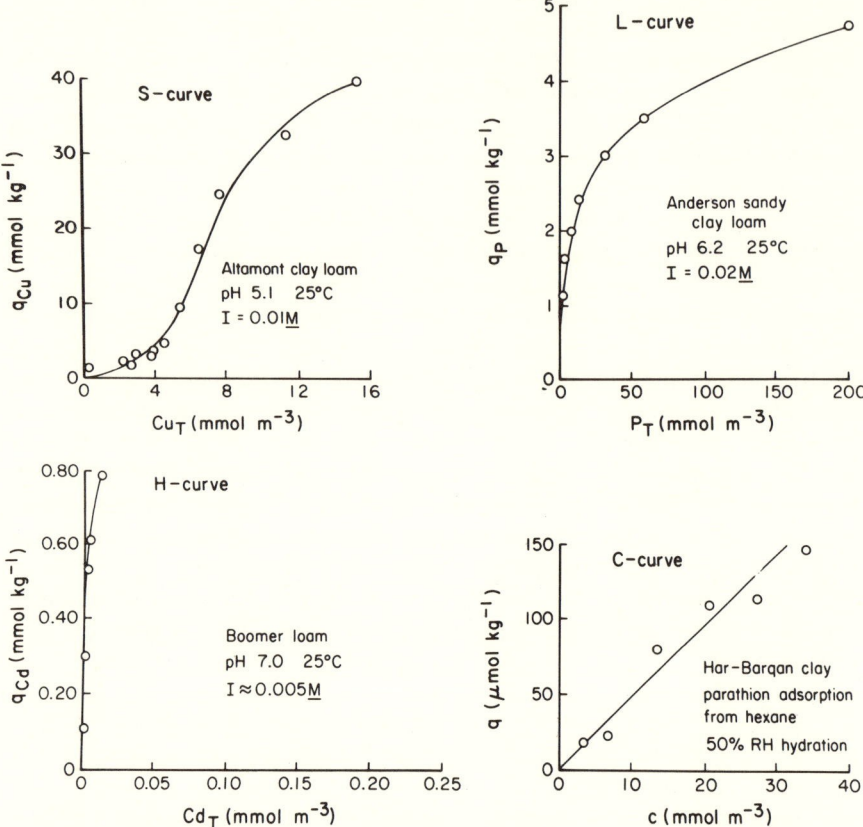

FIG. 8.1 The four general categories of adsorption isotherm. (Reprinted with permission from G. Sposito, *The Surface Chemistry of Soils*, Oxford Univ. Press, New York, 1984.)

isotherm is the resultant effect of a high relative affinity of the soil particles for the adsorbate at low surface coverage coupled with a decreasing amount of adsorbing surface remaining as the surface excess of the adsorbate increases. The example of phosphate adsorption in Fig. 8.1 illustrates a universal L-curve feature: the isotherm is concave to the concentration axis because of the combination of affinity and steric factors.

The *H-curve* isotherm is an extreme version of the L-curve isotherm. Its characteristic large initial slope (in comparison with the L-curve isotherm) suggests a very high relative affinity of the soil for an adsorbate. This condition is usually produced either by inner-sphere surface complexation or by significant van der Waals interactions in the adsorption process (Sections 3.4 and 3.5). The example of cadmium adsorption at very low concentrations by a kaolinitic soil, shown in Fig. 8.1, illustrates an H-curve isotherm caused by specific adsorption. Large organic molecules and inorganic polymers (e.g., aluminum hydroxy-polymers) provide examples of H-curve isotherms resulting from van der Waals interactions.

The *C-curve* isotherm is characterized by an initial slope that remains independent of adsorptive concentration until the maximum possible adsorption is achieved. This kind of isotherm can be produced either by a constant partitioning of an adsorptive between the interfacial region and the soil solution, or by a proportionate increase in the amount of adsorbing surface as the surface excess of an adsorbate increases. The example of parathion (diethyl *p*-nitrophenyl monothiophosphate) adsorption in Fig. 8.1 shows constant partitioning of this compound between hexane and the layers of water on a soil at 50% relative humidity (RH). The adsorption of amino acids by Ca-montmorillonite also exhibits a C-curve isotherm because the adsorptive can penetrate into the interlayer region, thereby creating new adsorbing surface for itself.

The L-curve isotherm is by far the one most commonly encountered in soil chemistry. The mathematical description of this isotherm almost invariably has involved either the Langmuir equation or the van Bemmelen–Freundlich equation. The *Langmuir equation* has the form:

$$q_i = \frac{bKc_i}{1 + Kc_i} \tag{8.6}$$

where b and K are adjustable parameters. The parameter b represents the value of q_i that is approached asymptotically as c_i becomes arbitrarily large. The parameter K determines the magnitude of the initial slope of the isotherm. The most precise way to calculate these two parameters with experimental data is to plot the *distribution coefficient* (Eq. 3.9),

$$K_d \equiv q_i/c_i \tag{8.7}$$

against the surface excess. After multiplying both sides of Eq. 8.6 by $(1/c_i + K)$ and solving for K_d, one finds that the Langmuir equation is equivalent to the linear expression:

$$K_d = bK - Kq_i \tag{8.8}$$

Thus a graph of K_d against q_i should be a straight line with slope equal to $-K$ and an x intercept equal to b, if the Langmuir equation is applicable.

The *van Bemmelen–Freundlich isotherm equation* has the form:

$$q_i = Ac_i^{\beta} \tag{8.9}$$

where A and β are positive-valued adjustable parameters, with β constrained to lie between 0 and 1. These two parameters can be estimated by plotting log q_i against log c_i for the range of adsorptive concentrations over which Eq. 8.9

applies. Then $\log A$ and β are calculated as the y intercept and slope, respectively, of the resulting straight line.

Like empirical rate laws, adsorption isotherm equations cannot be interpreted to indicate any particular adsorption mechanism, *or even if adsorption, as opposed to precipitation, actually has occurred.* On strictly mathematical grounds, it can be shown that a sum of two Langmuir equations with four adjustable parameters will fit *any* L-curve isotherm, regardless of the underlying adsorption mechanism. Thus adsorption isotherm equations should be regarded as curve-fitting models without particular molecular significance, but with predictive capability under limited conditions.

8.3 Metal Cation Adsorption

Metal cations adsorb onto soil particle surfaces via the three mechanisms illustrated in Fig. 7.3. The relative affinity that a given metal cation has for a soil adsorbent depends in a complicated way on soil solution composition. But to a first approximation, the selectivity of a soil for an adsorptive metal cation can be rationalized in terms of inner-sphere and outer-sphere surface complexation and diffuse-ion swarm concepts. As discussed in Section 7.2, the relative order of decreasing interaction strength among the three adsorption mechanism is: inner-sphere complex > outer-sphere complex > diffuse-ion swarm. For the inner-sphere surface complex, the electronic structures of the metal cation and surface functional group are important, whereas for the diffuse-ion swarm only the metal cation valence and surface charge should be critical to determining adsorption affinity. The outer-sphere surface complex is intermediate, in that valence is probably the most important factor, but the stereochemical-enhancement effect of immobilizing a cation in a well-defined complex must also play a role in determining affinity.

As a rule of thumb, the relative affinity of a soil adsorbent for a *free* metal cation will increase with the tendency of the cation to form inner-sphere surface complexes. For a series of metal cations of a given valence, this tendency is correlated positively with the ionic radius (Table 2.1). The reason for this correlation is twofold. First, for a given valence Z, the ionic potential Z/R (Section 7.5) decreases with increasing ionic radius R. This trend implies that metal cations with larger ionic radii will create a smaller electric field and will be less likely to remain solvated (see Problem 3 in Chapter 4) in the face of competition for complexation by a surface functional group. Second, a larger R implies a larger spread of the electron configuration in space and a greater tendency for a metal cation to polarize (distort) in response to the electric field of a charged surface functional group. This polarization is the necessary prerequisite for the distortion of the electron configuration leading to covalent bonding (Section 2.1). It follows from these considerations that relative adsorption affinity series ("selectivity sequences") can be established on the basis of ionic radius (Table 2.1):

$$Cs^+ > Rb^+ > K^+ > Na^+ > Li^+$$
$$Ba^{2+} > Sr^{2+} > Ca^{2+} > Mg^{2+}$$
$$Hg^{2+} > Cd^{2+} > Zn^{2+}$$

These selectivity sequences conform to what has been observed often in soil adsorption experiments. In respect to transition metal cations, however, ionic radius is not adequate as a single predictor of adsorption affinity, since electron configuration plays a very important role in the complexes of these cations (e.g., Mn^{2+}, Fe^{2+}, Ni^{2+}). Their relative affinities tend to follow the *Irving–Williams order*:

$$Cu^{2+} > Ni^{2+} > Co^{2+} > Fe^{2+} > Mn^{2+}$$

The molecular basis for this ordering is discussed in Section 13.1.

The effect of pH on metal cation adsorption is principally the result of changes in the net proton charge on soil particles. As pH increases, σ_H decreases toward negative values (see Fig. 7.4), and the electrostatic attraction of a soil adsorbent for a metal cation is enhanced. If a soil is reacted with a series of aqueous solutions containing a metal cation at the same initial concentration but having an increasing pH value, the amount of metal cation adsorbed will increase with pH, unless ligands in the soil solution compete overwhelmingly for the metal against surface functional groups. In the absence of significant ligand competition, a graph of metal cation adsorbed, q_M, vs. pH will have a characteristic sigmoid shape known as an *adsorption edge*. An adsorption edge of Na^+ is shown on the right in Fig. 7.4 and again in Fig. 8.2

FIG. 8.2 An adsorption edge for Ca^{2+} on an Oxisol. (Data from L. Charlet, Adsorption of some macronutrient ions on an Oxisol. An application of the Triple Layer model, Ph.D. dissertation, Univ. of California, Riverside, 1986.)

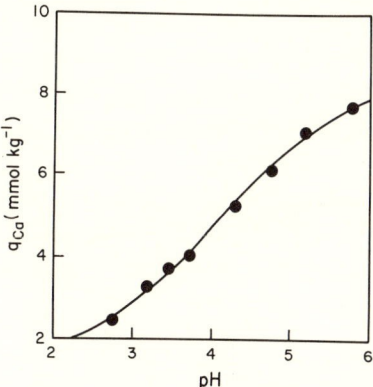

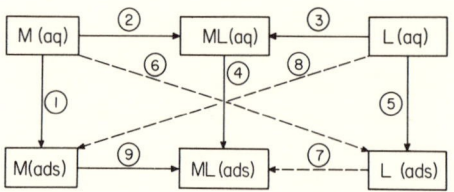

FIG. 8.3 Schematic diagram of the effects of soluble ligands on metal cation adsorption.

for Ca^{2+}. Often these curves are characterized numerically by the value of pH_{50}, the pH value at which one-half the maximum value of q_M is achieved. It is observed typically that pH_{50} correlates *negatively* with the relative affinity of the soil for the metal cation; for example, pH_{50} is larger for Mn^{2+} than Cu^{2+}, and larger for Mg^{2+} than Ba^{2+}. The effect of metal cation hydrolysis (Eq. 7.10) on the adsorption edge is a controversial matter. Nearly always, pH_{50} is well below the pH value at which significant hydrolysis occurs in aqueous solution. On the other hand, it is possible that hydrolytic species, like $MOH^{(m-1)+}$ in Eq. 7.10 are adsorbed strongly (because they are easier to desolvate than free metal cations) and contribute to the rapid ascent of q_M near pH_{50}. The typically low concentration of hydrolytic species is not a problem, since the complexes can be produced via Eq. 7.10 until the supply of free metal cation is depleted.

The presence of complex-forming ligands in the soil solution complicates the prediction of relative metal cation adsorption affinity. For example, Cu^{2+} usually adsorbs strongly on soil to produce an H-curve isotherm. But if organic ligands are available to form nonadsorbing soluble complexes with Cu^{2+}, its adsorption will be less strong and an S-curve isotherm can result (Fig. 8.1). The general effects of metal-complexing ligands in the soil solution on the adsorption of metal cations by soils can be classified as follows (Fig. 8.3).

1. The ligand has a low affinity for the metal and for the adsorbent (step 1).

2. The ligand has a high affinity for the metal and forms a soluble complex with it, and this complex has a low affinity for the adsorbent (steps 2 and 3).

3. The ligand has a high affinity for the metal and forms a soluble complex with it, and this complex has a high affinity for the adsorbent (steps 2–4).

4. The ligand has a high affinity for the adsorbent, and the adsorbed ligand has a low affinity for the metal (step 5).

5. The ligand has a high affinity for the adsorbent, and the adsorbed ligand has a high affinity for the metal (steps 5–7).

6. The metal has a high affinity for the adsorbent, and the adsorbed metal has a high affinity for the ligand (steps 1, 8, and 9).

Note that categories 3 and 5 result directly in enhanced metal adsorption from the presence of ligands, whereas category 4 can result indirectly in enhanced metal adsorption if the adsorbed ligand causes the surface charge to become more negative. Categories 3, 5, and 6 produce the same kind of surface species (adsorbed metal–ligand complex) and therefore cannot usually be identified separately on the basis of adsorption experiments alone. It is apparent in Fig. 8.3 that the prediction of selectivity sequences for metal cations in soil solutions will depend on quantitative information about the five different pathways to adsorption for each metal–ligand combination.

8.4 Anion Adsorption

Some important adsorptive, nonpolymeric anions in the soil solution are $B(OH)_4^-$, CO_3^{2-} and HCO_3^-, COO^-, NO_3^-, aromatic ring—O^- (phenolate), $H_3SiO_4^-$, PO_4^{3-}, HPO_4^{2-}, and $H_2PO_4^-$, SO_4^{2-}, Cl^-, and MoO_4^{2-} and $HMoO_4^-$. To this list may be added F^-, HS^-, AsO_4^{3-}, $HAsO_4^{7-}$, $H_2AsO_4^-$, SeO_4^{2-}, $HSeO_3^-$, and SeO_3^{2-} in special circumstances. The mechanisms by which these small anions adsorb are, as for cations, surface complexation and diffuse-ion swarm association. Outer-sphere surface complexation of anions involves coordination to a protonated hydroxyl or amino group or to a surface metal cation (e.g., the water-bridging mechanism in Eq. 3.14). If $A^{\ell-}$ is an anion adsorptive, then outer-sphere complexation can be depicted by the reactions:

$$SOH_2^+(s) + A^{\ell-}(aq) = SOH_2^+ A^{\ell-}(s) \qquad (8.10a)$$

$$SNH_3^+(s) + A^{\ell-}(aq) = SNH_3^+ A^{\ell-}(s) \qquad (8.10b)$$

$$SM^{m+}(s) + A^{\ell-}(aq) = SM^{m+} A^{\ell-}(s) \qquad (8.10c)$$

where S refers to the soil adsorbent, and each product on the right side is understood to have a water molecule interposed between the positive surface group and the complexed anion. Inner-sphere surface complexation of anions involves coordination to created or native Lewis acid sites (Section 7.1 and Figs. 7.1 and 7.2). Almost always, the mechanism of this coordination is hydroxyl ligand exchange, which was illustrated in Eqs. 3.15 and 3.16 for COO^-, and discussed in Section 7.1. For an anion $A^{\ell-}$ reacting with a created Lewis acid site, the reaction scheme is:

$$SOH(s) + H^+(aq) = SOH_2^+(s) \qquad (8.11a)$$

$$SOH_2^+(s) + A^{\ell-}(aq) = SA^{(\ell-1)-}(s) + H_2O(\ell) \qquad (8.11b)$$

If the Lewis acid site is present already, or if the concentration of A is very large, the protonation step in Eq. 8.11a is not required. In general, ligand exchange is favored by pH < PZNPC (Section 7.4).

The anions Cl^-, NO_3^-, and SeO_4^{2-}—and to some extent HS^-, SO_4^{2-}, HCO_3^-, and CO_3^{2-}—are considered to adsorb mainly as diffuse-ion swarm and outer-sphere complex species. The principal evidence for this conclusion is the observed, readily exchangeable character of these ions, and the fact that they often exhibit a *negative* surface excess in permanent-charge soils. Negative adsorption can occur only for species in the diffuse-ion swarm. On the molecular scale, it can be interpreted through the definitions:

$$n_i \equiv \int_{Su} c_i(\mathbf{x}) dV / m_s \qquad (8.12a)$$

$$M_w \equiv \rho_w \int_{Su} dV / m_s \qquad (8.12b)$$

where $c_i(\mathbf{x})$ is the concentration (moles per unit volume) of anion i at a point $\mathbf{x}$ in the aqueous solution portion of a suspension containing m_s kilograms of soil, ρ_w is the mass density of water in the suspension, and the integrals extend over the entire suspension volume. In dilute solutions, $\rho_w m_i = c_{0i}$, the concentration of ion i in the supernatant solution. Thus, Eqs. 8.1 and 8.12 can be combined to produce the expression:

$$q_i = \frac{1}{m_s} \int_{Su} [c_i(\mathbf{x}) - c_{0i}] dV \qquad (8.13)$$

Equation 8.13 actually applies to *any* ion in the diffuse swarm (Compare Eq. 7.5).

If negative adsorption occurs, $c_i(\mathbf{x}) < c_{0i}$ and $q_i < 0$. This condition is produced by electrostatic repulsion of the ion i away from a surface of like charge sign (i.e., an anion in a typical permanent-charge soil). An estimate of the size of the interfacial region over which this repulsion is effective can be made by defining the *exclusion volume*:

$$V_{ex} \equiv \int \left[1 - \frac{c_i(\mathbf{x})}{c_{0i}} \right] dV / m_s = -q_i / c_{0i} \qquad (8.14)$$

For example, in the chloride negative adsorption measurement discussed in Section 8.1, $V_{ex} = 0.006$ mol kg^{-1}/20 mol m$^{-3} = 3 \times 10^{-4}$ m^3 kg^{-1}. (Here $c_{0i} = 0.02$ molal ≈ 20 mol m^{-3}.) In suspensions of montmorillonite clay, this figure could be an order of magnitude larger for the same chloride concentration. In general, V_{ex} is the total volume in the soil solution (per kilogram of dry soil) in which $c_i(\mathbf{x})$ is smaller than its "bulk" value, c_{0i}. The observation of negative q_i and an appreciable V_{ex} is compelling evidence for significant diffuse-swarm-species of the anion i.

The anions remaining the the list presented at the beginning of this section, most notably borate, phosphate, and carboxylate, are considered to adsorb principally as inner-sphere complex species. Several different kinds of experimental evidence support this conclusion. Perhaps the most direct is the observed difficulty in desorbing anions like phosphate by leaching with anions like chloride. Another comparative type of evidence is the persistence of, for example, phosphate adsorption at pH > PZNPC, whereas chloride adsorption diminishes rapidly to zero at these pH values (Fig. 7.4). Finally, infrared and other spectroscopic methods have led to surface species concepts for phosphate, selenite, borate, silicate, and molybdate ions like that for biphosphate shown in Fig. 7.2. Although none of these pieces of evidence may be definitive when taken alone, when combined they make a very strong case for ligand exchange as the principal mode of fluoride and oxyanion (except nitrate, selenate, and possibly sulfate or carbonate) adsorption by soils.

The effect of pH on anion adsorption is the result of changes in the net proton charge on soil particles, if the adsorptive anion does not protonate significantly (e.g., Cl^-, NO_3^-, SO_4^{2-}, and SeO_4^{2-}). The decrease in σ_H with increasing pH produces a repulsion of the adsorptive anion from soil particle surfaces that becomes dominant at pH > PZNPC. Therefore, as shown on the right in Fig. 7.4 for Cl^-, the surface excess will decrease uniformly with pH. This kind of graph of anion adsorbed, q_A, vs. pH is termed an *adsorption envelope*. Typically, it is measured by reacting a soil with a series of aqueous solutions containing an adsorptive anion at the same initial concentration, but having an increasing pH value. For anions that protonate, the adsorption envelope is the resultant of a competition for H^+ by the adsorptive and the adsorbent, which is coupled by the reactions in Eq. 8.10 or 8.11. This coupled competition can be understood most readily by consideration of a monoprotic adsorptive, like fluoride or borate (Fig. 8.4).

The ligand exchange reaction in Eq. 8.11 occurs on variable-charge surfaces in soils. As indicated in Table 7.2, these surfaces usually become negatively charged in the alkaline pH range. Thus the essential reactant, SOH_2^+, in Eq. 8.11a will gradually decrease in population as the pH value increases to the PZNPC and above. Consider now the adsorptive F^-, which is protonated significantly at pH values < 3.2. As the pH increases from, say, 3 to 3.5, ever-increasing numbers of F^- ions are created, with little change in the SOH_2^+ population. The adsorption envelope for F^- will consequently show a rapid rise in response to a growing concentration of F^- on the left side of Eq. 8.11b. As the pH increases above 4 and into the alkaline range, the F^- population changes little, but the SOH_2^+ population drops, with the result that the ligand exchange reaction is disfavored and adsorption decreases. Somewhere between the rise and the fall, a maximum will occur—in this case near pH 3–4 because the PZNPC is much larger than minus the log protonation constant ($= 3.17$) of F^-. For the adsorptive $B(OH)_4^-$, protonation is significant at pH values < 9.2. As the pH increases from, say, 5 into the alkaline range, there will be a gradual increase in the population of borate ions reflected in a gradual rise of the adsorption envelope. Above pH 9.2, the envelope will drop sharply because the concentration of SOH_2^+ is small and that of $B(OH)_4^-$ is steady. In between, there will be a broad maximum in the envelope reflecting the pH range in

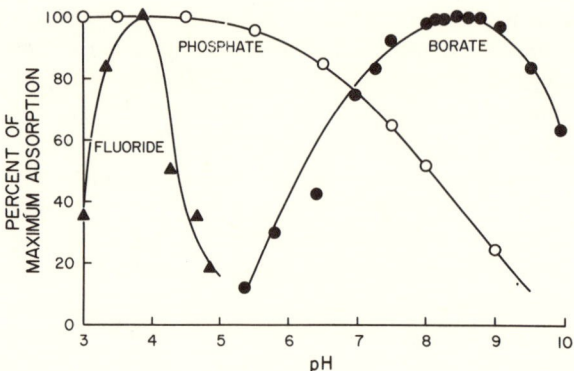

FIG. 8.4 Typical adsorption envelopes on soils for fluoride, phosphate, and borate ions.

which the rise in $B(OH)_4^-$ tends to compensate the drop in SOH_2^+ on the left side of Eq. 8.11b. For multiprotic adsorptives like phosphate and arsenate, the adsorption envelope is essentially a combination of the fluoride and borate scenarios (Fig. 8.4).

8.5 Molecular Adsorption Models

Isotherm data like those in Fig. 8.1 are often interpreted with the help of molecular adsorption models. These models are mathematical representations of q_i in Eq. 8.1 based on hypotheses about the interactions between an adsorptive and an adsorbent that result in a particular molecular arrangement of an adsorbate on a surface. Ideally, the hypotheses underlying a molecular adsorption model are developed from spectroscopic information about the adsorptive–adsorbent interaction and the structure of the adsorbate. Usually, however, the underlying hypotheses come from less complete information that is derived from experimental data on the effects of pH, ionic strength, competitive adsorption, and aqueous solution composition on the adsorption isotherm. The basic features of molecular adsorption models are perhaps best appreciated by a detailed consideration of two examples with very different foundational hypotheses.

The *diffuse double-layer model*—the oldest molecular adsorption model— develops from the following assumptions.

1. The adsorbent surface is a uniform plane of charge density σ (proportional to the intrinsic surface charge density in Section 7.3).

2. The adsorptive ions are point species that interact mutually and with the adsorbent through the coulomb force. Their only mechanism of adsorption is the diffuse-ion swarm.

3. The aqueous solution phase is a uniform continuum of dielectric constant D in which the point-ion adsorptive is immersed. The concentration of adsorptive species i at a point $\mathbf{x}$ in this continuum is expressed:

$$c_i(\mathbf{x}) = c_{0i} \exp(-Z_i F \phi(\mathbf{x})/RT) \tag{8.15}$$

where Z_i is the species valence, F is the Faraday constant, R is the molar gas constant, T is absolute temperature (see the Appendix), and $\phi(\mathbf{x})$ is the average electric potential at $\mathbf{x}$. Equation 8.15 is used in Eq. 8.13 to calculate q_i, which may be positive, zero, or negative.

4. The electric potential $\phi(\mathbf{x})$ is related to $c_i(\mathbf{x})$ through the *Poisson–Boltzmann equation*, which is a differential equation describing the net coulomb force in an ion swarm immersed in a continuum liquid. In one spatial dimension that is measured perpendicularly out from the adsorbent surface, this differential equation has the form:

$$\frac{d\phi}{dx} = \pm \left[\frac{2RT}{\varepsilon_0 D} \Sigma_k (c_k(x) - c_{0k}) \right]^{\frac{1}{2}} \tag{8.16}$$

where ε_0 is the permittivity of vacuum (see the Appendix), the sum is over each kind of ion in aqueous solution, and the plus (minus) sign is chosen if ϕ is negative (positive). Equations 8.15 and 8.16 are combined and solved for $\phi(x)$ subject to the boundary condition:

$$\sigma = \pm \frac{S}{F} \{2\varepsilon_0 DRT\Sigma_k c_{0k}[\exp(-Z_k F\phi_0/RT) - 1]\}^{\frac{1}{2}} \tag{8.17}$$

where S is the specific surface area of the adsorbent and ϕ_0 is the electric potential at the adsorbent surface.

As a concrete example, suppose that a permanent-charge soil with $\sigma = \sigma_0 < 0$ is suspended in a solution of NaCl at concentration c_1 mixed with KCl at concentration c_2. Then $\phi(x)$ will be negative, and Eqs. 8.15 and 8.17 become:

$$c_{Na}(x) = c_1 \exp(-F\phi(x)/RT) \qquad c_K(x) = c_2 \exp(-F\phi(x)/RT) \tag{8.18a}$$

$$c_{Cl}(x) = (c_1 + c_2) \exp(F\phi(x)/RT) \tag{8.18b}$$

$$\sigma_0 = -\frac{S}{F} \{2\varepsilon_0 DRT[c_1(\exp(-F\phi_0/RT) + \exp(F\phi_0/RT) - 2)$$
$$+ c_2(\exp(-F\phi_0/RT) + \exp(F\phi_0/RT) - 2)]\}^{\frac{1}{2}} \tag{8.18c}$$

In advanced textbooks of soil chemistry, it is shown mathematically that Eq. 8.18, in combination with the appropriate form of Eq. 8.16, can be substituted into Eq. 8.13 to produce the expressions:

$$q_{Na} = |\sigma_0| f(\phi_0) \frac{c_1}{c_1 + c_2} \qquad q_K = |\sigma_0| f(\phi_0) \frac{c_2}{c_1 + c_2} \qquad q_{Cl} = \sigma_0 f(-\phi_0) \quad (8.19)$$

where $|\sigma_0|$ is the absolute value of σ_0 and $f(\phi_0) = [\exp(-F\phi_0/2RT) - 1]/[\exp(-F\phi_0/2RT) - \exp(F\phi_0/2RT)]$. Equations 8.19 predict that Na^+ and K^+ will adsorb in proportion to their aqueous solution concentrations, and that the negative adsorption of Cl^- will be proportional to the surface charge density. Adsorption experiments with soils have shown that these predictions are only rough approximations, precisely because they neglect surface complex formation. In particular, the diffuse double-layer model cannot describe adsorption selectivity differences among metal cations of the same valence.

On a more fundamental level, the accuracy of Eq. 8.15 has been tested recently in precise computer simulations of ion distributions near charged planes. The simulations apply to a set of "hard-sphere" cations and anions immersed in a dielectric continuum and interacting with a charged plane and among themselves through the coulomb force. Because of the finite size of the ions, their centers cannot be closer to the charged plane than a distance equal to their hard-sphere radius, and the results of a computer simulation of their molecular behavior must be compared with Eq. 8.15 restricted to x values larger than this radius. Figure 8.5 shows the x dependence of $c_i(x)$ as deduced from computer simulations of 1:1 and 2:1 electrolytes near a negatively charged plane. The simulations were made for cations and anions with a radius of 0.213 nm immersed in a continuum with a dielectric constant equal to 78.5 (that of liquid water at 298 K). The concentration of the 1:1 electrolyte was 100 mol m^{-3}, and the ions confronted a plane whose surface charge density σ equaled -0.266 C m^{-2} (e.g., a vermiculite siloxane surface). The concentration of the 2:1 electrolyte was 50 mol m^{-3}, and it confronted a plane of charge density -0.177 C m^{-2} (e.g., a smectite surface). For the 1:1 electrolyte, there is excellent agreement between the data points provided by the computer simulation and the values of the ion concentrations, $\rho_+(x) = c_i(x)$ for cations and $\rho_-(x) = c_i(x)$ for anions, calculated for $c_0 = 100$ mol m^{-3} at 298 K with Eq. 8.15. At this low concentration, the errors inherent in Eq. 8.15 appear to be mutually compensating. At higher concentrations (e.g., $c_0 = 10^3$ mol m^{-3}), the computer simulation results deviated significantly from the predictions of Eq. 8.15, in that theory underestimated the extent of negative anion adsorption and failed to reproduce the oscillation in $\rho_+(x)$ produced by fluctuations in the true electric potential about its mean value. For the 2:1 electrolyte, these inadequacies of theory are apparent even at $c_0 = 50$ mol m^{-3}, as shown in Fig. 8.5. In this case, the model predictions of ion distribution have only qualitative significance. The results of computer simulation indicate that the Poisson–Boltzmann equation does not provide an accurate description of ion swarms containing bivalent species.

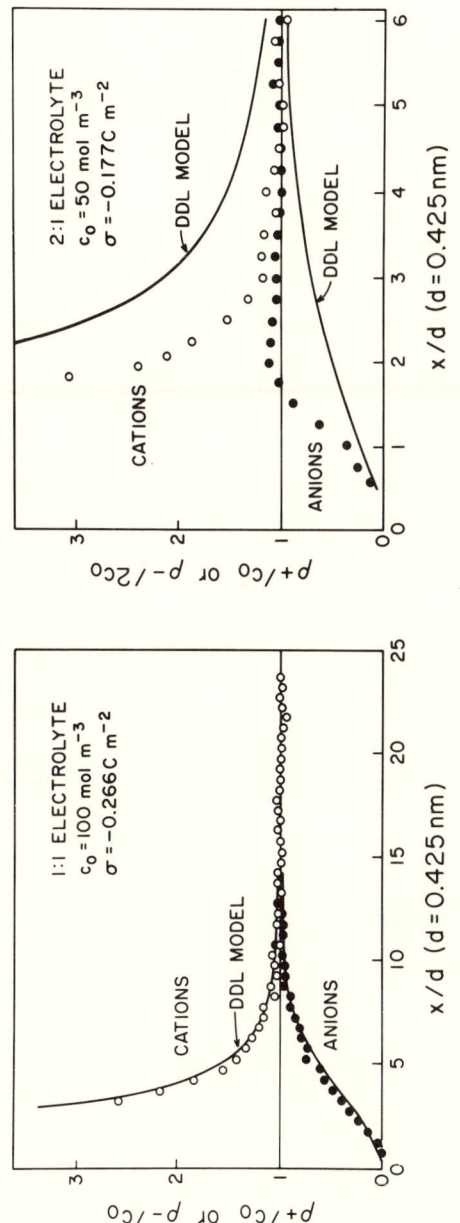

FIG. 8.5 Comparison of ion distributions near a charged surface as predicted by diffuse double-layer (DDL) theory (lines) and accurate computer simulation (circles). (Reprinted with permission from G. Sposito, *The Surface Chemistry of Soils*, Oxford Univ. Press, New York, 1984.)

The *constant capacitance model* takes a perspective completely opposite to that in the diffuse double-layer model in respect to the mechanism of adsorption. The basic assumptions of the model are as follows:

1. The adsorbent surface is a uniform plane of charge.
2. The adsorptive ions are point species that interact with the adsorbent to form only inner-sphere surface complexes.
3. The thermodynamic activity of a surface complex whose *overall* valence is Z is related to the surface excess by the equation:

$$(SOM) = [SOM] \exp(ZF\phi_0/RT) \qquad (8.20a)$$

$$(SA) = [SA] \exp(ZF\phi_0/RT) \qquad (8.20b)$$

where [ ] is a surface excess in moles per kilogram, M is a metal cation that forms the surface complex SOM, and A is an anion that forms the surface complex SA.

4. The electric potential at the adsorbent surface is related linearly to the net total particle charge:

$$\sigma_P = \frac{S}{F} C\phi_0 \qquad (8.21)$$

where C is a capacitance density in units of farads per square meter (see the Appendix).

As Eqs. 8.20a and 8.20b imply, the constant capacitance model is applied in conjunction with equilibrium constants formulated for surface complexation reactions. Suppose, for example, that Cu^{2+} is adsorbed by a permanent-charge soil according to the reaction:

$$SO^-(s) + Cu^{2+}(aq) = SOCu^+(s) \qquad (8.22)$$

The equilibrium constant for this reaction is:

$$K = \frac{(SOCu^+)}{(SO^-)(Cu^{2+})} = \frac{[SOCu^+]}{[SO^-](Cu^{2+})} \frac{\exp(F\phi_0/RT)}{\exp(-F\phi_0/RT)}$$

$$= \frac{[SOCu^+]}{[SO^-](Cu^{2+})} \exp(2\sigma_P/SCRT)$$

$$\equiv {}^cK \exp(2\sigma_p/SCRT) \qquad (8.23)$$

where SO^- is described according to Eq. 8.20b, Eq. 8.21 has been used to replace ϕ_0 with σ_p, and a convenient definition of a conditional equilibrium constant ${}^cK$ for the reaction has been applied. In the present example,

$$\sigma_P = \sigma_0 + \sigma_{IS} = \sigma_0 + 2[SOCu^+] = -[SO^-] + [SOCu^+] \qquad (8.24)$$

since

$$\sigma_0 = -\{[\text{SO}^-] + [\text{SOCu}^+]\} \qquad (8.25)$$

according to Eq. 7.4, with the surface species assumed.

Equation 8.23 can be understood chemically by analogy with the interpretation of conditional stability constants for complexes in aqueous solution, given in Section 4.5. The parameter ^cK in Eq. 8.23 is convenient to measure in an adsorption experiment, since σ_0 is presumed known and $q_{Cu} \equiv [\text{SOCu}^+]$. This conditional constant, however, will be composition-dependent, because it contains surface species concentrations only and does not correct for the electrostatic interactions among the species as their surface excess (and σ_P) change. In the limit of zero particle charge, these interactions cancel, and the extrapolated value of ^cK represents the chemical equilibrium of an "ideal" system wherein surface species interactions, other than those involved in complex formation, are unimportant. When σ_P is not zero, there are species interactions and ^cK differs from K. The activity coefficient, $\exp(ZF\phi_0/RT)$, then is introduced to "correct" the surface species concentrations in ^cK for nonideal behavior and thereby restore the value of K via Eq. 8.23. In the constant capacitance model, this correction is assumed to depend only on the net total particle charge, which thus plays the same role as ionic strength does for aqueous solution species.

In applications, Eq. 8.23 is written in logarithmic form:

$$\log {}^cK = \log K - (2/2.303 SCRT)\sigma_P \qquad (8.26)$$

and a graph of measured values of $\log {}^cK$ against σ_P yields $\log K$ from the y intercept and C from the slope (or SC if the specific surface area is unknown) of the resulting straight line. Equation 8.26, and similar expressions for other ions that are adsorbed specifically, often provide a good fit to experimental data, but with values of C that differ for different adsorbate ions on the same adsorbent. Thus C is effectively an adjustable parameter in applications.

These two examples illustrate the general features of molecular adsorption models. They begin with *molecular hypotheses* about the mechanisms of adsorption; they introduce *constraint equations* (like Eqs. 8.16, 8.17, 8.21, and 8.23) that serve to relate model parameters to measurable properties, and they provide *testable predictions* of surface excesses or quantities like ^cK in Eq. 8.26 that are closely related to surface excesses. The adherence of data to the model predictions, of course, does *not* prove that a given model is a correct molecular description of adsorption. It only shows that the model results are consistent with experimental measurements after adjustment of the model parameters to provide the best fit to data. The accuracy of a molecular model of adsorption can be ascertained only through experiments designed to verify directly the mechanisms of adsorption on which it is based.

FOR FURTHER READING

N. J. Barrow, Reactions of anions and cations with variable-charge soils, *Adv. Agron.* **38**:183–230 (1985). A thoughtful and provocative review of the development of molecular adsorption models for variable-charge soils.

R. D. Harter, *Adsorption Phenomena*, van Nostrand Reinhold, New York, 1986. This carefully edited collection of classic papers on adsorption reactions in soils is well worth the time of anyone interested in the historical roots of soil chemistry.

C. J. B. Mott, Anion and ligand exchange, in D. J. Greenland and M. H. B. Hayes (eds.), *The Chemistry of Soil Processes*, pp. 179–219. Wiley, Chichester, U.K., 1981. This explicative survey of anion adsorption by soils should be read as a companion to Section 8.4.

R. L. Parfitt, Anion adsorption by soils and soil materials, *Adv. Agron.* **30**:1–50 (1978). A careful review of the molecular aspects of ligand exchange.

D. L. Sparks, Kinetics of ionic reactions in clay minerals and soils, *Adv. Agron.* **38**:231–266 (1985). A useful review of the rate laws used in adsorption kinetics studies and their application to soils.

G. Sposito, Chemical models of inorganic pollutants in soils, *Crit. Rev. Environ. Control* **15**:1–24 (1985). A critical review of molecular adsorption models and their application.

PROBLEMS

The more difficult problems are indicated by an asterisk.

1. Dry soil (350 mg) is mixed with 20 mL of a solution containing 4.00 mol m^{-3} KNO_3 at pH 4. After equilibration for 24 h, a supernatant solution is collected and found to contain 3.96 mol m^{-3} KNO_3. Calculate q_K and q_{NO_3} for the soil in millimoles per kilogram.

2. The data in the accompanying table refer to Cu(II) adsorption by an Aridisol. Plot an adsorption isotherm with the data and classify it according to the criteria in Section 8.2.

q_{Cu} (mmol kg^{-1})	c_{Cu} (mmol m^{-3})	q_{Cu} (mmol kg^{-1})	c_{Cu} (mmol m^{-3})
6.87	1.87	25.39	28.11
10.64	3.06	34.14	77.68
18.05	9.19	37.34	155.1

3. The rate of adsorption of readily exchangeable K^+ onto an Entisol was observed to follow a first-order kinetics law with $k_f = 5.78 \times 10^{-4}$ s^{-1}. Calculate the half-life of the adsorption reaction.

4. Fit the data in Problem 2 to the Langmuir isotherm equation with a graph based on Eq. 8.8. Calculate the Langmuir parameters, b and K. If a linear regression program is used to fit Eq. 8.8, estimate the 95% confidence intervals for b and K.

5. Use the data in the accompanying table to apply the van Bemmelen–Freundlich isotherm equation to Cd(II) adsorption by an Alfisol. Calculate the parameters A and β.

q_{Cd} (mmol kg^{-1})	c_{Cd} (mmol m^{-3})	q_{Cd} (mmol kg^{-1})	c_{Cd} (mmol m^{-3})
0.11	0.89	0.61	4.45
0.30	1.78	0.79	12.5
0.53	3.56	1.14	17.8

*6. A common method used to ensure that an ion has been adsorbed by a soil instead of precipitated onto soil particle surfaces is to calculate ion activity products for likely solid phases to see if they are smaller than the solubility product constants for the solids. Discuss the possible error in this approach if IAP/K_{so} is between 0.1 and 1.0.
(*Hint:* Review Sections 1.3 and 5.1.)

7. Plot an adsorption edge for Mg(II) adsorption by an Oxisol based on the data given here. Calculate pH_{50} given a maximum adsorption of 8 mmol kg^{-1} at pH 6.

q_{Mg} (mmol kg^{-1})	pH	q_{Mg} (mmol kg^{-1})	pH
0.72	2.48	2.45	3.36
1.08	2.73	3.64	3.80
1.80	3.05	4.21	4.10
2.14	3.20	6.35	5.00

*8. A widely adopted procedure for measuring the concentration of organic complexes of a trace metal like Cu(II) or Al(III) is to percolate a soil solution through a strongly acidic cation exchange resin and then measure the concentration of the metal in the effluent solution. The metal concentration in the effluent solution is equated to that of organically complexed metal. Discuss critically the key assumptions underlying this procedure, and suggest experiments to validate them.
(*Hint:* Review Sections 4.2 and 8.3.)

9. Verify the statements in Section 8.5, that $\sigma_0 = -0.28$ C m^{-2} is a structural surface charge density characteristic of vermiculites and

$\sigma_0 = -0.18$ C m^{-2} is characteristic of smectites. Assume a specific surface area of 75 ha kg^{-1} for each clay mineral. (See the Appendix for a discussion of the units used here.)

10. In the accompanying table are values of the equilibrium constant K_{298} for the adsorption reaction:

$$SO^-Li^+(s) + M^+(aq) = SO^-M^+(s) + Li^+(aq)$$

on montmorillonite at 298 K. (The symbol SO$^-$ refers to 1 mol of negative montmorillonite surface charge.) Establish a relative selectivity sequence based on these data, and discuss a molecular mechanism for it.

M^+ :	Na^+	K^+	Rb^+	Cs^+
K_{298}:	1.26	4.05	26.4	66.4

11. Experimental measurements indicate that the surface species on a variable-charge soil suspended in $CaCl_2$ solution are SOH_2^+, SOH, SO$^-$, $SOCa^+$, and $SOH_2^+Cl^-$, where SOH refers to 1 mol of reactive surface OH groups and $SOCa^+$ is an inner-sphere complex. Give expressions for the surface charge components σ_H, σ_{IS}, σ_{OS}, and σ_D in terms of the concentrations of the surface species (e.g., $[SOCa^+]$ in moles per kilogram dry soil).

12. A soil suspended in $CaHPO_4$ solution contains the surface species SOH_2^+, SOH, SO$^-$, $(SO)_2Ca^0$, and $SHPO_4^-$, where SOH refers to 1 mol of reactive surface hydroxyls and $(SO)_2Ca^0$ is an inner-sphere complex. Give expressions for σ_H, σ_{IS}, σ_{OS}, and σ_D in terms of the concentrations of surface species (e.g., $[SOH_2^+]$ in moles per kilogram dry soil).

13. A variable-charge soil is known to contain 0.4 μmol m^{-2} reactive OH groups and 0.25 μmol m^{-2} native Lewis acid sites on a specific surface area of 0.8 ha kg^{-1}. If this soil adsorbs F$^-$ at pH < PZNPC, show that the molar ratio of H$^+$ adsorbed to F$^-$ adsorbed should be about 0.6 if all adsorption sites are equally accessible to the fluoride ion.

*14. Show that, according to diffuse double-layer theory,

$$V_{ex} = 2S[1 - \exp(F\phi_0/2RT)]/(\beta c_{Cl})^{\frac{1}{2}}$$

where $\beta = 2F^2/\varepsilon_0 DRT = 1.084 \times 10^{16}$ m mol^{-1} at 298 K (see the Appendix) and c_{Cl} is the concentration of Cl$^-$ in a supernatant solution (mol m^{-3}) of a soil suspension exhibiting the exclusion volume V_{ex}. (*Hint:* Combine Eqs. 8.18c and 8.19, noting that $c_1 + c_2 = c_{Cl}$.)

15. Apply the equation in Problem 14 to calculate the specific surface area of the soil whose exclusion volumes as a function of chloride concentration are given here. You may assume that $-F\phi_0/RT \gg 1$ in doing the calculation. (Note that 10^{-3} m^3 = 1 dm^3.)

V_{ex} $(10^{-3} \text{ m}^3 \text{ kg}^{-1})$	c_{Cl} (mol m^{-3})	V_{ex} $(10^{-3} \text{ m}^3 \text{ kg}^{-1})$	c_{Cl} (mol m^{-3})
1.06	0.79	0.50	6.2
1.00	1.1	0.49	6.8
0.70	2.0	0.38	7.7
0.68	3.1	0.29	9.9
0.55	4.0	0.24	20.5

16. The exclusion distance d_{ex} is defined by the equation:

$$d_{ex} \equiv -q_i/Sc_{0i}$$

where i refers to an anion negatively adsorbed by a soil whose specific surface area is S. Calculate d_{ex} as a function of chloride concentration using the equation for V_{ex} in Problem 14. You may assume that $\exp(F\phi_0/2RT)$ is negligible.

*17. Most electrochemistry textbooks state that the surface electric potential ϕ_0 cannot be measured. Why is it not possible to use Eq. 8.17 (or Eq. 8.18c) to measure ϕ_0 in terms of measured values of σ_0, c_{0k}, S, and the other known parameters like D, R, T, etc.?
(*Hint:* Define "measure" and proceed from there.)

18. Show that, for any soil in which adsorption can be described accurately *solely* by the diffuse double-layer model, PZC $\leqslant$ PZNPC. Under what condition does PZC = PZNPC?

*19. Apply the constant capacitance model to the surface acid–base reactions:

$$\text{SO}^-(s) + \text{H}^+(aq) = \text{SOH}(s) \qquad K_1$$

$$\text{SO}^-(s) + 2\text{H}^+(aq) = \text{SOH}_2^+(s) \qquad K_2$$

Derive expressions like Eq. 8.26 for K_1 and K_2, then show that:

$$\text{PZNPC} = \tfrac{1}{2} \log K_2$$

(*Hint:* Note that $\sigma_P = \sigma_H$ in this example.)

*20. Given that adsorbed Cu(II) consists of the two surface species SOCu^+ and $(\text{SO})_2\text{Cu}^0$, derive an equation for q_{Cu} in terms of the constant capacitance model parameters ϕ_0, K_1, and K_2, where K_1 and K_2 are equilibrium constants for the reactions:

$$\text{SO}^-(s) + \text{Cu}^{2+}(aq) = \text{SOCu}^+(s) \qquad K_1$$

$$2\text{SO}^-(s) + \text{Cu}^{2+}(aq) = (\text{SO})_2\text{Cu}^0(s) \qquad K_2$$

(*Hint:* Apply Eqs. 8.20 and 8.23, neglecting Eq. 8.21.)

9

Exchangeable Ions

9.1 Soil Exchange Capacities

The *ion exchange capacity* of a soil is the number of moles of adsorbed ion charge that can be desorbed from unit mass of soil, under given conditions of temperature, pressure, soil solution composition, and soil–solution mass ratio. In Section 3.3, a similar definition of the cation exchange capacity of soil humus was stated and, in Chapter 8, the operational definition of the surface excess of an ion was related to the soil chemical factors that affect ion exchange capacities. In most applications, the ion exchange capacity refers to the maximum adsorption (positive surface excess) of *readily exchangeable ions*, as defined in Section 7.2. These ions adsorb on soil particle surfaces solely via the outer-sphere complex and diffuse-ion swarm mechanisms (see Fig. 7.3).

The measurement of an ion exchange capacity often involves the replacement of native, readily exchangeable ions by an "index" cation or anion (Section 7.4) whose surface excess then is determined following the principles discussed in Section 8.1. Detailed laboratory procedures for this measurement are described in *Methods of Soil Analysis* (see the "For Further Reading" section at the end of Chapter 4). For soils in which the readily exchangeable cations are solely monovalent or bivalent (e.g., Aridisols), the "index" cation can be Na^+, whereas for soils also bearing trivalent readily exchangeable cations (e.g., Spodosols), Ba^{2+} is the "index" cation of choice (see also Section 3.3). Often NH_4^+ has been used an an "index" cation. Since this cation forms inner-sphere surface complexes with 2:1 layer-type clay minerals, like that shown for K^+ in Fig. 7.2, and since it can even dislodge cations from easily weathered primary soil minerals, the use of NH_4^+ to measure the soil cation exchange capacity has significant potential for inaccuracy (see Problem 1 in

Chapter 7). With regard to the anion exchange capacity, the use of Cl^- as the "index" anion is widespread because of its nonspecific adsorption characteristics.

Much controversy exists over the surface chemical significance of ion exchange capacities, particularly those measured for cations. If an excess of the "index" cation is used in a solution at high pH (e.g., $\geqslant 8.2$), the measured surface excess of the "index" cation should approximate closely the maximum (negative) intrinsic surface charge of a soil (Section 7.3). This parameter figures directly in molecular adsorption models (see Section 8.5) because it is the very largest value expected for the adsorbed cation charge. On the other hand, if the pH value or some other chemical property of the solution containing the "index" cation is arranged such that the maximum (negative) intrinsic surface charge is not neutralized by the adsorption of the "index" cation, then the measured surface excess of the latter will simply reflect the chemical conditions chosen. An example of this latter situation appears in Fig. 7.4 for Na^+ adsorption by an Oxisol under varying pH. Thus it is evident that *both* the maximum and the less-than-maximum variety of cation exchange capacity measurement are useful in soil chemistry. The maximum intrinsic surface charge measurement indicates the *potential* capacity of a soil for adsorbing cations or anions, whereas a less-than-maximum intrinsic surface charge measurement indicates the *actual* capacity of a soil for adsorbing ions under given conditions.

Table 9.1 lists representative cation exchange capacity (*CEC*) values for the 10 soil orders, based primarily on measurements made for U.S. soils using NH_4^+ as the "index" cation in a solution at pH 7. High variability of the *CEC* within each soil order is evident, but the very low values for Ultisols and Oxisols and the high values for Histosols and Vertisols are significant trends. Detailed studies of the *CEC* show that it is correlated positively with the content of soil organic matter and with the pH of a soil if an unbuffered solution containing the "index" cation is used in the measurement. The correlation of *CEC* with organic matter is illustrated in Fig. 9.1 with data for

TABLE 9.1 Representative cation exchange capacities (in $mol_c \ kg^{-1}$) of surface soils[a]

Soil order	*CEC*	Soil order	*CEC*
Alfisols	0.12 ± 0.08	Mollisols	0.22 ± 0.10
Aridisols	0.16 ± 0.05	Oxisols	0.05 ± 0.03
Entisols	0.13 ± 0.06	Spodosols	0.11 ± 0.05
Histosols	1.4 ± 0.3	Ultisols	0.06 ± 0.06
Inceptisols	0.19 ± 0.17	Vertisols	0.37 ± 0.08

[a]Based primarily on data compiled by G. G. S. Holmgren, M. W. Meyer, R. B. Daniels, R. L. Chaney, and J. Kubota (used with the permission of Dr. R. L. Chaney).

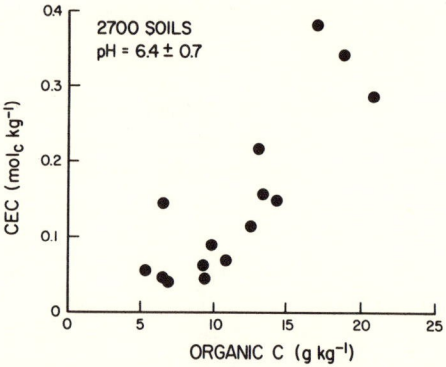

FIG. 9.1 Relationship between soil *CEC* measured by $NH_4C_2H_3O_2$ extraction and soil organic carbon content. (Data compiled by G. G. S. Holmgren, M. W. Meyer, R. B. Daniels, R. L. Chaney, and J. Kubota; used with the permission of Dr. R. L. Chaney.)

nearly 3000 mineral surface soils from the United States grouped according to textural class, irrespective of soil order. The basis for this correlation—which is also reflected dramatically in Table 9.1 by the *CEC* reported for Histosols—can be understood after comparison of the *CEC* values of humic substances $(4–9 \text{ mol}_c \text{ kg}^{-1}$, Section 3.3) to those for clay minerals like smectite and vermiculite $(0.7–2.5 \text{ mol}_c \text{ kg}^{-1}$, Section 7.3). The correlation with pH also is understandable after reviewing the pH dependence of the net proton charge in Figs. 3.4 and 7.4. Indeed, the pH dependence of a less-than-maximum intrinsic surface charge should parallel that of the adsorption edges in Figs. 7.4 and 8.2.

The composition of readily exchangeable ions in a soil can be determined by chemical analysis of the soil solution after reaction of the soil with an "index" ion, as outlined in Section 8.1. In alkaline soils, the readily exchangeable cations are Ca^{2+}, Mg^{2+}, Na^+, and K^+, decreasing in their contribution in the order shown. In acidic soils, the most important readily exchangeable metal cation is Al^{3+}, followed by Ca^{2+} and Mg^{2+}. Readily exchangeable Al(III)—which likely includes Al^{3+}, $AlOH^+$, and $Al(OH)_2^+$—can be measured by using K^+ as an "index" cation in an unbuffered KCl solution. The remaining nonhydrolyzable exchangeable metal cations can then be determined by replacement with Ba^{2+}.

Comprehensive data compilations like those in Table 9.1 are not available for the anion exchange capacity (*AEC*) of soils. The *AEC* tends to be negligible at the natural pH value of most soils, with the notable exception of the Spodosols, Ultisols, and Oxisols. Among these soil orders, *AEC* values in the range $1–10 \text{ mmol}_c \text{ kg}^{-1}$ are representative. Maximum *AEC* values for soil organic matter and metal oxides were discussed in Section 7.3; they range up to $1 \text{ mol}_c \text{ kg}^{-1}$.

9.2 Exchange Isotherms

Once the surface excess of each readily exchangeable ion in a soil has been measured under varying conditions of soil solution composition, an *exchange isotherm* can be constructed. An exchange isotherm is analogous to an adsorption isotherm (Section 8.2), except that the variables plotted are charge fractions instead of surface excesses and concentrations. The charge fraction of an adsorbed ion is:

$$E_i = |Z_i|q_i/Q \tag{9.1}$$

Where Z_i is the valence of ion i, q_i is its surface excess (Eq. 8.1), and Q is either the *CEC* or the *AEC* of a soil. The charge fraction of an ion in aqueous solution is:

$$\tilde{E}_i = |z_i|m_i/\tilde{Q} \tag{9.2}$$

where m_i is the molality (or other concentration variable) and

$$\tilde{Q} = \Sigma_k |Z_k|m_k \tag{9.3}$$

with the sum extending over all ions of the same valence sign (i.e., all cations or all anions) as ion i in Eq. 9.2. An exchange isotherm then is defined as a graph of E_i against $\tilde{E}_i$ at fixed temperature and applied pressure.

Exchange isotherms for Na → Ca, Na → Mg, and Ca → Mg exchange on an Aridisol are shown in Figs. 9.2 and 9.3. The maximum range of the charge fractions in each graph is from 0 to 1. In Fig. 9.2, and in Fig. 9.3 for the exchange experiment data represented by filled circles, only the pairs of cations

FIG. 9.2 Exchange isotherms for Na → Ca and Na → Mg exchange on an Aridisol.

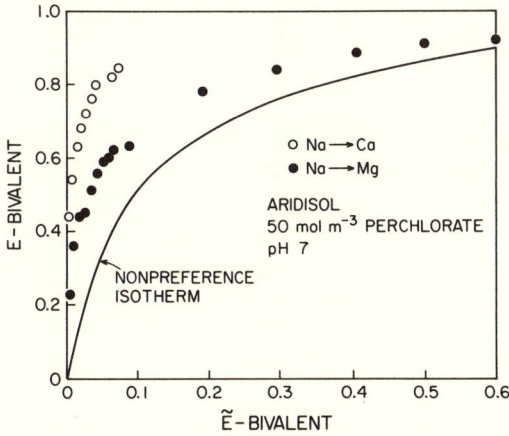

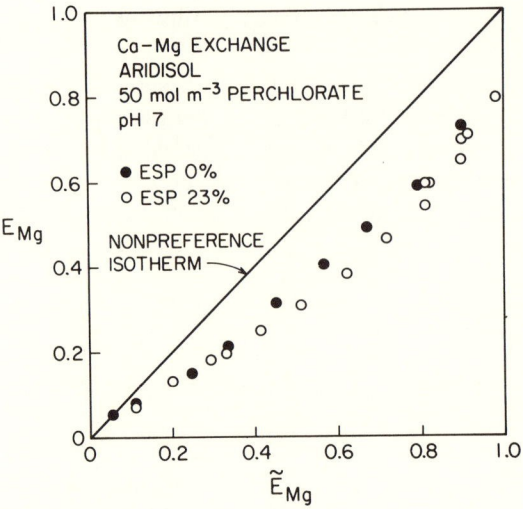

FIG. 9.3 Exchange isotherms for Ca → Mg exchange on an Aridisol at two exchangeable sodium percentages (*ESP*).

indicated explicitly were adsorbed by the soil, so the isotherms refer to *binary-exchange* systems. For the open-circle data in Fig. 9.3, adsorbed Na$^+$ was present during the Ca → Mg exchange experiment, and the resulting isotherm refers to a *ternary*-exchange system. In natural soils, ternary, quaternary, or even higher-order exchange systems, are the norm. The binary-exchange reaction is convenient for detailed laboratory study. In using it, one makes the critical assumption that naturally occurring, higher-order exchange systems can be understood in terms of "component" binary-exchange reactions. That this assumption may sometimes be true is indicated in Fig. 9.3, which shows Ca–Mg exchange reactions both in the absence and in the presence of adsorbed Na$^+$. The charge fractions of Mg^{2+} plotted for an *exchangeable sodium percentage* (*ESP*) of 23% are based on Eqs. 9.1 and 9.2, but with Q and $\tilde{Q}$ limited to contributions from Mg^{2+} and Ca^{2+} so as to allow direct comparison with the binary-exchange data (*ESP* = 0%). The closeness of the two exchange isotherms in Fig. 9.3 suggests that Ca–Mg exchange on the soil was largely independent of the presence of adsorbed Na$^+$ in the *ESP* range investigated.

The solid curves in Fig. 9.2 and 9.3 are the *thermodynamic nonpreference exchange isotherms* (monovalent–bivalent and bivalent–bivalent). For bivalent–bivalent exchange, and for any other exchange reaction involving ions of the same valence, the thermodynamic nonpreference isotherm is represented mathematically by the equation:

$$E_i = \tilde{E}_i \tag{9.4}$$

and will plot as a straight line making a 45° angle with the y and x axes. For monovalent–bivalent exchange, the nonpreference isotherm is expressed by the equation:

$$E_{biv} = 1 - \left[\frac{A(1 - \tilde{E}_{biv})^2}{\tilde{E}_{biv} + A(1 - \tilde{E}_{biv})^2} \right]^{\frac{1}{2}} \tag{9.5}$$

where E_{biv} is the charge fraction of the bivalent cation and $A = \tilde{Q}\gamma_{mon}^2/2\gamma_{biv}$, with the γ being single-ion activity coefficients given by Eq. 4.23. The curve resulting from Eq. 9.5 is shown in Fig. 9.2; it applies to any binary monovalent–bivalent binary–exchange system. In the case of Fig. 9.2, $\tilde{Q} = 0.05$ mol$_c$ L^{-1} and $\gamma_{Na}^2/\gamma_{biv} = 1.5$, so that $A = 0.037$. If $\tilde{E}_{biv} = 0.2$, then, by Eq. 9.5, $E_{biv} = 1 - (0.024/0.224)^{\frac{1}{2}} = 0.67$, which agrees with the solid-line curve in the figure. The derivations of Eqs. 9.4 and 9.5 are outlined in Section 9.4. Suffice it to say at present that the derivation is based on two chemical assumptions: (1) the thermodynamic equilibrium constant (Section 4.5) defined for the exchange reaction has unit value, and (2) the adsorbed ions behave as an "ideal mixture" in the soil (Section 8.5).

According to Fig. 9.2, the Ca and Mg exchange isotherms lie *above* the nonpreference isotherm, indicating selectivity of the soil for both bivalent cations relative to Na$^+$. Evidently, there is more of a selectivity difference between Ca^{2+} and Na$^+$ than Mg^{2+} and Na$^+$. This difference is reflected also in Fig. 9.3, where the Mg exchange isotherms lie *below* the nonpreference isotherm, indicating selectivity of the soil for Ca^{2+} over Mg^{2+}. Note, however, that these conclusions depend on the chemical conditions under which the exchange reactions occur, particularly for the monovalent–bivalent exchange reactions. The parameter A in Eq. 9.5 is related directly to the electrolyte concentration, such that E_{biv} increases as $\tilde{Q}$ decreases for a fixed $\tilde{E}_{biv}$. Therefore, a comparison like that in Fig. 9.2 for $\tilde{Q} = 0.05$ mol$_c$ L^{-1} may change completely as $\tilde{Q}$ changes to higher or lower values.

9.3 Ion Exchange Reactions

In its most general meaning, an ion exchange reaction involves the replacement of one ionic species in a solid compound by another ionic species taken from an aqueous solution in contact with the solid. Equation 1.5 shows a cation exchange reaction in which Ca^{2+} replaces K$^+$ in muscovite. Equation 2.3 has AlOH^{2+} replacing Ca^{2+} (see also Eq. 2.7a). Equation 3.13 shows cation exchange for organic cations, with a similar chemical equation possible for organic anions. Equation 3.16 describes the exchange of OH$^-$ for COO$^-$, and Eq. 8.11b gives this ligand exchange reaction for OH$^-$ and an arbitrary anion, A$^{\ell-}$. To these examples may be added the exchange of Na$^+$ for K$^+$ in feldspar,

$$KAlSi_3O_8(s) + Na^+(aq) = NaAlSi_3O_8(s) + K^+(aq) \tag{9.6}$$

and that of Mg^{2+} for Ca^{2+} in carbonate solids,

$$CaCO_3(s) + Mg^{2+}(aq) = MgCO_3(s) + Ca^{2+}(aq) \qquad (9.7)$$

The examples mentioned serve to point out two features of ion exchange reactions taken in their most general sense: (1) They may involve any kind of bonding mechanism (electrostatic, ionic, or covalent); and (2) they need not involve surface phenomena. Chloride replacement by nitrate on an Oxisol is as much an ion exchange reaction *in the general sense* as is hydroxyl replacement by phosphate. Potassium replacement by sodium inside the structure of a feldspar mineral is no less a cation exchange reaction *in the general sense* than is K^+ replacement by Na^+ on a smectite surface. As stated in Section 9.1, however, the usual meaning of "ion exchange reaction" in soil chemistry is the replacement of one *adsorbed, readily exchangeable* ion by another. On the molecular level, this means that ion exchange is a surface phenomenon involving charged species in outer-sphere complexes or in the diffuse-ion swarm. In practice, this conceptualization is adhered to only approximately. Cation exchange reactions on soil humus, for example, include protons (Section 3.3) that may be adsorbed in inner-sphere surface complexes. Common extracting solutions for soil cation exchange capacity measurements (e.g., $NH_4C_2H_3O_2$ and $BaCl_2$) may displace metal cations from inner-sphere surface complexes as well as readily exchangeable metal cations. The careful experimental methods that will quantitate only readily exchangeable ions or will partition adsorbed ions accurately into readily exchangeable and specifically adsorbed species remain the objective of future research.

Ion exchange reactions on whole soils or soil separates (e.g., the clay fraction) cannot be expressed as chemical equations like Eq. 2.7a, or even Eq. 3.16, which show details of the chemical composition of the adsorbent. Soil as an adsorbent is very heterogeneous, and, therefore, an approach similar to that used to describe cation exchange on soil humus (Section 3.3) must be adopted. The symbol X will denote the soil adsorbent in the same way that S was used to denote the humus adsorbent. This representation is meant to depict the ion exchange characteristics of soil only in some *average* sense, with chemical equations for ion exchange written *by analogy* with expressions like Eq. 2.7a.

Consider, for example, the $Na^+ \rightarrow Ca^{2+}$ exchange reaction underlying the data in Fig. 9.2. This reaction can be expressed:

$$Na_2X(s) + Ca^{2+}(aq) = CaX(s) + 2Na^+(aq) \qquad (9.8)$$

$$2NaX(x) + Ca^{2+}(aq) = CaX_2(aq) + 2Na^+(aq) \qquad (9.9)$$

$$2NaX(s) + Ca^{2+}(aq) = 2Ca_{\frac{1}{2}}X(s) + 2Na^+(aq) \qquad (9.10)$$

Equations 9.8–9.10 are three alternate ways to represent the *same* cation exchange reaction. In Eq. 9.8, X^{2-} denotes an amount of soil bearing 2 mol of intrinsic negative charge, whereas in Eqs. 9.9 and 9.10, X^- denotes an amount of soil bearing 1 mol of intrinsic negative charge. As long as the charge property is made clear, X can be used in either case as the symbol for the soil adsorbent, and in neither does the "valence" -1 or -2 have any molecular structural significance. Equation 9.10 differs from Eq. 9.9 by emphasizing that 1 mol of Ca *charge* reacts with 1 mol of soil adsorbent charge. Thus Eq. 9.9 is expressed in terms of the moles of Ca and Na that react with 1 mol of X, whereas Eq. 9.10 is expressed in terms of reacting moles of Ca and Na charge. The choice of which equation to use is solely a matter of personal preference, since both satisfy the requirements of mass and charge balance. The same conclusion applies to Eqs. 9.8 and 9.9 as well.

As indicated in Section 8.2, it is expected that the rate of ion exchange reactions involving readily exchangeable ions will be governed by a film diffusion mechanism (Special Topic 2). Equation 8.4 can be taken as a point of departure for representing these rates mathematically. The detailed specification of the forward and backward rate functions, however, depends on the way the ion exchange reaction itself is represented. Consider the reaction in Eq. 9.10. In this case, Eqs. 8.3, 8.4, and s2.1 (Special Topic 2) can be applied to each cation involved:

$$dq_1/dt = Sj_1 = (SD_1/\delta)(c_1 - c_1') \tag{9.11a}$$

$$dq_2/dt = 2SJ_2 = (2SD_2/\delta)(c_2 - c_2') \tag{9.11b}$$

where 1 refers to Na^{2+}, 2 refers to Ca^{2+}, q_i is moles of *charge* of ion i adsorbed, and all other symbols are defined in Special Topic 2. After multiplying Eq. 9.11a by $2D_2q_1c_2'/c_1'$ and Eq. 9.11b by D_1q_1, then subtracting the two equations, one derives the result:

$$(2D_2q_1c_2'/c_1')\frac{dq_1}{dt} - D_1q_1\frac{dq_2}{dt} = (2SD_1D_2/\delta)[q_1c_2'c_1/c_1')$$

$$-q_1c_2' - q_1c_2 + q_1c_2'] = (2SD_1D_2/\delta)[(q_1c_2', c_1/c_1') - q_1c_2]$$

This equation can be simplified on noting that, for a binary exchange reaction,

$$\frac{dq_1}{dt} + \frac{dq_2}{dt} = 0 \tag{9.12}$$

since $q_1 + q_2 = Q$, the soil *CEC*. Thus one can substitute for dq_2/dt and derive the equation:

$$\frac{dq_1}{dt} = (2SD_1D_2/\delta)\frac{(q_1c_2'c_1/c_1') - q_1c_2}{(2D_2q_1c_2'/c_1') + D_1q_1} \tag{9.13}$$

Equation 9.13 is put into a more compact form by defining the *exchange separation factor*, α_{21}:

$$\alpha_{21} \equiv 2q_1c_2'/q_2c_1' \tag{9.14}$$

which is simply the ratio of two distribution coefficients defined as in Eq. s2.4 (Special Topic 2). Then Eq. 9.13 becomes:

$$\frac{dq_1}{dt} = (SD_1D_2/\delta)\frac{(\alpha_{21}q_2c_1 - 2q_1c_2)}{(\alpha_{21}D_2q_2 + D_1q_1)} \tag{9.15}$$

Equation 9.15 describes the rate at which the moles of adsorbed Na^+ charge changes with time during Na^+–Ca^{2+} exchange. The prefactor in the equation reflects the film diffusion mechanism. The numerator in the equation is a measure of how much "out of equilibrium" the cation exchange reaction is at a given instant. The left side of Eq. 9.15 will vanish at equilibrium, and this condition is reached when

$$\alpha_{21} = 2q_1c_2/q_2c_1 \equiv \alpha_{21}^{eq} \tag{9.16}$$

which is the same as the condition $c_i = c_i'$ ($i = 1, 2$), discussed in Special Topic 2. Therefore, at equilibrium, film diffusion ceases (see Eq. s2.1) and the separation factor can be expressed in terms of the *bulk* aqueous solution concentrations, c_i ($i = 1, 2$), instead of the film–soil interface concentrations, $c_i' = (i = 1, 2)$. Note that, under the initial condition of a Na-saturated soil mixed with a Ca solution, $q_2 = 0$ and Eq. 9.15 reduces to the expression:

$$\left(\frac{dq_1}{dt}\right)_{t=0} = -(2SD_2/\delta)c_2^0 \tag{9.17}$$

where $c_2^0 = c_2$ at "time zero." The initial rate of Na^+ desorption depends on the initial Ca^{2+} bulk concentration and its diffusion coefficient in the film; large values of either parameter will hasten Na^+ desorption. Similar analyses can be carried out for dq_2/dt after the substitution of Eq. 9.12 into Eq. 9.15.

9.4 Cation Exchange Equilibria

Cation exchange reactions like those in Eqs. 9.8–9.10 can be described in terms of conditional and thermodynamic equilibrium constants by analogy with the concepts developed in Sections 4.2 and 4.5 for reactions in aqueous solution. The approach is quite similar in spirit to that taken in Section 5.1 for solid dissolution, Section 6.2 for redox reactions, and Section 8.5 for surface complexation. The thermodynamic equilibrium constants for the reactions in Eq. 9.8–9.10, for example, are defined by the equations:

$$K_{ex} \equiv (CaX)(Na^+)^2/(Na_2X)(Ca^{2+}) \tag{9.18a}$$

$$K_{ex} \equiv (CaX_2)(Na^+)^2/(NaX)^2(Ca^{2+}) \tag{9.18b}$$

$$K_{ex} \equiv (Ca_{0.5}X)^2(Na^+)^2/(NaX)^2(Ca^{2+}) \tag{9.18c}$$

Although the activity expressions on the right side of Eq. 9.18 differ, the numerical value of the exchange equilibrium constant K_{ex} is *the same* in each case because the same cation exchange reaction is described.

As with aqueous species, the activities of the solid species in Eq. 9.18 can be factored into an activity coefficient and a concentration variable. The concentration variable used conventionally with solid phases is the *mole fraction* (see also Section 2.5 and Problem 7 in Chapter 5). This parameter is the molar ratio of a species, whose chemical formula contains only integer stoichiometric coefficients, to the total number of moles of all such species in the solid phase. For example, in the case of Eq. 9.18a, the two species mole fractions are:

$$x_{CaX} = n_{CaX}/(n_{CaX} + n_{Na_2X}) \qquad (9.19a)$$

$$x_{Na_2X} = n_{Na_2X}/(n_{CaX} + n_{Na_2X}) \qquad (9.19b)$$

where x is mole fraction, and n is mole number. Similarly, for the solid species in Eq. 9.18b one writes:

$$x_{CaX_2} = n_{CaX_2}/(n_{CaX_2} + n_{NaX}) \qquad (9.20a)$$

$$x_{NaX} = n_{NaX}/(n_{CaX_2} + n_{NaX}) \qquad (9.20b)$$

Since Eq. 9.18c is formulated in moles of charge and contains a species with fractional stoichiometric coefficients ($Ca_{0.5}X$), the activity of this species cannot properly be factored into an activity coefficient and mole fraction. The reason for this difficulty is primarily that $n_{Ca_{0.5}X}$ cannot be related to an Avogadro number of calcium ions; only n_{CaX} or n_{CaX_2} can be so related.

The activity–concentration relationship analogous to Eq. 4.19 for aqueous species is:

$$(i) = f_i x_i \qquad (9.21)$$

where i is a species like $CaX_2(s)$, and f is a *rational activity coefficient*. All solid-species activity coefficients are required to approach the value 1.0 as the corresponding mole fractions approach 1.0. With Eq. 9.21, one can reformulate K_{ex} in Eq. 9.18b, for example, as the expression:

$$K_{ex} \equiv f_2 x_2 (Na^+)^2/f_1^2 x_1^2 (Ca^{2+})$$

$$\equiv f_2 {}^c K_{12}/f_1^2 \qquad (9.22)$$

where 1 refers to NaX, 2 refers to CaX_2, and

$$^c K_{12} \equiv x_2 (Na^+)^2/x_1^2 (Ca^{2+}) \qquad (9.23)$$

is defined to be the *conditional exchange constant* for the reaction in Eq. 9.9. An analogous relationship applies to Eq. 9.18a and the reaction in Eq. 9.8.

The chemical significance of the rational activity coefficient is parallel to that developed for the aqueous-species activity coefficient in Section 4.5 (see also the discussion following Eq. 8.23). The conditional constant $^cK_{12}$ is convenient to measure using composition data for the soil adsorbent and soil solution (using also Eq. 4.23). But $^cK_{12}$ is usually very dependent on the exchangeable-cation composition. This point is illustrated in Fig. 9.4 for the data in Fig. 9.3 at $ESP = 0\%$, corresponding to the exchange reaction:

$$MgX_2(s) + Ca^{2+}(aq) = CaX_2(s) + Mg^{2+}(aq) \qquad (9.24)$$

for which (with $1 = MgX_2$ and $2 = CaX_2$)

$$^cK_{12} = x_2(Mg^{2+})/x_1(Ca^{2+}) \qquad (9.25)$$

The cause of the composition dependence of $^cK_{12}$ is the interaction between adsorbed species (and often, changes in soil particle arrangements associated with colloidal phenomena) that occur as the adsorbate composition changes. If the mole fractions alone were adequate to reflect the effects of composition changes on K_{ex}, the solid-phase mixture of $NaX(s)$ and $CaX_2(s)$ would be termed "ideal" according to thermodynamic conventions. Real mixtures are not usually ideal, so the rational activity coefficient is introduced to "correct" the mole fractions in $^cK_{12}$ for this effect and thereby maintain a constant value of K_{ex} via Eq. 9.22. This correction is expected to be larger for exchanging cations of differing valence and solvated size (e.g., Na^+ vs. Ca^{2+}, or Li^+ vs. K^+).

If $^cK_{12}$ in Eq. 9.25 were equal to 1.0, irrespective of the values of x_{CaX_2} and x_{MgX_2}, then the mole fractions of the adsorbed exchangeable cations would simply be proportional to their concentrations in the soil solution (given in Eq.

FIG. 9.4   A graph of $\ln{^cK_{12}}$ against E_{CaX_2} for the reaction in Eq. 9.24.

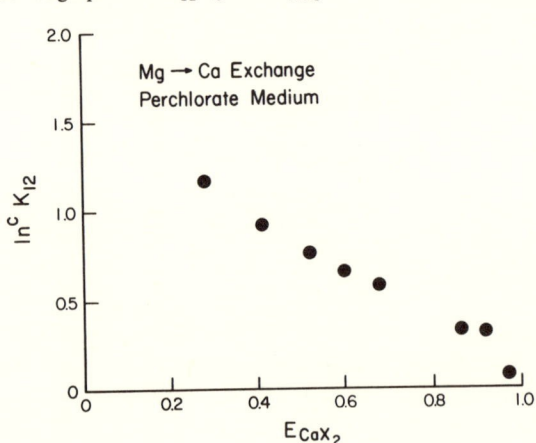

4.23). In this case, Eq. 9.25 would become:

$$1 = x_2(Mg^{2+})/x_1(Ca^{2+})$$

$$= n_{CaX_2}[Mg^{2+}]/n_{MgX_2}[Ca^{2+}]$$

$$= E_2\tilde{E}_1/E_1\tilde{E}_2$$

$$= E_2(1 - \tilde{E}_2)/(1 - E_2)\tilde{E}_2 \qquad (9.26)$$

where the second step comes from Eqs. 9.20 and 4.23, the third step from Eqs. 9.1 and 9.2, and the last step from the fact that $\Sigma_i E_i = \Sigma_i \tilde{E}_i = 1$ when all species charge fractions are summed. Equation 9.4, which defines the nonpreference exchange isotherm, is a direct result of rearranging Eq. 9.26. Similarly, if $^cK_{12} = 1.0$ in Eq. 9.23 irrespective of the mole fractions, then

$$1 = x_2(Na^+)^2/x_1^2(Ca^{2+}) \qquad (9.27)$$

and the relationships

$$x_1 = 2E_{NaX}/(E_{CaX_2} + 2E_{NaX}) \qquad x_2 = E_{CaX_2}/(E_{CaX_2} + 2E_{NaX}) \qquad (9.28)$$

that follow from Eqs. 9.20 and 9.1 lead to Eq. 9.5 for the nonpreference exchange isotherm. Equations 9.26 and 9.27 are based on the assumptions that $K_{ex} = 1.0$ and $f_1 = f_2 = 1$ ("ideal" mixture), as stated in Section 9.2.

An expression having the simplicity and generality of the Davies equation (Eq. 4.23) for the activity coefficients of aqueous ions has not been discovered yet for the rational activity coefficients of exchangeable cations. If the conditional exchange constant has been determined as a function of exchangeable-cation composition, however, chemical thermodynamic methods can be applied to derive equations for the rational activity coefficients as functions of composition. For the exchange reaction in Eq. 9.24 (or any reaction between cations of the same valence), the resulting expressions are $(1 = MgX_2, 2 = CaX_2)$:

$$\ln f_1 = E_2 \ln {^cK_{12}} - \int_0^{E_2} \ln {^cK_{12}} dE_2' \qquad (9.29a)$$

$$\ln f_2 = -(1 - E_2) \ln {^cK_{12}} + \int_{E_2}^1 \ln {^cK_{12}} dE_2' \qquad (9.29b)$$

where ln is natural logarithm. For the exchange reaction in Eq. 9.9 the result is $(1 = NaX, 2 = CaX_2)$:

$$2 \ln f_1 = E_2 \ln {^cK_{12}} - \int_0^{E_2} \ln {^cK_{12}} dE_2' \qquad (9.30a)$$

$$\ln f_2 = -(1 - E_2) \ln {^cK_{12}} + \int_{E_2}^1 \ln {^cK_{12}} dE_2' \qquad (9.30b)$$

Equation 9.22 then produces with Eq. 9.30 the important result:

$$\ln K_{ex} = \int_0^1 \ln {}^cK_{12} dE_2 \qquad (9.31)$$

In practice, data such as those in Fig. 9.4 are fitted to power series in a charge fraction by statistical regression techniques. The simplest case (approximately valid for the data in Fig. 9.4) is a linear relationship:

$$\ln {}^cK_{12} = a + bE_2 \qquad (9.32)$$

where a and b are empirical constants. With Eq. 9.32 introduced, Eq. 9.29 becomes:

$$\ln f_1 = E_2(a + bE_2) - \int_0^{E_2} (a + bE_2')dE_2' = \tfrac{1}{2}bE_2^2 \qquad (9.33a)$$

$$\ln f_2 = -(1 - E_2)(a + bE_2) + \int_{E_2}^1 (a + bE_2')dE_1'$$

$$= \tfrac{1}{2}b(1 - E_2)^2 = \tfrac{1}{2}bE_1^2 \qquad (9.33b)$$

$$\ln K_{ex} = a + \tfrac{1}{2}b \qquad (9.33c)$$

In this example, the natural logarithm of the rational activity coefficient depends on the square of the charge fraction, and the numerical value of K_{ex} can be found by calculating ${}^cK_{12}$ at $E_2 = E_{CaX_2} = 0.5$ (compare Eqs. 9.32 and 9.33c).

9.5 Cation Exchange Models

Equations 9.33 permit calculations of the rational activity coefficients and the thermodynamic cation exchange constant based on composition data for exchangeable cations. This information then can be applied to predict the speciation of soil particle surfaces by methods analogous to those used to speciate soil solutions (Section 4.5). Most soil chemistry research on cation exchange, however, has focused instead on the use of model equations to predict K_{ex} without the use of Eq. 9.33. These cation exchange models are based on assumptions about the composition dependence of the exchangeable-cation activities in Eq. 9.18. Nearly all of the models are special cases of two general activity–composition models:

$$(MX_{Z_M}) = x_1 \exp[a_1 a_2^2 x_2^2 / (a_1 x_1 + a_2 x_2)^2] \qquad (9.34)$$

$$(MX_{Z_M}) = E_1^{Z_M/\beta} \qquad (9.35)$$

where M is a cation of valence Z_M, 1 refers to MX_{Z_M}, 2 refers to the exchangeable cation with which M is exchanged, and a_1, a_2, and β are empirical parameters that characterize a given binary-cation exchange reaction. Equations 9.34 and 9.35 will be discussed for the cation exchange reaction in Eq. 9.9. Note that X always has the "valence" -1 in these applications.

Equation 9.34 is the *van Laar model*. The parameters a_1 and a_2 can be shown to have the chemical meaning:

$$a_1 = \ln f_1(x_1 \rightarrow 0) \qquad a_2 = \ln f_2(x_2 \rightarrow 0) \qquad (9.36)$$

by reference to Eq. 9.21. (To get the result for a_2, interchange 1 and 2 in Eq. 9.34.) If $a_1 = a_2 = b$, Eq. 9.34 becomes the *regular solution model*:

$$(MX_{Z_M}) = x_1 \exp(bx_2^2) \qquad (9.37)$$

This model leads to the same result for $f_1[=\exp(bx_2^2)]$ as in Eq. 9.33a when applied to the reaction in Eq. 9.24. Applied to Eq. 9.9, the regular solution model is:

$$(NaX) = x_1 \exp(bx_2^2) \qquad (CaX_2) = x_2 \exp(bx_1^2) \qquad (9.38)$$

If $a_1 = a_2 = 0$, Eq. 9.34 becomes the *Vanselow model*:

$$(MX_{Z_M}) = x_1 \qquad (9.39)$$

In this model, the exchangeable cations form an "ideal" mixture, by definition (Section 9.4). The application to Eq. 9.9 is:

$$(NaX) = x_1 \qquad (CaX_2) = x_2 \qquad (9.40)$$

Both the Vanselow and the regular solution models actually have been found to be more accurate for reactions involving cations of the same valence, like Eq. 9.24. Their respective predictions of K_{ex} for the reaction in Eq. 9.9 follow from Eqs. 9.18b and 9.38 or 9.40:

$$K_{ex}^{RS} = [x_2(Na^+)^2/x_1^2(Ca^{2+})] \exp[b(x_1^2 - 2x_2^2)] \qquad (9.41)$$

$$K_{ex}^{V} = x_2(Na^+)^2/x_1^2(Ca^{2+}) \qquad (9.42)$$

Since K_{ex}^{V} is the same as $^cK_{12}$ (Eq. 9.23), it is not usually observed to be constant as the exchangeable-cation composition varies. For this reason, K_{ex}^{V} and the other model exchange constants are called *cation exchange selectivity coefficients*. They show selectivity trends in cation exchange reactions, but they usually do not meet the requirements for a valid thermodynamic exchange constant.

Equation 9.35 is the *Rothmund–Kornfeld model*. When applied to Eq. 9.9, it produces the activity relations:

$$(NaX) = E_1^{1/\beta} \qquad (CaX_2) = E_2^{2/\beta} \tag{9.43}$$

where E is a charge fraction (Eq. 9.1). With the adjustable parameter β, the Rothmund–Kornfeld model K_{ex},

$$K_{ex}^{RK} = E_2^{2/\beta}(Na^+)^2/E_1^{2/\beta}(Ca^{2+}) \tag{9.44}$$

often works well for the monovalent–bivalent cation exchange reaction in Eq. 9.9. The special case, $\beta = 1$, is the *Gapon model*, where

$$K_{ex}^{G} \equiv E_2(Na^+)/E_1(Ca^{2+})^{1/2} = \sqrt{K_{ex}^{RK}} \tag{9.45}$$

This exchange selectivity coefficient has been observed to remain constant over a moderate variation in E_1 ($\leqslant 0.4$) in some arid-zone soils.

FOR FURTHER READING

R. D. Harter (ed.), *Adsorption Phenomena*, van Nostrand Reinhold, New York, 1986. Classic papers on ion exchange in soils and the fine historical review of ion exchange experiments by G. W. Thomas are collected in this anthology.

R. Levy (ed.), *Chemistry of Irrigated Soils*, van Nostrand Reinhold, New York, 1984. The original papers of Vanselow, Gapon, and Schofield on cation exchange theory may be found in Part III of this excellent anthology.

G. Sposito, *The Thermodynamics of Soil Solutions*, Clarendon Press, Oxford, U.K., 1981. Chapter 5 of this monograph is a detailed thermodynamic discussion of cation exchange equilibria.

O. Talibudeen, Cation exchange in soils, in D. J. Greenland and M. H. B. Hayes (eds.), *The Chemistry of Soil Processes*, pp. 115–177. Wiley, Chichester, U.K., 1981. A thorough review of the kinetics, equilibria, and mechanisms of cation exchange.

G. W. Thomas and W. L. Hargrove, The chemistry of soil acidity, in F. Adams (ed.), *Soil Acidity and Liming*, pp. 3–56. American Society of Agronomy, Madison, WI, 1984. A thought-provoking, careful review of the measurement of soil exchange capacities and their relation to soil pH.

PROBLEMS

The more difficult problems are indicated by an asterisk.

1. After consulting Chapters 8, 9, and 30 in *Methods of Soil Analysis* (listed in "For Further Reading" at the end of Chapter 4), compare the $BaCl_2$ – TEA, Baryta adsorption (see also Section 3.3), and $NH_4C_2H_3O_2$ methods of measuring *CEC* as applied to a variable-charge soil (e.g., Spodosol or Oxisol).

*2. For soils whose mineralogy reflects the advanced stage of Jackson–Sherman weathering (Table 1.7), the measurement of *CEC* as the sum of Na, K, Ca, and Mg extracted by $NH_4C_2H_3O_2$ and Al and H extracted by KCl has been proposed. Compare this method with the $BaCl_2$–TEA method as to advantages and disadvantages.

(*Hint:* Consult Chapters 8 and 9 of *Methods of Soil Analysis.*)

*3. Discuss critically the possibility of using data like those in Fig. 7.4 for the PZNC to determine the *CEC* and *AEC* of a variable-charge soil. Under what conditions would it be valid to state that *CEC = AEC* at the PZNC?

(*Hint:* Consult *Characterization of Soils,* listed in the "For Further Reading" section at the end of Chapter 7.)

4. Consult Chapters 8 and 9 of *Methods of Soil Analysis* to obtain details of the $NaC_2H_3O_2/NaCl$ and $NH_4C_2H_3O_2$ methods of measuring *CEC* in soils. Compare the advantages and disadvantages of each method as applied to a soil whose mineralogy reflects the early stage of Jackson–Sherman weathering (Table 1.7).

5. Plot an exchange isotherm like those in Fig. 9.2 using the data in the accompanying table. Show that the isotherm is essentially a nonpreference isotherm, as described in Section 9.2. (Take $\tilde{Q} = 0.05 \text{ mol}_c \text{ kg}^{-1}$ and calculate the single-ion activity coefficients with Eq. 4.23 for $I = 0.05$ mol kg^{-1}.)

m_{Na}	m_{Mg}	q_{Na}	q_{Mg}	m_{Na}	m_{Mg}	q_{Na}	q_{Mg}
(mol kg^{-1})		(mol$_c$ kg^{-1})		(mol kg^{-1})		(mol$_c$ kg^{-1})	
0.0495	0.00117	0.53	0.28	0.0340	0.0124	0.10	0.74
0.0474	0.00234	0.30	0.45	0.0291	0.0149	0.08	0.74
0.0440	0.00700	0.22	0.70	0.0237	0.0174	0.06	0.78
0.0383	0.00940	0.23	0.86	0.0185	0.0197	0.06	0.95

*6. Calculate the exchange separation factor as a function of $\tilde{E}_{Mg}$ for the data in Problem 5. (Take 1 = Na, 2 = Mg in α_{21}.) Sometimes the condition $\alpha_{21}^{eq} = 1.0$ is used as the criterion for a nonpreference isotherm. Compare this criterion with the conclusion implied in Problem 5 regarding nonpreference in the Na–Mg exchange data.

(*Hint:* Show that $\alpha_{21}^{eq} = 1.0$ leads to Eq. 9.4 as the nonpreference isotherm instead of Eq. 9.5.)

7. Plot an exchange isotherm like those in Fig. 9.2 using the composition data in the accompanying table. Include the nonpreference isotherm based on Eq. 9.5. (Take $\tilde{Q} = 0.05 \text{ mol}_c \text{ kg}^{-1}$ and calculate the single-ion activity coefficients with Eq. 4.23 for an ionic strength of 0.05 mol kg^{-1}.)

m_{Na}	m_{Ca}	q_{Na}	q_{Ca}	m_{Na}	m_{Ca}	q_{Na}	q_{Ca}
(mol kg^{-1})		(mol$_c$ kg^{-1})		(mol kg^{-1})		(mol$_c$ kg^{-1})	
0.0480	0.000136	0.100	0.037	0.0450	0.00164	0.047	0.103
0.0474	0.000320	0.074	0.056	0.0441	0.00212	0.039	0.108
0.0469	0.000717	0.060	0.081	0.0397	0.00475	0.016	0.121
0.0457	0.000118	0.049	0.095	0.0302	0.00976	0.009	0.134

8. Show that the rate of Na^+ adsorption by a Ca-saturated soil is determined only by properties of Na^+ in the soil solution and the soil property S/δ. Calculate this rate for the condition $c_{Na} = 3$ mol m^{-3} and the suggested parameter values in Special Topic 2.

9. Calculate the rate of K^+ adsorption by a soil containing 0.001 mol K kg^{-1} and 0.089 mol Ca kg^{-1} as exchangeable cations, with soil solution concentrations of 1 mol K m^{-3} and 3 mol Ca m^{-3}, under the condition $\alpha_{21} = 0.034$. The specific surface area, diffusion coefficients and the film thickness δ may be estimated as in Special Topic 2.

10. Calculate the conditional exchange equilibrium constant for the reaction in Eq. 9.24 based on the data in the table below. Plot $^cK_{12}$ against E_{CaX_2}. (Assume that single-ion activity coefficients can be calculated with the Davies equation.)

c_{Mg}	c_{Ca}	q_{Mg}	q_{Ca}	c_{Mg}	c_{Ca}	q_{Mg}	q_{Ca}
(mol m^{-3})		(mol$_c$ kg^{-1})		(mol m^{-3})		(mol$_c$ kg^{-1})	
20.4	2.4	0.21	0.078	10.1	12.3	0.093	0.21
17.8	4.8	0.17	0.11	4.9	17.0	0.053	0.26
14.8	7.2	0.14	0.15	2.4	19.6	0.027	0.31
12.4	9.7	0.12	0.17	1.2	20.9	0.016	0.30

11. Fit the values of $^cK_{12}$ obtained in Problem 10 to Eq. 9.32 and calculate K_{ex} for the reaction in Eq. 9.24. Write down the corresponding expressions for $\ln f_{NaX}$ and $\ln f_{CaX_2}$.

12. Calculate the Vanselow and Gapon model exchange equilibrium constants for the data in Problem 5. Which model appears to perform better?

*13. Calculate the Rothmund–Kornfeld model exchange equilibrium constant for the data in Problem 7 based on the reaction in Eq. 9.9.
(Hint: Under what condition is K_{ex}^{RK} dependent only on the activities of Na^+ and Ca^{2+} in aqueous solution?)

14. Test the Rothmund–Kornfeld model with the data in Problem 10 by plotting $\log(\tilde{E}_{Mg}/\tilde{E}_{Ca})$ against $\log(E_{CaX_2}/E_{MgX_2})$. According to the model as applied to the reaction in Eq. 9.24, this graph should be a straight line. How are K_{ex} and β calculated from the y intercept and slope of this line?

15. In the table below are values of the rational activity coefficients for $Na \rightarrow K$ exchange on a soil. Use these data to estimate the parameters a_1 and a_2 in the van Laar model. How good an approximation for the data would the regular solution model be?

E_K	f_{Na}	f_K	E_k	f_{Na}	f_K
0.1	1.000	0.693[a]	0.6	0.837	0.902
0.1	0.999	0.703	0.7	0.748	0.958
0.2	0.995	0.719	0.8	0.693	0.982
0.3	0.982	0.749	0.9	0.545	0.995
0.4	0.955	0.788	1.0	0.586[a]	1.000
0.5	0.895	0.854			

[a]Extrapolated values.

10

Soil Colloidal Phenomena

10.1 Colloidal Suspensions

Colloids are solid particles of very low water solubility with a diameter ranging between 0.01 and 10 μm. Therefore, in soils, the clay-sized and fine silt-sized insoluble particles are classified as colloids. The chemical composition of these particles may vary from that of a clay mineral or a metal hydrous oxide to that of soil humus or a combination of inorganic and organic material. Regardless of their composition, the characteristic property of colloids is that they do not dissolve in water to form solutions, but instead remain as an identifiable solid phase in *suspension*.

Colloidal suspensions are said to be *stable* (and the particles in them *dispersed*) if no measurable gravitational settling of the suspension occurs over practical time periods (e.g., 2–24 h). Stable suspensions of soil colloids lead to soil erosion and clay illuviation because the particles entrained by the soil solution remain highly mobile. Stable suspensions also have a secondary effect on the mobility of inorganic and organic adsorptives (e.g., trace metals, phosphate anions, or pesticides) that may become strongly bound to soil colloids. Thus colloidal stability is connected closely with particle and chemical transport in soils.

The process by which a colloidal suspension becomes unstable and undergoes gravitational settling is *coagulation*. Coagulation that produces bulky, settled masses of particles with high water contents is called *flocculation*. If dense, organized masses of settled particles are formed, the process is *aggregation*. The two kinds of coagulation are very important to soil structure formation, but are complex phenomena dependent on the surface chemistry and geometry of the colloids involved, as well as on the pH, ionic strength, and composition of the soil solution.

A basic understanding of coagulation can be had by considering in detail the microscopic nature of a colloidal suspension. In a quiescent soil solution, the motions of colloidal particles are incessant and chaotic because of the thermal energy the particles possess. These *Brownian motions* of colloids in suspension are analogous to the diffusive motions of molecules in solution. They can be described mathematically by a form of Fick's law (see Eq. s2.1 in Special Topic 2), with the diffusion coefficient expressed by the *Stokes–Einstein relation*:

$$D = k_B T / 6\pi\eta R \tag{10.1}$$

In Eq. 10.1, k_B is the Boltzmann constant (see the Appendix), η is the coefficient of viscosity of water, and R is the radius of the colloidal particle (assumed effectively spherical). The Stokes–Einstein relation indicates that a colloid will diffuse more rapidly if the absolute temperature T is high, if the viscosity η is low, or if the colloid is very small.

In the absence of interference from interparticle forces (e.g., coulomb forces), the coagulation of colloidal particles is produced by simple collisions of the particles as a result of their Brownian motions. The rate at which these collisions occur depends on how rapidly the particles can diffuse and how large a "target" they are; it will also depend on how many particles are in suspension. The rate at which the number of particles per unit volume of suspension, ρ, decreases because of a coagulation process can be described by the equation:

$$\frac{d\rho}{dt} = -8\pi R D \rho^2 \tag{10.2}$$

where ρ is the number of colloidal particles (of radius R and diffusion coefficient D) per cubic meter. Equation 10.2, known as the *von Smoluchowski equation*, contains the square of the number density ρ on the right side because *two* particles are involved in a collision, so the likelihood of coagulation depends on the number density of each.

The von Smoluchowski equation describes the kinetics of the onset of coagulation in the absence of suspension flow and interparticle forces. [It also implies tacitly that *all* the colliding particles stick to one another permanently; if this assumption is not correct, a "sticking fraction" parameter α ($0 \leqslant \alpha \leqslant 1$) can be included on the right side of Eq. 10.2.] As a kinetics expression, Eq. 10.2 is second-order in ρ (Section 4.2). Therefore, according to Table 4.2, a graph of $1/\rho$ against time should be a straight line with a slope that is equal to the rate constant,

$$K = 8\pi R D = 4k_B T / 3\eta \tag{10.3}$$

At 298.15 K, this rate constant equals 6.2×10^{-18} m^3 s^{-1}, given $\eta = 8.9 \times 10^{-4}$ kg m^{-1} s^{-1} for liquid water. The corresponding half-life is:

$$t_{\frac{1}{2}} = 1/K\rho_0 = 1.62 \times 10^{17}/\rho_0 \tag{10.4}$$

where ρ_0 is the initial number density (see Table 4.2). For example, if a suspension contains initially 1 kg m^{-3} colloidal particles, each of which has a mass density of 2.65×10^3 kg m^{-3} and a 1 μm radius, then

$$\rho_0 = 1 \text{ kg m}^{-3} \bigg/ \left[2.65 \times 10^3 \text{ kg m}^{-3} \times \frac{4\pi}{3} (10^{-6} \text{ m})^3 \right]$$

$$= 9 \times 10^{13} \text{ m}^{-3}$$

where spherical particles are assumed in order to ca'culate the particle volume, $4\pi R^3/3$. It follows from Eq. 10.4 that $t_{\frac{1}{2}} \approx 1800$ s for the coagulation of these particles.

Often the number density in a colloidal suspension is measured by light scattering. According to the Lambert–Beer law, the intensity of light transmitted (i.e., not scattered) by a colloidal suspension is described by the equation:

$$I = I_0 \exp(-A\rho m_p^2) \tag{10.5}$$

where I_0 is the intensity of light incident on the suspension; A is a parameter dependent on the wavelength of light used, the light-scattering properties of the colloidal particles, and the optical path length in the suspension; ρ is the suspension number density; and m_p is the mass of a colloidal particle. If A and m_p do not depend on the suspension concentration, then measurements of I/I_0 can be used to infer the number density. Usually the absorbance, $\ln(I_0/I)$, is measured directly, since this variable is proportional to ρ, according to Eq. 10.5.

10.2 Soil Colloids

The colloids suspended in soil solutions will exhibit shapes and sizes that reflect both chemical composition and the effects of weathering processes. Kaolinite particles, for example, are seen in electron micrographs as roughly hexagonal plates comprising perhaps 50 unit layers (each layer a wafer the thickness of a unit cell), which are stacked irregularly and interconnected through hydrogen bonding between the OH groups of the octahedral sheet and the oxygens of the tetrahedral sheet (Fig. 2.4). In the soil environment, weathering reactions produce a rounding of the corners of the kaolinite hexagons, as well as surface coatings of iron hydrous oxides and humus. Fracturing of the plates also is apparent along with a "stair-step" surface

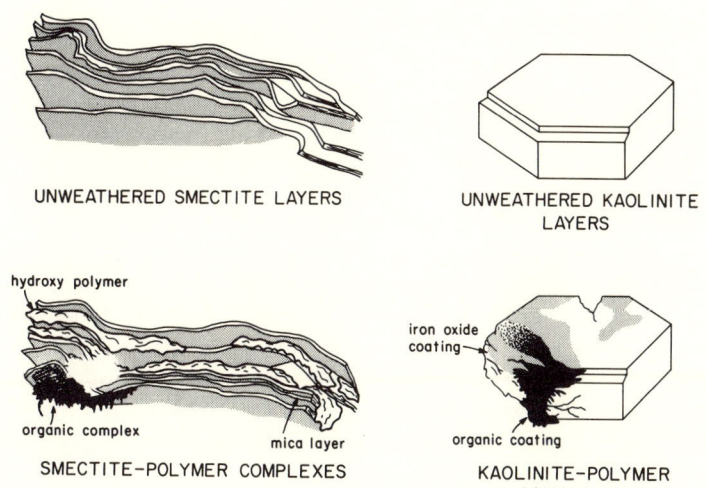

UNWEATHERED SMECTITE LAYERS

UNWEATHERED KAOLINITE LAYERS

hydroxy polymer

iron oxide coating

organic complex

mica layer

organic coating

SMECTITE–POLYMER COMPLEXES

KAOLINITE–POLYMER COMPLEXES

FIG. 10.1 Effects of weathering on the shapes and surface features of clay minerals. (Reprinted with permission from G. Sposito, *The Surface Chemistry of Soils*, Oxford Univ. Press, New York, 1984.)

topography caused by the stacking of unit layers of differing lateral dimensions (see Fig. 10.1). These heterogeneous features lead to flocculation products ("floccules") that are not well organized. The fabric of these floccules consists of many stair-stepped clusters of stacked plates, interspersed with plates in edge–face contact, that are arranged in a chaotic, porous, three-dimensional network.

Similar statements can be made for 2:1 layer-type clay minerals. Illite, for example, is seen in electron micrographs as platelike particles stacked irregularly [although the interlayer bonding mechanism is the inner-sphere surface complex of K^+ (Fig. 7.2), not hydrogen bonding]. These particles exhibit a stair-step surface topography and frayed edges produced by cation exchange and hydration weathering reactions (Section 1.5). Polymeric coatings of Al-hydroxy and humus compounds may occur on these surfaces. These features are made even more heterogeneous by the possibility of a nonuniform distribution of isomorphic substitution in illite, with regions of layer charge approaching 2.0 grading to regions of layer charge near 0.5 (see Table 2.3). These characteristics and the slight flexibility of the illite plates (probably caused by strains associated with isomorphic substitution) lead to floccule structures that are like those for kaolinite particles, but with even greater porosity and chaotic arrangement.

Smectites and vermiculites have a lesser tendency to flocculate in stacks because their layer charge is smaller than that of illite. They are also more flexibile particles, probably because of more extensive isomorphic substitution in the octahedral sheet. Floccule structures comprise very irregularly shaped plates in a random framework of high porosity. Surface heterogeneities brought on by nonuniform layer charge and the adsorption of Al-hydroxy or humus polymers add to the chaos.

Aggregate structures in coagulated clay minerals have been observed for the 2:1 layer types. These consist of a fairly regular stacking of crystal units to form organized arrangements called *domains* in the case of illite and *quasi crystals* in the case of smectite. A domain comprises parallel alignments of interleaved illite packets containing stacks of five to seven unit layers each. The interleaved structure extends laterally over distances about 50 times the dimension of an illite plate. A quasi crystal comprises parallel alignments of unit smectite layers (see Fig. 10.1, upper left). This alignment of unit layers can be observed in thick suspensions of Na-smectite and as a microaggregate in any kind of suspension of bivalent cation-saturated smectite. Quasi crystals of Na-smectite are important in arid-zone soils because their ordered structure prevents the development of large pore spaces essential to soil permeability. They are created by the dewatering of suspensions originally containing dispersed unit layers (i.e., initially stable suspensions).

Quasi crystals of Ca-smectite (or any M^{2+}-saturated smectite), on the other hand, are observed in dilute suspensions. The key organization unit in this microaggregate is an outer-sphere complex of Ca^{2+} with a pair of opposing siloxane cavities. The octahedral solvation complex, $Ca(H_2O)_6^{2+}$, is arranged in the interlayer region with its principal symmetry axis perpendicular to the siloxane surface. Four of the solvating water molecules lie in a central plane parallel with the opposing siloxane surfaces, while the remaining two water molecules reside in planes between the siloxane surfaces and the central plane to give an interlayer spacing of 1.91 nm (see Fig. 10.2). An outer-sphere surface

FIG. 10.2 "Exploded" view of the outer-sphere surface complex of Ca^{2+} in the smectite quasi crystal. The solvating water molecules form a three-layer structure of height 0.95 nm. (Reprinted with permission from G. Sposito, *The Surface Chemistry of Soils*, Oxford Univ. Press, New York, 1984.)

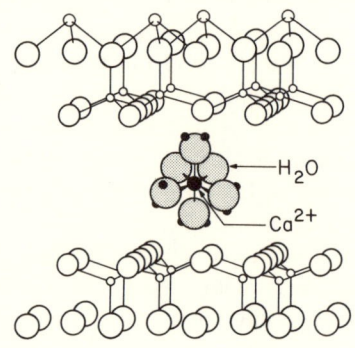

complex of this kind, which brings together two clay layers separated by three water layers, is a characteristic structure in suspensions of smectite bearing bivalent exchangeable cations.

A variety of experimental methods has been applied to determine the number of unit layers and other properties of the montmorillonite quasi crystal in dilute suspensions. Of these, perhaps light scattering is the most straightforward to interpret. As implied by Eq. 10.5, the light scattered by an aggregate in suspension depends on the refractive-index characteristics of the aggregate, the concentration of clay in the suspension, and the number of unit layers in face-to-face association within the aggregate. Under the assumption that the refractive-index characteristics and the mass density of the clay mineral do not depend on the type of exchangeable cation, absorbance measurements made for a series of homoionic montmorillonites at the same suspension concentration will be proportional to the number of unit layers per aggregate. With the convention that Li-montmorillonite suspensions contain only dispersed unit layers, *relative* estimates, as given in Table 10.1, can be made.

To see this point in detail, one can express m_p in Eq. 10.5 as the product $m_c N$, where m_c is the mass of a unit layer of montmorillonite and N is the number of unit layers per quasi crystal. It follows that

$$\ln(I_0/I) = A\rho m_p m_c N \tag{10.6}$$

Therefore, from absorbance measurements made on stable suspensions of the same *mass* concentration, ρm_p, an estimate of N is obtained. For Li-montmorillonite, diffraction data and viscosity measurements suggest $N = 1$ in dilute (<2 kg m^{-3}) suspensions.

The data in Table 10.1 indicate that, in relatively dilute stable suspensions, unweathered Na-montmorillonite particles will have a different structure from unweathered Ca-montmorillonite particles. One would infer also that there should be some kind of continuous transition from quasi crystals to more or

TABLE 10.1 Light-scattering estimates of the average number of unit layers in a specimen montmorillonite quasi crystal

Exchangeable cation	N		Exchangeable cation	N	
Li$^+$	1.0[a]	1.0[b]	Cs$^+$	2.9[a]	3.0[b]
Na$^+$	1.2	1.7	Mg^{2+}	5.5	4.2
K$^+$	1.5	2.7	Ca^{2+}	6.2	7.0

[a] A. Banin and N. Lahav, Particle size and optical properties of montmorillonite in suspension, *Israel J. Chem.* **6**:235–250 (1968). Data recalculated by Schramm and Kwak.[b]
[b] L. L. Schramm and J. C. T. Kwak, Influence of exchangeable cation composition on the size and shape of montmorillonite particles in dilute suspension, *Clays and Clay minerals* **30**:40–48 (1982).

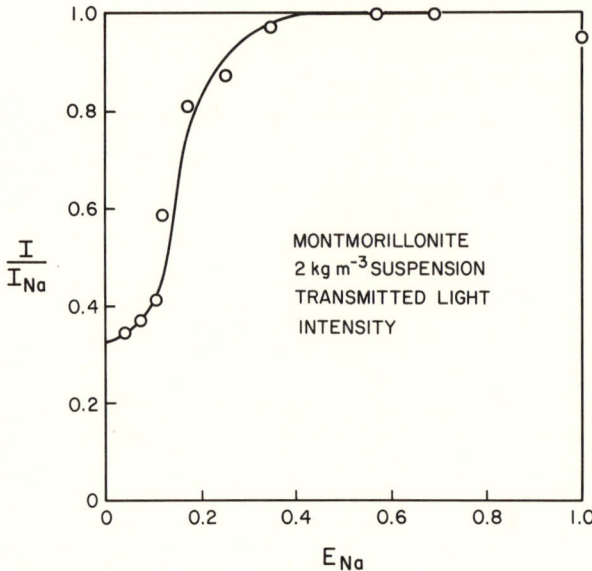

FIG. 10.3 The ratio of transmitted light intensity for a mixed Na/Ca-montmorillonite suspension to the transmitted light intensity for a Na-montmorillonite suspension. [Data from L. L. Schramm and J. C. T. Kwak, Influence of exchangeable cation composition on the size and shape of montmorillonite particles in dilute suspension, *Clays and Clay Minerals* **30**:40–48 (1982).]

less single-unit layer particles as the fraction of exchangeable Na^+ increases in stable suspensions of montmorillonite bearing both Na^+ and Ca^{2+} on the basal planes. This hypothesis is verified experimentally by the light-scattering data graphed in Fig. 10.3. The ordinate of the graph represents the ratio of the transmitted light intensity for suspensions of the mixed-ion clay to its value for a Na-montmorillonite suspension, whereas the abscissa refers to values of the charge fraction of exchangeable Na^+ on the clay. Sharp increases in I/I_{Na} are observed when the charge fraction of Na^+ on the clay increases from 0.15 to 0.30. In this range, the quasi crystals of Ca-montmorillonite are being broken up in favor of more or less single-unit layer particles. Evidently, when $E_{Na} < 0.3$, the quasi crystals are stable entities, with the residual exchangeable Na^+ residing principally on the external surfaces.

Light-scattering data can also be used to obtain kinetics information about quasi crystal formation and breakdown. Measurements of the intensity of light transmitted after the mixing of Na- and Ca-montmorillonite suspensions to produce an overall charge fraction of Na^+ on the clay particles <0.1 have indicated a very rapid (<1 min) formation of quasi crystals from conversion of the Na-montmorillonite particles. This rapid conversion involves a redistribution of the exchangeable cations, such that Na^+ ions are displaced to the external surfaces of already-formed quasi crystals that contain Ca^{2+} ions on their internal surfaces.

On the other hand, if the mixing together of two homoionic montmorillonite suspensions produces an overall charge fraction of $Na^+ > 0.6$, then it is observed that a gradual breakdown of the initial quasi crystals occurs, with the final product being a suspension with light-scattering properties resembling closely those of a collection of dispersed clay particles. This slower process of quasi crystal conversion to dispersed unit layers reflects both a lower mobility of Ca^{2+} in the interlayer region and the difficulty Na^+ has in replacing these bivalent cations in the initial step of quasi crystal destruction. The decomposition process is aided by maintaining a low electrolyte concentration, since interlayer swelling is thereby enhanced (see Section 10.3).

These kinetics aspects of quasi crystal formation and breakdown are consistent with the trends in Fig. 10.3. When $E_{Na} < 0.1$, for example, the properties of a mixed Na/Ca-montmorillonite do not differ much from those of Ca-montmorillonite and the quasi crystal should remain a stable entity. When $E_{Na} > 0.6$, however, the mixed Na/Ca-montmorillonite exhibits light-scattering properties that are indistinguishable from those of a Na-montmorillonite, and a quasi crystal should be an inherently unstable structure.

10.3 Interparticle Forces

Regardless of how complicated the structure and shape of a soil colloid may be, it still is subject to basic forces brought on by its fundamental properties of mass and electric charge. The property of mass gives rise to the *gravitational force* and the *van der Waals force*. The property of charge gives rise to the *electrostatic force* and, indirectly, the *solvation force*. The first two forces cause a colloidal suspension to be unstable, whereas the second two cause it to remain stable. The gravitational force (corrected for the effect of bouyancy) initiates and sustains the coagulation of a suspension via particle settling. This force is created simply by the gravity field of the earth. The other three forces are properly *interparticle* forces: they act between the colloids either to attract them or to repel them.

The particles in a stable soil suspension can be envisioned, in an ideal geometric sense, to be roughly spherical (hydrous oxides) or to comprise one or more unit layers stacked together (clay minerals). If the spherical particles are large relative to the dimension of their adsorbed swarm of ions or—in the case of clay minerals—if the stacking is not extensive or highly irregular, one can imagine the forces the particles exert on one another as coming from roughly planar surfaces. The nature of interparticle forces will be discussed on the basis of this geometric simplification. Actual interparticle forces are more complicated, but their essential physical features can be represented adequately in the planar-surface approximation.

The van der Waals interaction between soil colloids in suspension is exactly analogous to that between soil humus compounds and organic polymers or clay minerals (Sections 3.4 and 3.5). Over a time interval that is much longer than 10^{-16} s, the distribution of electronic charge in a nonpolar molecule is geometrically spherical. However, on a time scale comparable with or $< 10^{-16}$ s (approximately the period of an ultraviolet light wave), the charge distribution of a nonpolar molecule will exhibit significant deviations from spherical symmetry, taking on a flickering dipolar character. These deviations fluctuate rapidly enough to average to zero when observed over, say, 10^{-14} s (the period of an infrared light wave), but they persist long enough to induce distortions in the charge distributions of other neighboring molecules. If two nonpolar molecules are brought close together, each will induce in the other a fluctuating dipolar character and the *correlations* between these induced dipole charge distributions will not average to zero, even though the individual dipole distributions themselves will average to zero. The correlations between the two instantaneous dipole moments produces an attractive interaction whose potential energy is proportional to the inverse sixth power of the distance of separation, according to classical electrostatics. The corresponding attractive force is known as the *van der Waals dispersion force*. At small values of the separation distance, this interaction can be strong enough to cause particles to stick together and flocculate.

Suppose that a nonpolar molecule confronts the planar surface of a solid comprising N atoms per unit volume. The van der Waals dispersion energy for the attractive interaction between the single molecule and the solid can be calculated and shown to vary as the inverse third power of their distance apart. This inverse power is smaller than six because of the *additive* effect of the van der Waals forces between the N atoms of the solid and the nonpolar molecule. Now suppose that the nonpolar molecule is embedded in a solid just like the one it confronts. A calculation of the van der Waals attraction per unit area of solid surface then gives the equation:

$$V_{vw}(d) = -\frac{A}{12\pi d^2} \qquad (10.7)$$

where d is the distance separating the solid surfaces and A is called the *Hamaker constant*. Equation 10.7 shows that the van der Waals dispersion energy (per unit area) falls off as the inverse square of the distance separating two opposing planar surfaces. Because it is additive for all of the molecules in the two solids, this attractive interaction (hence the negative sign in Eq. 10.7) decreases with distance of separation much more slowly than the interaction between an isolated pair of molecules. The Hamaker constant, which gives a measure of the magnitude of the van der Waals energy at any separation distance, is thought to have a value near 2×10^{-20} J for soil colloids. Experiments have shown this value to be independent of the type of electrolyte solution that may lie between interacting planar surfaces.

The van der Waals force is the cause of particle coalescence ("sticking") after a collision induced by Brownian motion, as described in Section 10.1. If the particles each bear a net electric charge, however, their tendency to collide and stick together is affected by the electrostatic force between them. A repulsive force arises if the sign of the charge on each particle is the same; it manifests itself as a pressure exerted on the particles, forcing them apart. This pressure is equal to the osmotic pressure of the electrolyte solution between the particle surfaces, given approximately by the standard equation for an ideal electrolyte solution:

$$P = RT \, \Sigma_i c_i \tag{10.8}$$

where the sum is over each ionic species present. Equation 10.8 usually is evaluated by applying the diffuse double-layer model to the ion swarm between the particle surfaces (Section 8.5).

Consider two identical, opposing planes of charge arranged with the origin of coordinates set in one of them and with the other located at the point $x = d$ along a perpendicular line from the first. Each plane is to represent a particle surface to which σ in Eq. 8.17 refers. Within the aqueous solution between the planes, the concentration of ionic species i is, according to diffuse double-layer theory, given by Eq. 8.15. It follows that the osmotic pressure at a point midway between the planes ($x = d/2$) can be expressed:

$$P_m = P_0 + RT \, \Sigma_i c_{0i}[\exp(-Z_i F \phi_m / RT) - 1] \tag{10.9}$$

where P_m is the osmotic pressure at $x = d/2$, P_0 is that at $x = \infty$, and all of the symbols in the second term on the right side have been defined following Eq. 8.15. A further specification of P_m as a function of position between the planes cannot be made without solving Eq. 8.16 subject to the conditions:

$$\phi(d/2) = \phi_m \qquad \left(\frac{d\phi}{dx}\right)_{x=d/2} = 0 \tag{10.10}$$

If the electrolyte solution is $1:1$ and dilute (e.g., 1 mol m^{-3} NaCl), then Eq. 8.16 can be solved readily to yield a result for the osmotic pressure:

$$P_m = P_0 + 64a^2 c_0 RT \exp(-\kappa d) \tag{10.11}$$

where $a = \tanh(F\phi(0)/4RT)$, c_0 is the concentration of the electrolyte far from the planar surfaces, and

$$\kappa = (2F^2 c_0 / \varepsilon_0 DRT)^{1/2} \tag{10.12}$$

is a parameter whose inverse gives a measure of the spatial extent of the diffuse double layer. (See Section 8.5 for the definitions of ε_0 and D in Eq. 10.12.)

Calculated values of P_m based on Eq. 10.11 have been found to agree semiquantitatively with experimentally determined, reversible curves of the *swelling* pressure of Na-montmorillonite expressed as a function of the interlayer separation. These data, which apply to the clay mineral suspended in dilute solutions of NaCl, agree fairly well with theoretical prediction obtained using $\sigma = -0.118$ C m^{-2}. A more direct test of the diffuse double-layer model of the electrostatic force has been made by measuring the repulsive force between two clean muscovite surfaces contacting a dilute solution of KNO_3. The force between the two mica planes was measured through the deflection of a spring attached to one of them, and the surface separation distance was measured by optical techniques. Excellent agreement between experiment and theory was found, and this is convincing evidence for the applicability of diffuse double-layer theory in $1:1$ electrolyte solutions at low concentrations. However, at higher concentrations of KNO_3 (>10 mol m^{-3}) and even at low concentrations (0.1 mol m^{-3}) in $2:1$ electrolyte solutions [$Ca(NO_3)_2$ and $BaCl_2$], the diffuse double-layer model predictions fail to describe the observed force curves. This failure can be traced to the inherent inadequacy of Eq. 8.15 for clay minerals suspended in moderately concentrated $1:1$ electrolyte solutions and in any $2:1$ electrolyte solution, as discussed in Section 8.5 (see Fig. 8.5).

The absence of parameters referring directly to the molecular nature of liquid water is a common feature of the electrostatic and van der Waals dispersion forces. But this simplicity must cease to be realistic at some point as two planar surfaces are brought closer and closer together. For example, during the process of quasi crystal formation by unit layers of montmorillonite bearing exchangeable Ca^{2+} cations (Section 10.2), one can imagine that the competition between the repulsive electrostatic force and the attractive van der Waals force will, along with random thermal motions, largely determine the behavior of two siloxane surfaces approaching each other from a distance of separation >10 nm. However, at the separation distance of 0.95 nm, which is characteristic of the outer-sphere surface complex in Fig. 10.2, it can be expected that the force required to bring the particle surfaces together must have a component that reflects the effort necessary to desolvate the exchangeable Ca^{2+} cations. Indeed, between 10 and 0.95 nm, the force bringing the siloxane surfaces into close proximity must displace all of the water molecules from the second solvation shell of the Ca^{2+} cation. When the two surfaces coalesce into the quasi crystal configuration, these outermost solvating water molecules will have been ejected from the interlayer region, and the force required to accomplish this task must to some extent depend on the structure of the cation solvation complex at the molecular level.

It is not possible to derive an explicit relationship between the additional force required to desolvate exchangeable cations at a given separation distance based only on qualitative arguments. Since most of the solvated metal cations encountered on the surfaces of natural soil colloids would tend to form octahedral solvation complexes in aqueous solution, it is likely that the additional force would increase markedly as the separation distance decreased below 1 nm, the characteristic diameter of octahedral solvation complexes. For separations >1 nm, it is possible that the additional force would drop rapidly to zero if the van der Waals dispersion force were strong enough to overcome the electrostatic force and itself displace water molecules from secondary solvation shells of the cations. The effect of the finite diameter (0.29 nm) of the water molecules on the decay of the additional force with distance thus should be to superimpose an oscillatory feature that exhibits relative maxima at the mean positions of the molecules.

Qualitative evidence for the existence of this kind of *solvation force* between siloxane surfaces has been obtained from experiments on the compression of Na-montmorillonite particles in a NaCl background solution. In a solution of 0.1 mol m^{-3} NaCl, the reversible experimental compression curve for Na-montmorillonite is observed to lie significantly above the theoretical curve based on Eqs. 10.7 and 10.11, when the average interlayer distance is decreased from 5.0 to 1.8 nm. This positive deviation of the experimental compression curve suggests that an additional force must be significant at separations <5 nm.

As with the electrostatic and van der Waals dispersion forces, a more precise characterization of the solvation force has come from direct experimental measurements of the net force between opposing muscovite surfaces. Figure 10.4 shows this net force as a function of interplanar separation for two clean

FIG. 10.4 The measured net force between mica surfaces contacting a KBr solution compared with the net force calculated as the sum of van der Waals and electrostatic contributions. (Reprinted with permission from G. Sposito, *The Surface Chemistry of Soils*, Oxford Univ. Press, New York, 1984.)

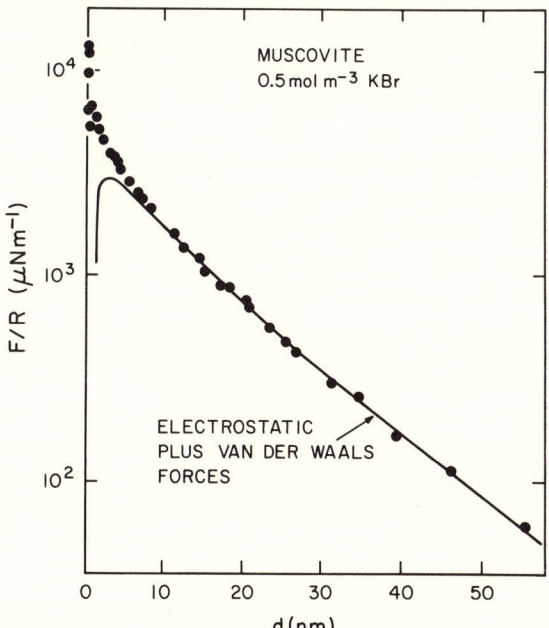

mica surfaces contacting a solution of 0.5 mol m^{-3} KBr at pH 6.2. The solid line in the figure represents the combination of the electrostatic and van der Waals forces with $\phi(0) = -0.1$ V taken as a fixed surface potential. For separations of the planes smaller than about 6 nm, there is a marked positive deviation of the experimental data from the theoretical curve. The difference between the data and the curve can be fit by statistical regression analysis to the equation:

$$V_{solv}(d) = \frac{\alpha}{2\pi} \exp(-d/\delta) \tag{10.13}$$

where α and δ are empirical constants. From this kind of curve fitting, it is found that $\alpha \approx 0.03-0.05$ N m^{-1} and $\delta \approx 0.3-1.0$ nm. The solvation force thus decays approximately exponentially with a "decay length," δ, near 1 nm in agreement with the qualitative description outlined previously.

10.4 The DLVO Model

The smallest concentration of electrolyte, in moles per cubic meter, at which a soil colloidal suspension becomes unstable and begins to undergo rapid coagulation is called the *critical coagulation concentration* (*ccc*). The value of the *ccc* will, in general, depend on the nature of the colloid particles, the composition of the aqueous solution in which they are suspended, and the time allowed for settling. The measurement of the *ccc* entails the preparation of dilute (<3 kg m^{-3} solids concentration) suspensions in a series of electrolyte solutions of increasing concentration. After about 1 h of shaking and a subsequent standing period of 2–24 h, the coagulated suspensions will show a clear boundary separating the settled solid mass from an aqueous-solution phase, and the *ccc* can be bracketed between two values determined by the largest electrolyte concentration at which coagulation did not occur and the smallest at which it did.

The study of soil colloidal stability has not yet produced exact, quantitative theories, but general relationships between stability, interparticle forces, and surface chemistry have been developed that are of predictive value. One of these general relationships is the *Schulze–Hardy Rule*. This empirical relationship concerning the *ccc*, first suggested by H. Schulze about 100 years ago and generalized by W. B. Hardy in 1900, can be stated:

> The critical coagulation concentration for a colloid suspended in an aqueous electrolyte solution is determined by the ions with a charge opposite in sign to that on the colloid and is proportional to an inverse power of the valence of the ions.

The Schulze–Hardy Rule is illustrated in Table 10.2 for five important soil minerals. Mean values of the *ccc* extracted from published studies, together

TABLE 10.2 Critical coagulation concentrations for colloidal suspensions of soil minerals

Soil mineral	$ccc(\lvert Z\rvert = 1)$ (mol m^{-3})	$ccc(\lvert Z\rvert = 2)$ (mol m^{-3})	$\dfrac{ccc(\lvert Z\rvert = 2)}{ccc(\lvert Z\rvert = 1)}$
Al hydrous oxide	50 ± 9	0.5 ± 0.2	0.010
Fe hydrous oxide	11 ± 2	0.21 ± 0.01	0.019
Illite	48 ± 11	0.14 ± 0.02	0.003
Kaolinite	10 ± 4	0.3 ± 0.2	0.030
Montmorillonite	8 ± 6	0.12 ± 0.02	0.015
		DLVO theory:	0.0156

with standard deviations reflecting the range of *ccc* values reported, are presented in the table for ions whose absolute valence, $\lvert Z\rvert$, is equal to 1 or 2. For metal oxides, the particle charge is positive and the coagulating ions are anions, whereas for the clay minerals, the particle charge is negative and the coagulating ions are cations. The ratio of *ccc* values for $\lvert Z\rvert = 2$ and $\lvert Z\rvert = 1$ is given in the fourth column of the table. These ratios may be compared with a theoretical value, $2^{-6} = 1/64 = 0.0156$, which will be derived later. It is evident from Table 10.2 that the Schulze–Hardy Rule provides a semiquantitative prediction of relative *ccc* values for soil colloidal suspensions. The inverse sixth power of the coagulating ion valence appears to be a reasonable factor by which the *ccc* values can be related.

A theoretical derivation of the Schulze–Hardy Rule can be developed on the basis of the interparticle forces described in Section 10.3. Each of the three forces is associated with a potential energy that contributes additively to the total potential energy between two planar particle surfaces that are a distance d apart. If $\phi(d)$ is the total potential energy per unit area of planar surface, then:

$$\phi(d) = V_m(d) + V_{vw}(d) + V_{solv}(d)$$

$$\approx \frac{64a^2}{\kappa} c_0 RT \exp(-\kappa d) - \frac{A}{12\pi d^2} + \frac{\alpha}{2\pi} \exp(-d/\delta) \tag{10.14}$$

according to Eqs. 10.7, 10.11, and 10.13, along with the force-potential relationship, that P_m is minus the derivative of V_m with respect to d. Since $\phi(d)$ contains both positive and negative terms, it is expected to show a relative maximum at some d value for many possible choices of the experiental parameters κ, a, A, α, and δ. This kind of mathematical behavior is illustrated in Fig. 10.5 for the case $\kappa = 0.329$ nm^{-1}, $a = 0.462$, $A = 2.2 \times 10^{-20}$ J, $\alpha = 0.05$ J m^{-2}, and $\delta = 1$ nm. These parameter values would be appropriate for a stable suspension of illite in 10 mol m^{-3} NaCl at 298 K. Figure 10.5 shows the characteristic behavior of $\phi(d)$: a pronounced minimum for very small values of d, followed by a maximum and *at least* one more relative

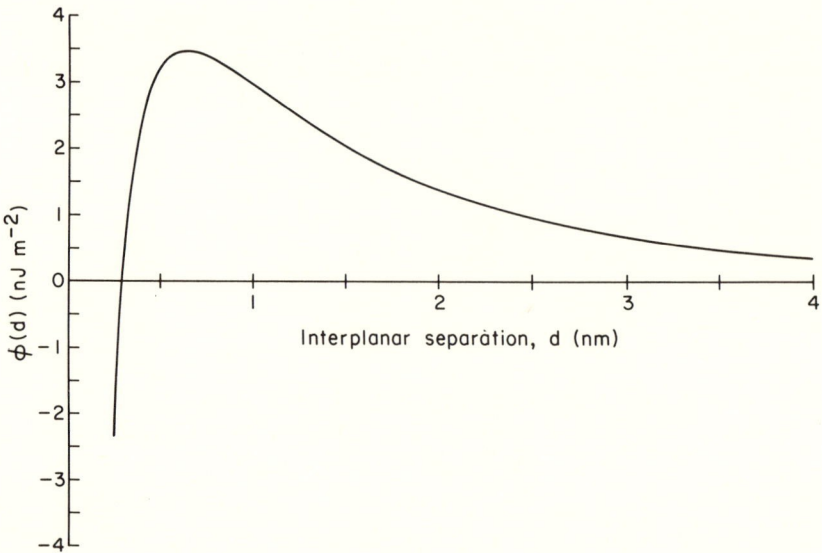

FIG. 10.5 The total interparticle potential energy (per unit area) in a stable suspension calculated according to Eq. 10.14. (Reprinted with permission from G. Sposito, *The Surface Chemistry of Soils*, Oxford Univ. Press, New York, 1984.)

extremum. If two more extrema in fact occur, the one at a finite d value will be a shallow minimum termed the *secondary minimum*. The *primary minimum* is the very deep one at small d values. Sometimes flocculation is associated with the secondary minimum, and aggregation is associated with the primary minimum.

The hypothesis was developed by B. V. Derjaguin, L. D. Landau, E. J. W. Verwey, and J. Th. G. Overbeek that a colloidal suspension will become rapidly unstable if, at its maximum value, $\phi(d)$ is very small as compared with the random thermal energy of colloidal particles. This hypothesis forms the basis of the *DLVO model* of colloidal stability. In the DLVO model, the solvation force is neglected, and Eq. 10.14 is solved for c_0 subject to the condition that $\phi(d)$ and its first derivative with respect to d vanish when $c_0 = ccc$. To make the equation apply to an ion of arbitrary valence Z, κ is replaced everywhere by $|Z|\kappa$. The result of this calculation is:

$$ccc = \left(\frac{3072\pi}{e^2} \times \frac{a^2 RT}{\beta^{\frac{3}{2}} A} \right)^2 Z^{-6} \qquad (10.15)$$

where $e = 2.7183$ and $\beta = \kappa^2/c_0 = 1.084 \times 10^{16}$ m mol^{-1} for water at 298 K (see Problem 14 in Chapter 8). Equation 10.15 represents the Schulze–Hardy Rule according to the DLVO model. It follows from this equation that

$ccc(Z = 2)/ccc(Z = 1) = 2^{-6} = 0.0156$, in reasonable agreement with the trend in the fourth column of Table 10.2. However, this agreement must be considered in large measure the result of a fortuitous cancellation of the effects of the factors neglected when Eq. 10.14 is applied without the terms in $V_{solv}(d)$. Although the competition of electrostatic and van der Waals forces is sufficient to derive a proportionality between the ccc and a power of the ionic valence via the DLVO hypothesis, the detailed numerical relationship found in Eq. 10.15 is not accurate: it predicts impossibly large ccc values. The derivation of an equation to predict accurately the ccc values in Table 10.2 awaits future research on both the parameters and the distance dependence in the component potentials of $\phi(d)$.

10.5 Adsorption Effects on Colloidal Stability

According to the DLVO model, the rapid coagulation of a colloidal suspension is induced by a reduction in both the magnitude and the range of the repulsive electrostatic force as the concentation of background electrolyte increases. The essential soundness of this conceptual view was demonstrated in Table 10.2. When the coagulating ions have a valence different from one or, in general, when these ions engage significantly in complexation reactions with surface functional groups, the nature of the coagulation process changes and the DLVO model becomes less appropriate. The formation of surface complexes alters the total particle charge density, σ_P (Eq. 7.4) directly, whereas the formation of a compact diffuse double-layer, inherent to the DLVO model, only "screens" the total particle charge density indirectly to reduce its long-range effect.

If the surface functional group is the siloxane cavity, inner-sphere complex formation with monovalent metal cations can reduce the total particle charge sharply. Since the likelihood of this type of surface complex formation increases in direct proportion to the ionic radius of the metal cation (Section 8.3), Cs^+ should be much more effective than Na^+, for example, at reducing σ_P and facilitating coagulation. Experimental evidence supporting this hypothesis comes, for example, in the observation that the ccc of a dilute Na-montmorillonite suspension is 2.1 mol m^{-3} in $NaNO_3$, whereas the ccc of a Cs-montmorillonite suspension at the same clay concentration and pH value is 0.79 mol m^{-3} in $CsNO_3$. The lower ccc for Cs-montmorillonite evidently occurs because a larger number of inner-sphere surface complexes are formed per unit area, producing a smaller value of σ_P to screen.

In the case of bivalent metal cations, siloxane cavities tend to form outer-sphere surface complexes leading to quasi crystals, as described in Section 10.2. Since two cavities are involved, quasi crystal formation also produces an electrically neutral surface, but it is the internal surface of a quasi crystal that has its total charge density reduced. Experiments on the coagulation of Ca-montmorillonite suggest that outer-sphere complexation of Ca^{2+} is an aggregation process that follows a flocculation process induced by Ca^{2+} cations participating in the surface charge-"screening" mechanism hypothesized in the DLVO model. In this case, outer-sphere complexation is a gradual particle rearrangement phenomenon that may require several days' time. When initiated from a stable suspension, however, outer-sphere complexation and quasi crystal formation can be very rapid, as pointed out in Section 10.2.

When the principal surface functional group is the inorganic or organic hydroxyl group, colloidal stability can be affected strongly by the pH value, since inner-sphere complexes with protons can form. A soil colloidal suspension containing particles bearing surface hydroxyl groups (e.g., hydrous oxides or kaolinite) will tend to coagulate at the PZC (Section 7.4), regardless of the background electrolyte concentration. In the absence of surface complexes involving ions other than H^+ or OH^-, the PZC will coincide with the PZNPC (Table 7.1), and coagulation will be dependent on the balance of surface charge between protonated and dissociated functional groups as expressed through σ_H.

In the presence of surface complex-forming ions, σ_P is the determining property for rapid coagulation. As a general rule, the existence of ions that can form surface complexes significantly can be detected by examining the *ccc* as a function of the colloid concentration in the suspension. If the DLVO mechanism is the principal cause of coagulation, the *ccc* will be essentially independent of the colloid concentration—at least over a severalfold change—whereas if surface complexation is the principal cause, the *ccc* will tend to increase with the colloid concentration, since the surface complexation capacity is also thereby increased.

If the surface-complexing ion is multivalent, like Al^{3+}, its adsorption can result in a reversal of the sign of σ_P. When this happens, the ions in aqueous solution that previously were of the same charge sign as the colloidal particles now become coagulating ions! The mechanism of any subsequent coagulation induced by these ions can be either surface charge "screening" or adsorption.

When polymer ions (e.g., Al hydroxy polymers or soil humus polymers) form surface complexes with soil colloidal particles, stability depends on stereochemistry as well as surface charge density. If the extent of polymer adsorption is small, a soil colloidal suspension may be coagulated at a lower concentration of an added, noncomplexing electrolyte (like NaCl) than in the absence of the polymer. In this situation, the addition of electrolyte brings the colloidal particles closer together until the polymer chains can form bridges among them, thereby inducing coagulation. Since the polymer bridging can occur with the particles farther apart than the separation required to make van der Waals forces effective, coagulation can occur at lower electrolyte concentrations than in the absence of the polymer. If polymer adsorption is extensive, coagulation by the polymer bridging mechanism may take place at extremely

TABLE 10.3 Factors that affect the stability of soil colloidal suspensions

Factor	Effects	Promotes coagulation	Promotes stability
Electrolyte concentration	Extent of diffuse double layer	When increased	When decreased
pH value	Changes σ_P	pH = PZC	pH $\neq$ PZC
Adsorption of small ions	Changes σ_P	$\sigma_P = 0$	$\sigma_P \neq 0$
Adsorption of polymer ions	Changes σ_P and/or particle association	With polymer bridges	By electrostatic repulsion

low electrolyte concentrations. Alternatively, the colloidal suspension may be *stabilized* by the repulsive electrostatic force between the coatings of adsorbed polymers! Which phenomenon occurs depends on the pH value, the electrolyte concentration, and the configuration of the adsorbed polymer ion.

The principal surface chemical factors that determine the stability of soil colloidal suspensions are summarized in Table 10.3. Surface reactions affect colloid stability through changes in the strength of the repulsive electrostatic and solvation forces and, if adsorptive macromolecules are involved, through changes in particle association mechanisms. Coagulation is the result of a reduction in the repulsive electrostatic and solvation forces, whether through charge "screening," surface complexation, or stereochemically induced particle bridging.

FOR FURTHER READING

W. W. Emerson, Interparticle bonding, in *Soils: An Australian Viewpoint*, Academic Press, London, 1983, pp. 477–498. An excellent review of the mechanisms of particle interactions in colloidal suspensions and soil aggregates, including the roles played by humus and metal oxides.

P. C. Hiemenz, *Principles of Colloid and Surface Chemistry*, Marcel Dekker, New York, 1986. A fine introductory textbook on colloid chemistry with a useful discussion of flocculation in Chapter 11 and light-scattering techniques in Chapter 5.

J. M. Oades, Soil organic matter and structural stability: Mechanisms and implications for management, *Plant and Soil* 76:319–337 (1984). A comprehensive discussion of the role of humus in soil aggregate formation, from fundamental chemistry to soil structure management in the field.

G. Sposito, *The Surface Chemistry of Soils*, Oxford Univ. Press, New York, 1984. Chapter 6 of this monograph gives a detailed discussion of interparticle forces.

B. K. G. Theng, *Formation and Properties of Clay–Polymer Complexes*, Elsevier, Amsterdam, 1979. Chapters 2 and 6 of this excellent monograph can be read to amplify the discussion of polymer adsorption effects in Section 10.5.

PROBLEMS

The more difficult problems are indicated by an asterisk.

1. Calculate the diffusion coefficient of a colloid of radius $1 \, \mu m$ moving through water at 25°C. The time required to diffuse a distance Δx is $2(\Delta x)^2/3D$. Estimate the time required for the colloid to diffuse $10 \, \mu m$, and compare the result with the time required by an ion (see Special Topic 2 for an estimate value of an ionic diffusion coefficient).

2. A suspension consists of platelike particles $1 \, \mu m \times 1 \, \mu m \times 8 \, nm$ with a mass density of $2.5 \times 10^3 \, kg \, m^{-3}$. Calculate the half-life for Brownian coagulation at 25°C in a suspension whose initial density is $1 \, kg \, m^{-3}$. Is the coagulation rapid?

3. The data in the accompanying table give the number density in a kaolinite suspension during coagulation. Use these data to calculate the half-life for Brownian coagulation.

$\rho \times 10^{-14} \, (m^{-3})$	Time (s)	$\rho \times 10^{-14} \, (m^{-3})$	Time (s)
5.00	0	2.52	335
3.90	105	2.00	420
3.18	180	1.92	510
2.92	255	1.75	600

4. Equation 10.2 applies strictly to a suspension of colloids of uniform radius R ("monodisperse suspension"). If colloids of radii R_1 and R_2 are present, the equation changes to the approximate expression:

$$\frac{d\rho}{dt} = -8\pi D_1 \bar{R} \rho^2$$

where

$$\bar{R} = (R_1 + R_2)^2/4R_2$$

and D_1 is the diffusion coefficient of a colloid with radius R_1. Calculate the rate constant at 25°C for the coagulation of a mixture of 1-μm and 10-μm colloids, then compare the result with the rate constant for a monodisperse suspension of 1-μm colloids. Note the enhancement of the coagulation rate by addition of the 10-μm colloids.

5. The absorbance of two Ca-saturated soil colloidal suspensions was found to depend on the suspension mass concentration (in $kg \, m^{-3}$) according to the equations:

$$Absorbance = 1.005c_s$$

$$Absorbance = 1.453c_s$$

where c_s is the suspension mass density. What inference as to the relative mass of the colloidal particles can be made from these results? Be sure to indicate any assumptions made.

6. Go to the library and read the article by O. Onofiok and M. J. Singer, Scanning electron microscope studies of surface crusts formed by simulated rainfall, *Soil Sci. Soc. Am. J.* **48**:1137–1143 (1984). Examine Fig. 1 in the article and discuss the likely colloidal behavior leading to the soil structure in Fig. 1b, using Section 10.2 as a basis.

7. Light-scattering data plotted as in Fig. 10.3 for Li- and K-montmorillonite suspensions instead of Na-montmorillonite show the same kind of rapid increase near E_{Li} or $E_K = 0.2$. The maximum values of the transmitted light intensity, I_{Li}, I_{Na}, and I_K, however, are not the same, decreasing in the order: $I_{Li} = 0.81I_0$, $I_{Na} = 0.68I_0$, $I_K = 0.55I_0$, where I_0 is the intensity of incident light. Give an explanation for these results in terms of concepts discussed in Sections 8.3 and 10.2 (Table 10.1).

*8. Light-scattering data plotted as in Fig. 10.3 for Mg-montmorillonite (i.e., I/I_{Mg} against E_{Mg}) show an approximately linear increase from 0.55 to 1.00 as E_{Mg} increases from 0.0 to 1.0. The absolute value of I at $E_{Mg} = 0.0$ is $0.22I_0$, where I_0 is the intensity of incident light. Use these results to estimate the *ratio* of unit layers in Ca and Mg quasi crystals. (*Answer:* $N_{Mg}/N_{Ca} = 0.6$)

*9. Suspensions of Ca-montmorillonite in a chloride background show an anion exclusion volume (Section 8.4 and Problem 14 in Chapter 8) that decreases to a limiting value near 3×10^{-4} m^3 kg^{-1} as the chloride concentration increases. Show that this value is consistent with the complete exclusion of chloride ions from the interlayer region of a quasi crystal with opposing basal surfaces separated by about 1 nm. Take the specific surface area of the clay mineral equal to 75 ha kg^{-1}.
(*Hint:* Show that the interlayer volume per unit mass of clay equals one-half the specific surface area times the separation of the basal surfaces.)

10. Calculate the value of P_m at 25°C corresponding to the parameter values used to construct Fig. 10.5. Express the result in atmospheres. (1 atm $= 1.01325 \times 10^5$ N m^{-2}.) What is the spatial extent of the diffuse double layer in this case?

11. Calculate the *ccc* value at 25°C corresponding to the parameter values used to construct Fig. 10.5. Assume that the DLVO model applies and compare your result with the data in Table 10.2.

*12. A sample of an acidic soil (pH 6.4) with $E_{Na} = 0.2$ and $E_K + E_{Ca} + E_{Mg} = 0.75$ was found to disperse completely in water, whereas another sample of the soil taken from elsewhere in the profile (pH 5.0, $E_{Na} = 0.24$, $E_K + E_{Ca} + E_{Mg} = 0.61$) did not disperse. Use the concepts discussed in Section 10.5 to suggest a chemical explanation for these results.

(*Hint:* Review also Section 9.1.)

13. A suspension of birnessite colloids in distilled water showed no particle migration in an applied electric field at pH 1.40. A similar suspension of this Mn mineral (Table 2.4) coagulated at pH 1.55. Explain in surface chemical terms why the two pH values are essentially the same.

14. Use the concepts discussed in Sections 3.5 and 10.5 to explain the chemical basis for the statement: "Organic matter prevents the dispersion of dry soil aggregates; once the soil particles in the aggregates are forced apart (by shaking in suspension), however, the organic matter helps to stabilize the separated particles in suspension."

15. It is observed experimentally that the concentration of synthetic organic polymer cation required to coagulate suspensions of clay minerals is proportional to the initial solids concentration. Use this result to suggest a mechanism for the coagulation process.

11

Soil Acidity

11.1 Proton Cycling

A soil is acidic if the pH value of its aqueous-solution phase is < 7.0. This condition is found in many soils worldwide, particularly those weathering under intensive leaching by fresh water, which always contains free protons at concentrations > 1 mmol m^{-3}. Soils of the humid tropics are the most obvious examples of acid soils, as are soils of forested areas in the temperate zones of the earth.

The phenomena that combine to produce a given proton concentration in the soil solution and render it acidic are complex and interrelated. They are indicated schematically in Fig. 11.1. Among them, the field-scale transport processes involved are *wetfall* (in the form of rain, snow, fog drizzle, etc.), *dryfall* (in the form of deposited solid inorganic and organic particles, whether natural or anthropogenic), and *interflow* (i.e., the lateral movement of soil solution beneath the land surface on hill slopes). Each of these physical processes can carry protons and other solutes into an ambient soil solution from external sources. Their existence and the existence of the corresponding proton-exporting processes (e.g., volatilization, wind erosion) underscore the fact that the soil solution is an *open* natural water system (Section 1.1) subject, therefore, to anthropogenic and natural inputs and outputs that may surpass internal causes of its chemical composition. Industrial effluents (e.g., sulfur and nitrogen oxide gases) that produce acidic wetfall, and nitrogenous fertilizers that produce acidic soil conditions are common examples of anthropogenic inputs.

The important, strictly chemical processes that influence soil solution pH are carbonic acid ($H_2CO_3^*$) dissociation, the acid–base reactions of soil humus and aluminum hydroxy polymers, and mineral-weathering reactions.

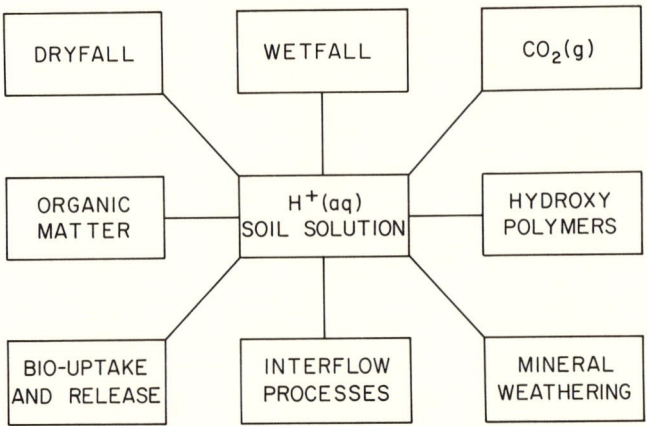

FIG. 11.1 The physical, chemical, and biological factors that influence the pH value of the soil solution.

The formation of carbonic acid and its dissociation reactions in the soil solution are discussed in Sections 1.4 and 2.5, and in Problems 6 and 7 of Chapter 4. The key mathematical relationship in respect to soil pH is presented in Problem 7 of Chapter 4:

$$P_{CO_2}/(H^+)(HCO_3^-) = 10^{7.8} \qquad (T = 298.15\,\text{K}) \qquad (11.1)$$

Equation 11.1 shows that the pressure of $CO_2(g)$ (in atmospheres) and the bicarbonate ion activity determine the pH of the soil solution. Numerical calculation is facilitated by writing the equation in logarithmic form:

$$\text{pH} \equiv -\log(H^+) = 7.8 + \log(HCO_3^-) - \log P_{CO_2} \qquad (11.2)$$

The pH value of a pure solution of $H_2CO_3^*$, for which $(H^+) = (HCO_3^-)$, can be calculated with this expression once P_{CO_2} is specified. For atmospheric air, $P_{CO_2} = 10^{-3.52}$ atm and pH = 5.7; for soil air in B horizons or near plant roots, $P_{CO_2} = 10^{-2}$ atm and pH = 4.9. Thus pH values near 5.0 can be expected in soil solutions if carbonic acid dissociation is the chemical reaction governing soil acidity.

The acid–base reactions of soil humus are discussed in Section 3.3 and illustrated in Problems 7–9 of Chapter 3. For soils in which there is an active cycling of organic matter (e.g., forest soils, peat and muck soils, and Mollisols under agriculture), these reactions are the primary influence on soil solution pH. Examples are given in Eqs. 3.3, 3.4, and 3.6. These proton exchange reactions can be described quantitatively with conditional exchange constants (Section 9.4), but their relation to soil acidity is seen more directly in terms of the concepts of *acid-neutralizing capacity* and its derivative with respect to pH, the *buffer intensity*, β_H.

The acid-neutralizing capacity (ANC) is the moles of protons per unit volume or mass required to change the pH value of an aqueous system to the pH at which the net charge from ions that do not react with OH^- or H^+ is zero. These latter ions are taken conventionally to be those that do not protonate or hydrolyze in the normal pH range of acidic soil solutions, 3.5–7.0 (Fig. 6.1). Thus leaving aside for now the possibility of complex formation, the ANC of a soil solution is expressed in terms of the *free-ion* concentrations:

$$ANC = [Na^+] + [K^+] + 2[Ca^{2+}] + 2[Mg^{2+}]$$
$$-[Cl^-] - 2[SO_4^{2-}] - [NO_3^-] \qquad (11.3)$$

with all concentrations in units of moles per cubic decimeter (liter). The ANC vanishes when charge balance in the soil solution is achieved solely with the ions on the right side of Eq. 11.3. In the presence of dissolved $CO_2(g)$, *overall* charge balance in the soil solution would require inclusion of the proton with the cations and OH^-, HCO_3^-, and CO_3^{2-} with the anions:

$$[Na^+] + [K^+] + 2[Ca^{2+}] + 2[Mg^{2+}] + [H^+] - [Cl^-] - 2[SO_4^{2-}]$$
$$-[NO_3^-] - [OH^-] - [HCO_3^-] - 2[CO_3^{2-}] = 0 \qquad (11.4)$$

It follows from Eqs. 11.3 and 11.4 that, in this case, the ANC can be expressed in the form:

$$ANC = [HCO_3^-] + 2[CO_3^{2-}] + [OH^-] - [H^+] \qquad (11.5)$$

In problem 8 of Chapter 4, the right side of Eq. 11.5 is defined as the *alkalinity*. The ANC or alkalinity of a pure solution of $H_2CO_3^*$ is zero because charge balance is determined solely by the four ionic species comprised in Eq. 11.5. When other ions are present, the ANC given by Eq. 11.3 will not vanish unless the pH has been adjusted to make the right side of Eq. 11.5 vanish. In a similar manner, one can use a charge balance expression like Eq. 11.4 to show that the ANC of a suspension of soil humus is given by Eq. 3.7 (with vanishing ANC for a pure humus suspension). Both Eqs. 3.7 and 11.5 can be expressed per unit mass instead of per unit volume after multiplying each term on the right side by the volume of aqueous solution per unit mass of solids present (equal to c_s^{-1} in Eq. 3.7).

The buffer intensity can be expressed operationally as the number of moles of proton charge per unit mass or volume that are dissociated from (complexed by) a soil or soil constituent when the pH value of the soil solution increases (decreases) by one pH unit. The buffer intensities of organic-rich surface horizons in temperate-zone acidic soils generally have maximal values in the range 0.1–1.5 $mol_c \ kg_{om}^{-1} \ pH^{-1}$ around pH 5, *when expressed per unit mass of soil organic matter*. Thus, for example, the addition of 20 mmol of proton charge per kilogram to a soil horizon whose organic-matter content is $0.4 \, kg_{om}$ kg^{-1}, and for which the buffer intensity is $0.2 \, mol_c \, kg_{om}^{-1} \, pH^{-1}$, would decrease the pH by: $0.02 \, mol_c \, kg^{-1}/(0.4 \, kg_{om} \, kg^{-1} \times 0.2 \, mol_c \, kg_{om}^{-1}$ $pH^{-1}) = 0.25$ pH units. The general relationship exemplified by this calculation is:

$$\Delta pH = \Delta n_A/\beta_H = \Delta n_A/f_{om}\beta_H^{om} \qquad (11.6)$$

where Δn_A is moles of proton charge added or removed per kilogram of *soil*, f_{om} is the soil organic-matter mass fraction, and β_H^{om} is the buffer intensity expressed per unit mass of soil organic matter. Note that, since β_H is usually pH-dependent (see Problem 9 in Chapter 3), ΔpH will be pH-dependent.

The roles that aluminum hydroxy polymers and weathering of aluminum-bearing minerals play in soil acidity are discussed in Section 11.3. Suffice it to say here that hydrolytic species of Al(III)—in aqueous solution, adsorbed on soil particles, or in solid phases—often govern the soil solution pH in mineral horizons of acid soils. The soil buffer intensity in this case, however, is typically an order of magnitude or more smaller than the values just quoted for β_H^{om}.

The biological processes important in soil acidity are ion uptake or release by plant roots and the microbial catalysis of redox reactions. Plants often take up more cations from soil than anions, with the result that protons are excreted to maintain charge balance. The rhizosphere (Section 3.1) may thereby become acidified relative to the soil in bulk. The same will follow if organic acids are excreted, particularly those having pH_{dis} values below the ambient rhizosphere pH (Table 3.1). Under controlled experimentation, rhizosphere pH values as much as two pH units below bulk soil values have been measured. The other, quite different biological influence on soil acidity, redox catalysis, is discussed in Section 11.4. It pertains especially to the transformations of C, N, and S (Section 6.2).

Like the chemical elements listed in Table 1.1 for which the cycle was illustrated in Fig. 1.1, proton transfers to and from the soil solution can be estimated (with some difficulty) in respect to the cycling implied in Fig. 11.1. Table 11.1 shows four such estimates for acid soils in the United States and Europe. The first and third rows of data represent the annual proton cycle in two soils underlying a rural forest, whereas the second and fourth rows represent the cycle in a woodland soil near agricultural land and in an agricultural soil, respectively. The soil pH values are in the range 4–6. Minus signs refer to net proton losses.

The data in Table 11.1 suggest that, overall, there is net soil acidity created by deposition, CO_2 and organic-matter production, and biocycling (ion uptake and redox reactions involving C, N, and S). The protons produced by these processes are consumed in mineral-weathering reactions (hydrolysis, complexation, and ion exchange) that release metal cations and silica to the soil solution. If there is no exportation of $H^+(aq)$ by interflow processes (or

TABLE 11.1 Proton transfer components ($kmol\ ha^{-1}\ yr^{-1}$) in acid soils of rural areas

Soil	Wetfall+dryfall	CO_2(aq) + organic matter	Mineral weathering	Bio-uptake and release	Interflow processes
Spodosol[a] (forested)	1.3	0.1	−2.1	0.9	−0.1
Inceptisol[a] (woodland)	3.1	0.0	−4.8	2.2	−0.1
Inceptisol[b] (forested)	0.90	0.24	−0.98	0.65	0.0
Inceptisol[b] (agricultural)	0.89	0.48	−3.86	4.24	0.0

[a]N. van Breemen, J. Mulder, and C. T. Driscoll, Acidification and alkalinization of soils, *Plant and Soil* **75**:283–308 (1983).
[b]T. Pačes, Sources of acidification in Central Europe estimated from elemental budgets in small basins, *Nature* **315**:31–36 (1985).

deep percolation), the net proton production should be zero. That it is not zero for the data in Table 11.1 (especially row 4) shows how difficult it is to estimate proton budgets in field soils.

11.2 Exchangeable Acidity

Methods to measure the cation exchange capacity of an acid soil are described in *Methods of Soil Analysis* (see the "For Further Reading" section at the end of Chapter 4). As mentioned in Section 9.1, these methods involve the use of Ba^{2+} as an "index" cation in exchange reactions like those in Eq. 3.4 or 9.24:

$$2HX(s) + Ba^{2+}(aq) = BaX_2(s) + 2H^+(aq) \qquad (11.7a)$$

$$CaX_2(s) + Ba^{2+}(aq) = BaX_2(s) + Ca^{2+}(aq) \qquad (11.7b)$$

where X represents 1 mol of negative intrinsic surface charge (Section 9.3). If an unbuffered solution of $BaCl_2$ is used to supply Ba^{2+} for the reactions in Eq. 11.7, the resulting *CEC* will usually represent a less-than-maximum intrinsic surface charge component (Section 9.1), whereas the use of a buffered $BaCl_2$ solution at pH 8.2 will lead to a *CEC* approximating a maximum intrinsic surface charge. As pointed out in Section 9.1, both kinds of *CEC* measurement are useful: that measured with a buffered $BaCl_2$ solution indicates the *potential* negative surface charge that can be balanced by H^+ and hydrolyzable metal cations (Al^{3+}) that acidify soils, whereas the *CEC* measured with an unbuffered $BaCl_2$ solution approximates the *actual* negative surface charge that can be balanced by acidifying cations.

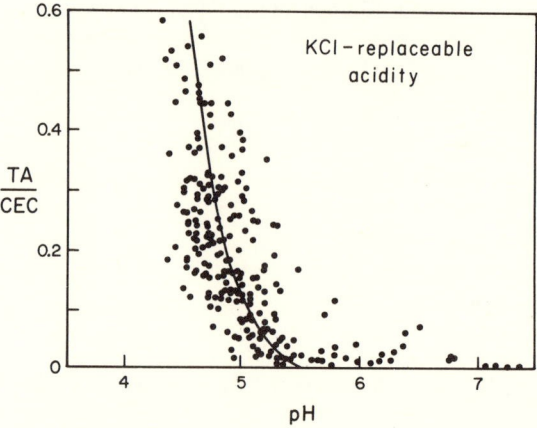

FIG. 11.2   Salt-replaceable acidity as a function of pH for some acidic soils of the eastern United States. (After G. W. Thomas and W. L. Hargrove, The chemistry of soil acidity, in F. Adams (ed.), *Soil Acidity and Liming*, American Society of Agronomy, Madison, WI, 1984).

The moles of surface charge per unit mass balanced by readily exchangeable Na^+, K^+, Ca^{2+}, or Mg^{2+} defines the *ANC* of the soil adsorbent. The difference between the *CEC* and the *ANC* is the *total acidity* (*TA*) of the soil adsorbent:

$$CEC = TA + ANC \qquad (11.8)$$

If the *CEC* is measured with a buffered $BaCl_2$ solution, the *TA* calculated as the difference between *CEC* and *ANC* is a *potential* total acidity. The *actual* total acidity is often measured as the moles of titratable protons per unit mass displaced by an unbuffered KCl solution, termed the *salt-replaceable acidity* or *exchangeable acidity*. Experiments with a variety of mineral acid soils have shown that the principal contribution to salt-replaceable acidity is made by readily exchangeable forms of Al(III): Al^{3+}, $AlOH^{2+}$, and $Al(OH)_2^+$. The protons released when these Al(III) species are displaced by K^+ and then hydrolyze in the soil solution are the titratable protons measured experimentally. For soil humus, the salt-replaceable acidity comprises mostly protons displaced from strongly acidic organic functional groups and from Al- or Fe-hydroxy species complexed by organic functional groups. The pH dependence of the salt-replaceable acidity is illustrated in Fig. 11.2 for subsurface horizons of some soils from the eastern United States. The *TA* values were measured using unbuffered KCl solutions, whereas the *CEC* values were measured using $BaCl_2$ buffered at pH 8.2. The ratio *TA/CEC* declines sharply to zero as the pH value increases from 4.5 to 5.5. This trend is typical of acid soils.

11.3 Aluminum Chemistry

The aqueous-phase, adsorbed, and solid-phase forms of Al(III) are of critical importance in acid soils. The principal solid phases that can control the activity of Al^{3+}(aq) in acidic soil solutions are introduced in Sections 2.3, 2.4, and 5.5. They are gibbsite, kaolinite, beidellite, and hydroxy-interlayer vermiculite (HIV). Dissolution reactions and their log K_{so} values at 298.15 K for these minerals are listed in Table 11.2. As discussed in Sections 5.4 and 5.5 (see also Problem 14 in Chapter 5), the degree of crystallinity in soil gibbsite or kaolinite varies, with a corresponding variation in the value of log K_{so}. For example, log K_{so} = 8.11, 8.77, or 9.35 for the reaction:

$$Al(OH)_3(s) + 3H^+(aq) = Al^{3+}(aq) + 3H_2O(\ell) \qquad (11.9)$$

depending on whether $Al(OH)_3(s)$ represents well-crystallized gibbsite, "soil gibbsite," or microcrystalline gibbsite, respectively. The difference between the extreme values of 9.35 and 8.11 for log K_{so} illustrates the gibbsite "window" in respect to Al^{3+}(aq) activity control by gibbsite in soils (Section 5.4). A similar "window" applies to kaolinite, for which log K_{so} ranges between 3.56 and 5.24 for the dissolution reaction given in Table 11.2 (Section 5.5). Beidellite and HIV "windows" also are expected, but are not known quantitatively at present.

An activity-ratio diagram based on the reactions, log K_{so} values, and expressions for log(Al^{3+}) in Table 11.2 is shown in Fig. 11.3. This fairly

TABLE 11.2 Dissolution reactions of aluminum-bearing minerals[a]

Reaction	log K_{298}
$Al(OH)_3(s) + 3H^+(aq) = Al^{3+}(aq) + 3H_2O(\ell)$ $\log(Al^{3+}) = \log K_{298} - 3\ pH$	8.11–10.8
$\frac{1}{2}Al_2Si_2O_5(OH)_4(s) + 3H^+(aq) = Al^{3+}(aq) + Si(OH)_4^0(aq) + \frac{1}{2}H_2O(\ell)$ $\log(Al^{3+}) = \log K_{298} - 3\ pH - \log(Si(OH)_4^0)$	3.56–5.24
$0.48Al_{0.218}[Si_{3.55}Al_{0.45}](Al_{1.41}Fe(III)_{0.385}Mg_{0.205})O_{10}(OH)_2(s)$ $\quad + 3.75H^+(aq) + 1.06H_2O(\ell)$ $= Al^{3+}(aq) + 0.19Fe^{3+}(aq) + 0.099Mg^{2+}(aq) + 1.71Si(OH)_4^0(aq)$ $\log(Al^{3+}) = 3.45 - 3.75\ pH - 0.19\ \log(Fe^{3+}) - 0.099\ \log(Mg^{2+})$ $\quad - 1.71\ \log(Si(OH)_4^0)$	3.45
$0.27K_{0.24}Ca_{0.08}(Al(OH)_{2.61})_{1.45}[Si_{3.24}Al_{0.76}](Al_{1.56}Fe(III)_{0.24}Mg_{0.20})$ $\quad + O_{10}(OH)_2(s) - 3.4H^+(aq)$ $= 0.064K^+(aq) + 0.021Ca^{2+}(aq) + Al^{3+}(aq) + 0.064Fe^{3+}(aq) + 0.053Mg^{2+}(aq)$ $\quad + 0.86Si(OH)_4^0(aq) + 0.75H_2O(\ell)$ $\log(Al^{3+}) = 5.33 - 3.4\ pH - 0.064\ \log(Fe^{3+}) - 0.064\ \log(K^+)$ $\quad - 0.021\ \log(Ca^{2+}) - 0.053\ \log(Mg^{2+}) - 0.86\ \log(Si(OH)_4^0)$	5.33

[a]G. Sposito, Chemical models of weathering in soils, in J. I. Drever (ed.), *The Chemistry of Weathering*, pp. 1–18. D. Reidel, Dordrecht, The Netherlands, 1985.

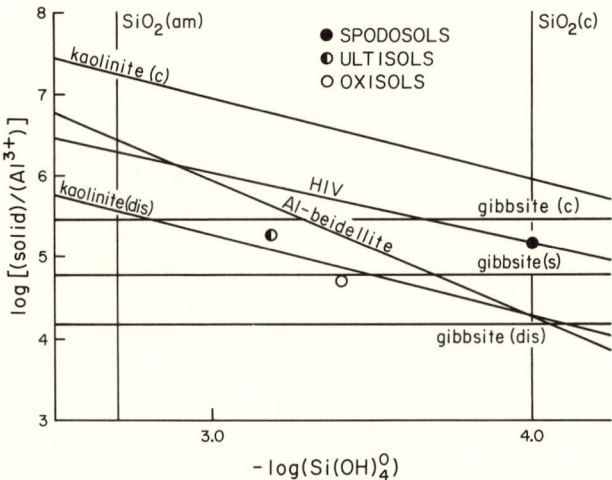

FIG. 11.3   An activity-ratio diagram for Al^{3+}(aq) based on the reactions in Table 11.2. The data points indicate representative Al^{3+}–$Si(OH)_4^0$ activity pairs for acidic soils.

complex diagram was constructed as described for Fig. 5.6 in Section 5.5. The pH value chosen was 4.5, and appropriate values for the activities of Fe^{3+}(aq), Ca^{2+}(aq), Mg^{2+}(aq), and K^+(aq) were introduced so as to produce linear relationships between $\log(Al^{3+})$ and the "master variable," $\log(Si(OH)_4^0)$. Three "windows" appear in Fig. 11.3: gibbsite, kaolinite, and silica (quartz to amorphous silica). Shown also are representative $[(Al^{3+}), (Si(OH)_4^0)]$ activity pairs for soil solutions in the surface horizons of Spodosols, Ultisols, and Oxisols under silviculture or agriculture. These representative points lie for the most part in the common intersection of the silica, gibbsite, and kaolinite "windows."

A strict interpretation of an activity-ratio diagram requires that the solid phase whose activity-ratio line lies highest be assigned control of the Al^{3+}(aq) activity (Section 5.2). If a $\log K_{so}$ "window" exists, however, the GLO Step Rule (Section 5.3 must be invoked to permit a range of metastable solid phases to control (Al^{3+}) in slow succession. Therefore, in respect to Fig. 11.3, unless very well-crystallized kaolinite has formed, the four Al-bearing solid phases represented in the diagram may coexist in acid soils that are weathering actively. This conclusion hinges very much on the likely presence of disordered kaolinite in most acid soils that sustain $Si(OH)_4^0$ activities $> 3 \times 10^{-5}$ (the value at which well-crystallized kaolinite and gibbsite are in equilibrium, according to the data in Table 11.2).

The aqueous-phase speciation of Al(III) can be calculated as described in Sections 4.3 and 4.4. Table 11.3 gives an example analogous to that in Table 4.4. The total concentration (C_T) data are representative of the soil solution in

TABLE 11.3 Speciation of an acidic soil solution (pH 4.7)

Constituent	C_T (mmol m^{-3})	Percentage speciation
Ca	20	Ca^{2+} (99%), CaSO$_4^0$ (1%)
Mg	6	Mg^{2+} (99%), MgSO$_4^0$ (1%)
K	3	K$^+$ (100%)
Na	20	Na$^+$ (100%)
Al	17	AlL (71%), AlF^{2+} (11%), Al^{3+} (11%), Al(OH)$^{2+}$ (5%), Al(OH)$_2^+$ (2%)
CO$_3$	10	H$_2$CO$_3^*$ (100%)
SO$_4$	54	SO$_4^{2-}$ (99%)
Cl	24	Cl$^-$ (100%)
F	2	AlF^{2+} (95%), F$^-$ (5%)
SiO$_2$	100	Si(OH)$_4^0$ (100%)
L^a	12	AlL (100%)

^{a}Organic ligands from soil humus.

the surface horizons of a Spodosol, with the ligand "L" being the average organic ligand (i.e., L is 1 mol of organic anion charge) in the sense described in Section 3.3. The computer calculation of the species distribution was performed as outlined in Fig. 4.1. The Al(III) speciation results, indicating that most of the metal is in organic complexes with significant fluoride complexes and free-ion species at pH < 5, are typical of acid soils.

Besides the monomeric Al species indicated in Table 11.3, evidence exists for relatively stable *polynuclear* Al species, particularly in complexes with OH$^-$(aq) and organic anions. Typical of the polynuclear Al-hydroxy species are Al$_2$(OH)$_2^{4+}$(aq), Al$_6$(OH)$_{12}^{6+}$(aq), and [AlO$_4$Al$_{12}$(OH)$_{24}$]$^{7+}$(aq). These species can engage in acid–base reactions as aqueous solutes or suspended colloids, or as adsorbed species on both soil humus and soil minerals. The formation of HIV and chloritized smectite involves the adsorption of Al-hydroxy polymers as the first step (Eq. 2.7a).

Given the kind of aqueous-phase speciation data listed in Table 11.3, the distribution of exchangeable cations can be calculated if conditional exchange constants have been measured (Section 9.4). Consider, for example, the analog of Eq. 11.7b for Ca–Al exchange:

$$3CaX_2(s) + 2Al^{3+}(aq) = 2AlX_3(s) + 3Ca^{2+}(aq) \tag{11.10}$$

The conditional exchange constant (Eq. 9.23) is:

$$^cK_{12} = x_2^2(Ca^{2+})^3/x_1^3(Al^{3+})^2 \tag{11.11}$$

where 1 refers to CaX$_2$, and 2 refers to AlX$_3$. If $^cK_{12}$ has been measured as a function of x_2 (or if one of the cation exchange models of Section 9.5 is assumed accurate), then Eq. 11.11 can be used to calculate the mole fraction of exchangeable Al^{3+} in equilibrium with the soil solution. Suppose, for example, that $^cK_{12} \approx 3.0$ independent of x_2 (i.e., the Vanselow cation exchange model applies). Then, with $(Al^{3+}) = 1.8 \times 10^{-6}$ and $(Ca^{2+}) = 1.98 \times 10^{-5}$, based on Table 11.3 (and Eq. 4.23 with $I = 227$ mmol m^{-3}), one calculates from Eq. 11.11:

$$3.0 = 2.4 \times 10^{-3}[x_2^2/(1 - x_2)^3]$$

which indicates that $x_2 \approx 0.91$. Thus under the conditions given, exchangeable Al^{3+} is predicted to dominate the soil adsorbent.

The dissolution reactions in Table 11.12 consume protons and bring metal cations and neutral silica into the soil solution, thereby contributing to the buffer intensity of the soil while depleting the acid-neutralizing capacity of the soil solids. The same effect accompanies the cation exchange reaction in Eq. 11.10 (understood to proceed from left to right), because Al^{3+}(aq) removal from the soil solution provokes the release of OH^- and decreases the ANC of the soil adsorbent (Eq. 11.8). [The loss of Al^{3+}(aq) through adsorption impies a conversion of $AlOH^{2+}$ to Al^{3+} to maintain species equilibrium, such that OH^- is released to the soil solution to neutralize protons there.] This kind of change in the ANC of the soil solids is also reflected in the fourth column of Table 11.1 by the negative values of the proton flux: the change in ANC of the soil solids is approximately proportional to the net proton flux caused by mineral weathering.

On the other hand, the release of nonhydrolyzing metal cations into the soil solution via mineral weathering *increases* its ANC if the cations are not leached from the soil profile. The conclusion follows directly from Eq. 11.3, as does the statement that the release of nonprotonating anions into the soil solution via weathering will *decrease* its ANC. These inferences remain unchanged if complexes form between the cations and anions in Eq. 11.3. In that case, each concentration term is replaced by a total concentration (e.g., $[Na^+] \to Na_T$, $[SO_4^{2-}] \to SO_{4T}$, and the ANC is redefined accordingly.

11.4 Redox Effects

In Section 6.2, it is emphasized that most of the reduction half-reactions that occur in soils result in proton consumption. Therefore, an important source of acid-neutralizing capacity and buffer intensity in soil solutions is the redox reactions that feature a net proton consumption and an effective microbial catalysis. Several of these reactions are listed in Table 11.4. In each example, a reduction half-reaction from Table 6.2 has been coupled with the oxidation half-reaction:

$$\tfrac{1}{4}CH_2O(aq) + \tfrac{1}{4}H_2O(\ell) = \tfrac{1}{4}CO_2(g) + H^+(aq) + e^-(aq) \qquad (11.12)$$

TABLE 11.4 Redox reactions important in soil acidity

Reaction	$\log K_{298}$
$\frac{1}{4}CH_2O(aq) + \frac{1}{8}NO_3^-(aq) + \frac{1}{4}H^+(aq) = \frac{1}{8}NH_4^+(aq) + \frac{1}{4}CO_2(g) + \frac{1}{8}H_2O(\ell)$	15.1
$\frac{1}{4}CH_2O(aq) + \frac{1}{2}NO_3^-(aq) + \frac{1}{5}H^+(aq) = \frac{1}{10}NO_2^-(aq) + \frac{1}{4}CO_2(g) + \frac{7}{20}H_2O(\ell)$	21.3
$\frac{1}{4}CH_2O(aq) + \frac{1}{2}MnO_2(s) + H^+(aq) = \frac{1}{2}Mn^{2+}(aq) + \frac{1}{4}CO_2(g) + \frac{3}{4}H_2O(\ell)$	20.9
$\frac{1}{4}CH_2O(aq) + FeOOH(s) + 2H^+(aq) = Fe^{2+}(aq) + \frac{1}{4}CO_2(g) + \frac{7}{4}H_2O(\ell)$	11.5
$\frac{1}{4}CH_2O(aq) + \frac{1}{8}SO_4^{2-}(aq) + \frac{7}{8}H^+(aq) = \frac{1}{8}HS^-(aq) + \frac{1}{4}CO_2(g) + \frac{1}{4}H_2O(\ell)$	4.5

which appears in the next-to-last row of Table 6.2. The reactant, CH_2O, replaces $C_6H_{12}O_6$ in Table 6.2 and is to represent an abbreviated form of the "Redfield formula" for the chemical composition of aquatic plant material susceptible to oxidation (Section 1.2).

The reactions in Table 11.4 indicate that typical reduction processes in soils will deplete the soil solution of protons if they couple effectively to the oxidation of labile organic matter. This loss of protons reduces the alkalinity of the soil solution (Eq. 11.5), since the dissolving of $CO_2(g)$ produced from the oxidation half-reaction cannot change the alkalinity. (Recall that the alkalinity of a pure solution of $H_2CO_3^*$ is zero.) In quantitative terms, the redox processes that produce reduced aqueous species lead to a modification of Eq. 11.3 for the soil solution ANC:

$$ANC = [Na^+] + [K^+] + 2[Ca^{2+}] + 2[Mg^{2+}] + 2[Fe^{2+}] + 2[Mn^{2+}]$$

$$+ [NH_4^+] - [Cl^-] - 2[SO_4^{2-}] - [NO_3^-] \qquad (11.13)$$

The inclusion of $NH_4^+(aq)$, $Fe^{2+}(aq)$, and $Mn^{2+}(aq)$ as nonhydrolyzing cations reflects a choice of zero ANC for aqueous systems in which only the redox reactions involving these cations produce aqueous species. In the case of $HS^-(aq)$, however, zero ANC is assigned to a pure solution of H_2S, analogous to the convention used for $H_2CO_3^*$. The combination of Eq. 11.13 with the overall condition of charge balance then leads to a generalization of Eq. 11.5:

$$ANC = [HCO_3^-] + 2[CO_3^-] + [HS^-] + [OH^-] - [H^+] \qquad (11.14)$$

where $S^{2-}(aq)$ has been neglected as an unstable species (Problem 1 in Chapter 6).

Experimental field studies have indicated the importance of the reactions in Table 11.4 to the generation of acid-neutralizing capacity in the soil environment. Indeed, the reactions involving Fe and Mn are examples of an increase in ANC caused by mineral weathering. Because C, N, and S are essential elements for the nutrition of green plants, however, the ultimate impact of their redox reactions cannot be estimated without a full specification of their biogeochemical cycles. For example, reduction processes parallel to those in Table 11.4

occur when nitrate and sulfate are taken up and become part of the plant biomass (note the complete "Redfield formula" in Section 1.2). These bio-uptakes also increase the soil solution alkalinity. Conversely, the mineralization of soil organic matter will reduce the alkalinity of the soil solution through the production of NO_3^- and SO_4^{2-}. This effect, on the other hand, can be diminished significantly by removal of the biomass before mineralization (harvesting of agricultural crops or forest clear-cutting). If N is supplied to plants as the species NH_4^+, the soil solution alkalinity decreases, both from NH_4^+ uptake and from NH_4^+ oxidation to NO_3^- in the soil solution (Eq. 11.13). Reduced-N fertilizers, like NH_4NO_3, $(NH_2)_2CO$ (urea), NH_3, and $(NH_4)_2SO_4$, can reduce the soil solution alkalinity via oxidation. Long-term field studies on fertilized plots have shown that NH_4NO_3 and $(NH_4)_2SO_4$ applications in particular can decrease the soil solution alkalinity greatly through nitrification. Ammonium sulfate and reduced-S species are often introduced into soil by dry deposition and produce decreases in alkalinity after oxidation. These processes may account for half the input of protons into soil from deposition sources. These examples illustrate the broad scope of redox effects on soil acidity as well as the interrelatedness of the proton transfer components identified in Fig. 11.1.

11.5 Neutralization of Soil Acidity

The natural processes that increase the pH value of the soil solution are mineral weathering, anion bio-uptake, protonation of anions or surface functional groups, adsorption of nonhydrolyzing metal cations, and reduction half-reactions (Fig. 11.1). In acid soils, these processes usually are not adequate to maintain soil pH in the range optimal for agriculture, especially if acidifying fertilizers are applied to the soil. When soil pH is such that the total acidity of the soil exceeds about 15% of its cation exchange capacity, a variety of serious problems for plant growth (e.g., Al and Mn toxicity, or Ca, Mg, and Mo deficiency) is expected. Under this condition, soil amendments to decrease the total acidity must be used.

The practice of neutralizing soil acidity is formalized in the concept of the *lime requirement*. This parameter is the moles of Ca^{2+} charge per kilogram of soil required to decrease the total acidity to a value deemed acceptable for agricultural use of the soil. Typically, the lime requirement is expressed in the convenient units of $cmol_c\ kg^{-1}$ and is found to have a value somewhere between that of the salt-replaceable acidity and the *CEC* of a soil as measured with a buffered solution of $BaCl_2$ (Section 11.2). The previous discussion in this chapter makes clear the fact that the line requirement of a soil will depend on its mineralogy, its content of clay and organic matter, and the extent of leaching with fresh water. Special consideration must be given also to proton inputs from acid deposition and acidifying fertilizers. Often the lime requirement is seen as a means to increase soil pH to around 6, where the salt-replaceable acidity is usually nil (Fig. 11.2).

Methods for measuring the lime requirement are described in Chapter 12 of *Methods of Soil Analysis* (see the "For Further Reading" section at the end of Chapter 4). The procedures that have been used range from field applications of $CaCO_3$ (involving years to achieve a steady state), to laboratory incubations of soil samples with $CaCO_3$ (involving months to complete), to soil titration with $Ca(OH)_2$ over several days, to rapid soil equilibrations with buffer solutions optimized for a given group of agricultural soils. Considerations such as the number of soil samples to be analyzed and the accuracy of the estimated lime requirement enter into the final choice of method.

The fundamental chemical reaction underlying the concept of the lime requirement appears in Eq. 11.10. This reaction, reversed so that $CaX_2(s)$ and $Al^{3+}(aq)$ are the products, can be coupled with the dissolution reactions of a Ca-bearing mineral added to the soil and an appropriate Al-bearing mineral that precipitates to produce an overall reaction that removes Al^{3+} from the soil solution. For example, if $CaCO_3(s)$ is added to the soil and $Al(OH)_3(s)$ precipitates in response, one can combine Eqs. 5.14, 11.9, and 11.10 to obtain the overall reaction:

$$2AlX_3(s) + 3CaCO_3(s) + 3H_2O(\ell)$$
$$= 3CaX_2(s) + 2Al(OH)_3(s) + 3CO_2(g) \qquad (11.15)$$

Similarly, if gypsum is added to the soil and jurbanite precipitates, one combines Eq. 5.2, 5.20a, and 11.10 to obtain the reaction:

$$2AlX_3(s) + 3CaSO_4 \cdot 2H_2O(s) + 6H_2O(\ell)$$
$$= 3CaX_2(s) + 2H^+(aq) + SO_4^{2-}(aq) + 2AlOHSO_4 \cdot 5H_2O(s) \qquad (11.16)$$

Note that the reaction in Eq. 11.15 will not change the acid-neutralizing capacity of the soil solution, although both reactions increase the ANC of the soil adsorbent. If the Al^{3+} that exchanges for Ca^{2+} does not precipitate, however, the ANC of the soil solution will decrease. This effect can be seen by adding the species $Al^{3+}(aq)$, $AlOH^{2+}(aq)$, and $Al(OH)_2^+(aq)$ to the left side of Eq. 11.4, then incorporating Eq. 11.3 to derive the expression:

$$ANC = [HCO_3^-] + 2[CO_3^{2-}] + [OH^-] - [H^+] - 3[Al^{3+}]$$
$$- 2[AlOH^{2+}] - [Al(OH)_2^+] \qquad (11.17)$$

If Eq. 11.15 occurs in a soil, the activities of $Al^{3+}(aq)$ and $Ca^{2+}(aq)$ in the soil solution are governed by Eq. 5.8 and the expression for the thermodynamic exchange constant (Section 9.4):

$$K_{ex} = (AlX_3)^2(Ca^{2+})^3/(CaX_2)^3(Al^{3+})^2 \qquad (11.18)$$

These equations can be combined to derive the relationship:

$$pH + \tfrac{1}{2} \log(Ca^{2+}) = \tfrac{1}{6} \log[*K_{so}{}^2 K_{ex}] + \tfrac{1}{3} \log[(CaX_2)^{3/2}/(AlX_3)] \qquad (11.19)$$

The left side of Eq. 11.19 is the *lime potential*, an activity variable equal to one-half the common logarithm of the *IAP* of $Ca(OH)_2(s)$ (Section 5.1). Soils that have adequate acid-neutralizing capacity exhibit lime potentials > 3.0. Equation 11.19 shows that the lime potential depends sensitively on the activities of $CaX_2(s)$ and $AlX_3(s)$ in the soil adsorbent.

The application of Eq. 11.19 to soil acidity problems can be illustrated by choosing the Rothmund–Kornfeld model (Section 9.5) to express the exchangeable-cation activities. The introduction of Eq. 9.35 (with $\beta = 1$) into Eq. 11.19 produces the expression:

$$pH + \tfrac{1}{2} \log(Ca^{2+}) = \tfrac{1}{6} \log[*K_{so}{}^2 K_{ex}] + \log[E_{CaX_2}/(1 - E_{CaX_2})] \qquad (11.20)$$

where E is a charge fraction (Section 9.2) and $E_{AlX_3} = 1 - E_{CaX_2}$. Reasonable values for $*K_{so}$ and K_{ex} are $10^{8.77}$ and $10^{0.3}$, respectively; this choice leads to the value 3.0 for the first term on the right side of Eq. 11.2. It then follows that the lime potential will be > 3.0 if the charge fraction of Ca^{2+} on the soil adsorbent is > 0.50.

FOR FURTHER READING

F. Adams (ed.), *Soil Acidity and Liming*, 2nd Ed., American Society of Agronomy, Madison, WI, 1984. The standard reference on the chemistry of soil acidity and liming practice in agriculture.

I. R. Kennedy, *Acid Soil and Acid Rain*, Wiley, New York, 1986. A useful introduction to the biological aspects of acid soils.

J. A. Kittrick, D. S. Fanning, and L. R. Hossner, *Acid Sulfate Weathering*, Soil Science Society of America, Madison, WI, 1982. A valuable collection of research papers devoted to the problem of acidity induced by sulfur oxidation in soils forming on coastal plains.

J. O. Reuss and D. W. Johnson, *Acid Deposition and the Acidification of Soils and Waters*, Springer-Verlag, New York, 1986. A fine introduction to the soil chemical processes involved in the acidification of natural waters.

The following four articles give valuable research-oriented discussions of the chemical processes in acidic soils and waters.

B. W. Bache, Aluminum mobilization in soils and waters, *J. Geol. Soc. (London)* **143**:699–706 (1986).

D. G. Kinniburgh, Towards more detailed methods for quantifying the acid susceptibility of rocks and soils, *J. Geol. Soc. (London)* **143**:679–690 (1986).

W. Stumm, L. Sigg, and J. L. Schnoor, Aquatic chemistry of acid deposition, *Environ. Sci. Technol.* **21**:8–13 (1987).

N. van Breemen, J. Mulder, and C. T. Driscoll, Acidification and alkalinization of soils, *Plant and Soil* **75**:283–308 (1983).

PROBLEMS

The more difficult problems are indicated by an asterisk.

1. Partial pressures of $CO_2(g)$ measured in the B horizons of some Spodosols ranged between 10^{-3} and $10^{-1.9}$ atm throughout the year. Calculate the corresponding range of soil solution pH if carbonic acid dissociation is the governing reaction.

*2. Use the concept of charge balance and the data in Problem 6 of Chapter 4 to show that $(H^+) = (HCO_3^-)$ in a pure solution of $H_2CO_3^*(aq)$. (*Hint*: Calculate (H^+) for the range of P_{CO_2} typical of soils using Eq. 11.2 and the assumption that $(H^+) = (HCO_3^-)$. Then show that, for the range of (H^+) calculated, $[OH^-]$ and $[HCO_3^{2-}]$ are negligible.)

3. State whether the addition of small amounts of each of the following compounds to a soil solution will increase, decrease, or not change its *ANC*. Explain your choice in each case.

   (a) CO_2 (c) H_4SiO_4 (e) Na_3PO_4

   (b) $NaNO_3$ (d) H_2SO_4 (f) CH_3COOH

4. The buffer intensity of a sample collected from an A horizon of a mineral soil is 0.015 mol_c kg^{-1} pH^{-1} at pH 4.0. Calculate the moles of proton charge per kilogram of soil that must be removed to raise the pH value by 0.5 units.

5. In the accompanying table are data on the buffer intensity of surface and subsurface horizons of Spodosols at pH 4.5. Plot β_H against f_{om} and fit a straight line through the data points. Calculate β_H^{om} from the slope of the line.

f_{om} (kg_{om} kg^{-1})	β_H (mol_c kg^{-1} pH^{-1})	f_{om} (kg_{om} kg^{-1})	β_H (mol_c kg^{-1} pH^{-1})
0.103	0.0088	0.054	0.0054
0.162	0.0166	0.102	0.0084
0.030	0.0068	0.038	0.0054
0.902	0.0738	0.401	0.0416
0.947	0.0824	0.870	0.1122

6. The pH dependence of the buffer intensity β_H^{om} was found to obey the empirical equation:

$$\beta_H^{om} = 3.27/[(pH - 3.5)(7.29 - pH)]$$

for a series of acidic agricultural soils. Plot a graph of β_H^{om} against pH in the range 4–7. At which pH values is the buffer intensity least?

7. The ratio of the *ANC* measured with unbuffered $BaCl_2$ solution to that

measured with buffered $BaCl_2$ solution at pH 8.2 ($CEC_{8.2}$) was observed to depend linearly on soil pH:

$$\frac{ANC}{CEC_{8.2}} = 0.33 \text{ pH} - 1.41$$

Calculate the buffer intensity of the soil if $CEC_{8.2} = 0.12 \text{ mol}_c \text{ kg}^{-1}$. Plot $TA/CEC_{8.2}$ against pH in the range 4.3–7.4, based on the empirical equation just given.

8. Given that NH_4^+ uptake by plants usually produces excess cation over anion uptake, and that NO_3^- uptake usually produces the opposite effect, what is the expected change in rhizosphere pH from the bio-uptake of each N species?

*9. Prepare an activity-ratio diagram like that in Fig. 11.3 for the soil solution described in Table 11.3. You may take $(Fe^{3+}) = 10^{-11}$ and ignore the difference between activity and concentration. Plot a point on the diagram representing (Al^{3+}) and $(Si(OH)_4^0)$.

10. Long-term field experiments indicate that acid soils receiving nitrogen fertilizer as $(NH_4)_2SO_4$ decrease in pH, whereas those receiving $NaNO_3$ increase in pH. Give an explanation for these results in terms of the ANC of the soil solution and all of the processes depicted in Fig. 11.1. (You may neglect the deposition processes.)

11. Manganese toxicity in acid soils is associated with (Mn^{2+}) equal to about $10^{-3.7}$ in the soil solution. Use the relevant information in Table 6.2 to estimate the pH below which Mn toxicity should occur as the pE ranges between 8 and 12.

*12. Ammonium sulfate is applied to an acid soil at the rate of 236 mg $(NH_4)_2SO_4$ per kilogram of soil. Calculate the lime requirement to neutralize the total acidity expected if complete nitrification occurred, without bio-uptake of NH_4, and the protons produced were adsorbed by the soil.
(*Answer:* 0.714 cmol$_c$ kg^{-1})

13. The relationship between ANC and pH for a soil was observed to follow the equation:

$$ANC = \frac{CEC}{1 + K\ 10^{-pH}}$$

where $CEC = 0.15 \text{ mol}_c \text{ kg}^{-1}$ and $K = 10^{6.0}$. Use this equation to calculate the lime requirement for the soil in order to increase its pH value from 4.8 to 6.0.

14. Prepare a graph of the lime potential vs. the charge fraction of exchan-

geable Ca^{2+}, based on Eq. 11.20. Let $0.05 < E_{CaX2} < 0.95$. Identify the range of E_{CaX_2} where the buffer intensity is largest.

*15. Calculate the lime potential of the soil solution described in Table 11.3, and estimate the corresponding ANC of the soil using Eq. 11.20 with $*K_{so} = 10^{9.35}$ and $K_{ex} = 2.0$. The ionic strength of the soil solution is 0.23 mol m^{-3}.

12

Soil Salinity

12.1 Saline Soil Solutions

A soil is *saline* if the electrolytic conductivity (EC_e) of its aqueous phase, obtained by extraction from a saturated paste (Section 4.1), has a value $>4\,\mathrm{dS\,m^{-1}}$. (For a discussion of these SI units, see the Appendix.) Values of $EC_e > 1\,\mathrm{dS\,m^{-1}}$ are encountered typically in *arid-zone soils*, for which the climatic regime produces evaporation rates that exceed precipitation rates on an annual basis. Ions released into the soil solution by mineral weathering, or introduced there by the intrusion of saline groundwater, tend to accumulate in secondary minerals formed as these soils dry. The secondary minerals include the clay minerals (Section 2.3), carbonates and sulfates (Section 2.5), and chlorides. Since Na, K, Ca, and Mg are relatively easily brought into solution—either as exchangeable ions on smectite and illite, or as structural ions in carbonates, sulfates, and chlorides—it is this set of the metals that contributes most to soil salinity. The corresponding set of ligands that contributes then would be CO_3, SO_4, and Cl. Thus arid-zone soil solutions are essentially electrolyte solutions containing chloride, sulfate, and carbonate salts of Group IA and IIA metals of the Periodic Table.

According to Eq. 4.22, an electrolytic conductivity of $4\,\mathrm{dS\,m^{-1}}$ corresponds to an ionic strength of $58\,\mathrm{mol\,m^{-3}}$ ($\log I = 1.159 + 1.009 \log 4 = 1.77$). This level of salinity is only 10% of that in seawater, but it is high enough in an agricultural context that only crops that are relatively salt-tolerant could withstand it. Moderately salt-sensitive crops are affected when the electrolytic conductivity of a soil extract approaches $2\,\mathrm{dS\,m^{-1}}$, corresponding to an ionic strength of $29\,\mathrm{mol\,m^{-3}}$. Salt-sensitive crops are affected even at $EC_e = 1\,\mathrm{dS}$

m^{-1} ($I = 14 \, mol \, m^{-3}$). Thus in respect to crop salinity tolerance, a soil can be "saline" at any ionic strength $>15 \, mol \, m^{-3}$ if the plants growing in it are stressed. The visual evidence of this is a reduction in crop growth and yield caused by a diversion of energy from normal physiological processes to those involved in the acquisition of water under osmotic stress.

The chemical speciation of a saline soil solution can be calculated as described in Sections 4.3 and 4.4. Total-concentration data and the percentage speciation that are representative of the saturation extract of an irrigated Aridisol are listed in Table 12.1. Notable in the table are the dominance of soluble Ca over Mg; the relatively complicated speciation of Ca, Mg, HCO_3^-, SO_4, and PO_4; the unimportance of organic complexes of the metal cations (this would not be so for the metals Fe, Cu, and Zn were they included); and the high free-ion percentages for Na, K, Cl, and NO_3. Neutral sulfate complexes reduce the contribution of SO_4 to the ionic strength—and, therefore, the electrolytic conductivity—by nearly one-fourth. The computed ionic strength for the soil solution in Table 12.1 is $24 \, mol \, m^{-3}$, which corresponds to an electrolytic conductivity of $1.66 \, dS \, m^{-1}$, high enough to affect salt-sensitive crops.

The alkalinity of a saline soil solution can be defined by Eq. 11.3, with the application of charge balance leading to the expression:

$$\text{Alkalinity} = [HCO_3^-] + 2[CO_3^{2-}] + [H_2PO_4^-] + 2[HPO_4^-] + 3[PO_4^{3-}]$$
$$+ [B(OH)_4^-] + [L^-] + [OH^-] - [H^+] \tag{12.1}$$

for a soil solution like that described in Table 12.1 (compare Eq. 11.5). As mentioned in Section 11.3, the concentrations of metal complexes of the anions on the right side of Eq. 12.1 are included implicitly. In the example of Table

TABLE 12.1 Speciation of an alkaline soil solution (pH 7.6)

Constituent	C_T (mol m^{-3})	Percentage speciation
Ca	5.9	Ca^{2+}(82%), $CaSO_4^0$(16%), $CaHCO_3^+$(1%)
Mg	1.3	Mg^{2+}(85%), $CaSO_4^0$(13%), $MgHCO_3^+$(1%)
Na	1.9	Na^+(99%), KSO_4^-(1%)
K	1.0	K^+(98%), KSO_4^-(1%)
HCO_3	3.0	HCO_3^-(91%), $H_2CO_3^0$(4%), $CaHCO_3^+$(3%), $MgHCO_3^+$(1%)
SO_4	4.4	SO_4^{2-}(74%), $CaSO_4^0$(21%), $MgSO_4^0$(4%)
Cl	5.0	Cl^-(99%), $CaCl^+$(1%)
NO_3	0.28	NO_3^-(99%), $CaNO_3^+$(1%)
PO_4	0.065	HPO_4^{2-}(36), $CaHPO_4^0$(35%), $MgHPO_4^0$(13%), $H_2PO_4^-$(8%), $CaPO_4^-$(4%), $MgPO_4^-$(1%)
B	0.038	$H_3BO_3^0$(96%), $B(OH)_4^-$(3%)
L^a	0.022	L^-(95%), CaL^+(4%), MgL^+(1%)

[a] Organic ligands from soil humus.

12.1, the alkalinity equals $3.0 \, \text{mol m}^{-3}$, with 95% derived from bicarbonate. Bicarbonate alkalinity values ranging from 1 to $4 \, \text{mol m}^{-3}$ are common in saline soils.

The pH value of a saline soil solution whose alkalinity comes mainly from bicarbonate is governed by Eq. 11.2:

$$pH = 7.8 + \log(HCO_3^-) - \log P_{CO_2} \qquad (12.2)$$

If the Davies equation (Eq. 4.23) is used to calculate the activity coefficient of $HCO_3^-(aq)$, and if the Marion–Babcock equation (Eq. 4.22) is used to relate ionic strength to electrolytic conductivity, then Eq. 12.2 takes the form:

$$pH = 7.8 + \Delta(\kappa) + \log[HCO_3^-] - \log P_{CO_2} \qquad (12.3)$$

where

$$\Delta(\kappa) \approx -0.512\left[\frac{0.12\sqrt{\kappa}}{1 + 0.12\sqrt{\kappa}} - 0.0043\kappa\right] \qquad (12.4)$$

is a small correction for ionic strength effects ($\Delta(\kappa) < -0.1$ for $\kappa < 4 \, \text{dS m}^{-1}$). Equation 12.3 shows that the soil solution pH is determined by the bicarbonate alkalinity and the $CO_2(g)$ pressure (in atmospheres). Conversely, the $CO_2(g)$ pressure at equilibrium can be calculated with Eq. 12.3 from measured pH and alkalinity values. Given $P_{CO_2} = 10^{-3.5} \, \text{atm}$, the range of $[HCO_3^-]$ quoted earlier leads to pH values in the range 8–9, according to Eq. 12.3. For the soil solution of Table 12.1, $pH = 7.6$, $[HCO_3^-] = 0.00273 \, \text{mol dm}^{-3}$, and a $CO_2(g)$ pressure of $10^{-2.4} \, \text{atm}$ is calculated. Note that Eq. 12.3 predicts increasing soil solution pH with increasing bicarbonate alkalinity or decreasing $CO_2(g)$ partial pressure. This relationship has often been observed in experiments on arid-zone soils.

12.2 Cation Exchange and Colloidal Phenomena

Exchange reactions among the cations Na^+, Ca^{2+}, and Mg^{2+} are of great importance in arid-zone soils. These reactions are described in Eqs. 9.8–9.10 and 9.24. In the convention of Eq. 9.9, they can be expressed:

$$2NaX(s) + Ca^{2+}(aq) = CaX_2(s) + 2Na^+(aq) \qquad (12.5a)$$

$$2NaX(s) + Mg^{2+}(aq) = MgX_2(s) + 2Na^+(aq) \qquad (12.5b)$$

$$MgX_2(s) + Ca^{2+}(aq) = CaX_2(s) + Mg^{2+}(aq) \qquad (12.5c)$$

where X represents 1 mol of negative exchanger charge. (Note that any one of these reactions can be obtained by combining the other two.) Exchange isotherms based on the reactions in Eq. 12.5 are shown in Figs. 9.2 and 9.3 for an Aridisol. They indicate, as observed typically for arid-zone soils, that Ca and Mg are preferred over Na and that Ca is preferred slightly over Mg by the soil exchanger.

The conditional exchange equilibrium constants for the reactions in Eqs. 12.5a and 12.5c are given in Eqs. 9.23 and 9.25. An expression analogous to Eq. 9.23 applies to the Na $\rightarrow$ Mg exchange reaction in Eq. 12.5b. The conditional constants usually are observed to vary with exchanger composition, as demonstrated in Fig. 9.4, but this variability is often small and is neglected in a first approximation. In the case of Na $\rightarrow$ Ca exchange, laboratory experiments on many arid-zone soils from the United States and the Middle East have shown that $^cK_{12}$ defined by Eq. 9.23 remains within the range 1.4–2.9 regardless of exchanger composition; a reasonable average value is 2.2. The use of an average value is equivalent to adopting the Vanselow model (Section 9.5) of the exchanger reaction, as in Eq. 9.42. Thus in a first approximation,

$$K_{ex}^V = x_2(Na^+)^2/x_1^2(Ca^{2+}) \approx 2.2 \tag{12.6}$$

describes Na $\rightarrow$ Ca exchange in arid-zone soils.

Equations 9.28, which relate the mole fractions in Eq. 12.6 to exchanger charge fractions, can be used to rewrite K_{ex}^V in the form:

$$K_{ex}^V = \frac{\Gamma[Na^+]^2/[Ca^{2+}]}{4E_{Na}^2}(1 - E_{Na}^2) \tag{12.7}$$

where $\Gamma = \gamma_{Na}^2/\gamma_{Ca}$ and γ is a single-ion activity coefficient (Eq. 4.23). Equation 12.7 contains two important chemical variables. The parameter

$$SAR = 10^{3/2}[Na^+]/[Ca^{2+}]^{1/2} \tag{12.8a}$$

is the *sodium adsorption ratio*. The quantity $100E_{Na} \equiv ESP$ is the *exchangeable sodium percentage* (Section 9.2). With these definitions, Eq. 12.7 becomes the expression:

$$K_{ex}^V = 2.5\Gamma\left(\frac{SAR}{ESP}\right)^2 [1 - (ESP/100)^2] \tag{12.9}$$

Given $K_{ex}^V \approx 2.2$, Eq. 12.9 provides a unique relationship between *ESP* and *SAR*. This relationship is plotted in Fig. 12.1 using also the typical value $\Gamma \approx 1.3$ ($I \approx 20 \text{ mol m}^{-3}$). The *ESP–SAR* relation is indistinguishable from a straight line of slope 1.2 for $SAR < 20 \text{ mol}^{1/2} \text{ m}^{-3/2}$.

The *SAR* defined by Eq. 12.8a is equivalent to utilizing concentrations in the SI units of moles per cubic meter instead of moles per liter:

$$SAR \equiv c_{Na}/\sqrt{c_{Ca}} \tag{12.8b}$$

where c is in SI units (see the Appendix). Since direct measurements of the free-ion concentrations of Na^+ and Ca^{2+} are not common in practical work, *SAR*

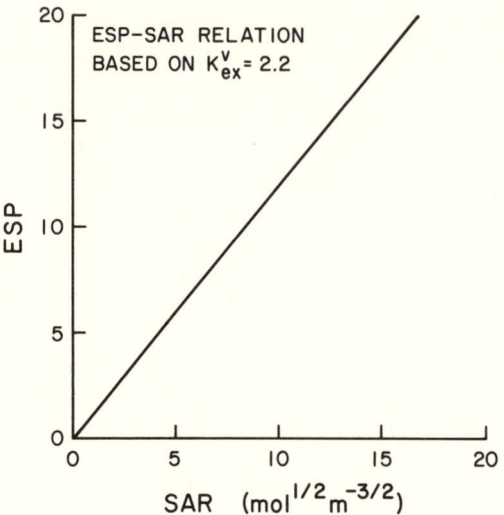

FIG. 12.1 A graph of the ESP–SAR relationship in Eq. 12.9 for $K_{ex}^v = 2.2$ and $\Gamma = 1.3$.

in Eq. 12.8a usually is replaced by the parameter SAR_p:

$$SAR_p \equiv Na_T/\sqrt{Ca_T} \qquad (12.8c)$$

where the subscript p indicates the "practical" SAR, and *total* concentrations in SI units are indicated on the right side. This parameter is smaller than SAR because of typically greater soluble-complex formation by Ca^{2+} than Na^+ (Table 12.1). Statistical analyses of the SAR_p–SAR relationship in saturation extracts indicate that SAR_p is about 12% smaller than SAR, on the average. This small difference and the small deviation of the graph in Fig. 12.1 from a 1:1 line are often neglected in applications of Eq. 12.9 to irrigation water quality evaluation, with the result that the graph in Fig. 12.1 simplifies to the expression:

$$ESP \approx SAR_p \qquad (12.10)$$

Equation 12.10 has been found to be a useful rule of thumb in field studies of ESP. Given the several approximations leading to this equation and its intended application to irrigation water quality (Section 12.5), the expedient assumption that $K_{ex}^v \approx 2.2$ for $Na \rightarrow Mg$ exchange also is made, such that SAR_p can be redefined by the formula:

$$SAR_p \equiv Na_T/(Ca_T + Mg_T)^{1/2} \qquad (12.8d)$$

and incorporated into Eq. 12.10. Although the chain of assumptions linking

Eq. 12.10 to Eq. 12.7 via Eq. 12.8d is long, it is also well defined enough to make the basis of the *ESP–SAR*$_p$ relation in cation exchange theory apparent.

The importance of exchangeable Na in maintaining the stability of soil clay particles in suspension is discussed in Section 10.2. Figure 10.3 provides evidence for the dispersion of smectite particles as *ESP* increases above 15%. Studies of the permeability characteristics and aggregate structures of arid-zone soils have substantiated that this effect of adsorbed Na$^+$ is quite general. These studies have led to the designation *sodic* for a soil in which the *SAR*$_p$ value of the saturation extract is > 13. Sodic soils correspondingly will have *ESP* values larger than about 15 (Fig. 12.1). In these soils, clay-sized particles will tend to disperse in the soil solution, and a reduction in pore diameters will occur because of clogging and swelling phenomena.

The dispersive effect of exchangeable Na$^+$, however, will be observed only if the electrolyte concentration in the soil solution is smaller than that required to flocculate clay particles (the critical coagulation concentration, defined in Section 10.4). As shown in Table 10.2, even a fully Na-saturated smectite suspension will flocculate if the electrolyte concentration is > 8 mol m^{-3}, and a suspension of Na-illite will do the same if the electrolyte concentration reaches about 50 mol m^{-3}. *Thus soil salinity tends to counteract the effect of exchangeable sodium on soil structure.*

FIG. 12.2 A Quirk–Schofield diagram for montmorillonite suspended in NaCl–CaCl$_2$ solution. [Data from J. D. Oster, I. Shainberg, and J. D. Wood, Flocculation value and gel structure of sodium/calcium montmorillonite and illite suspensions, *Soil Sci. Soc. Am. J.* **44**:955–959 (1980).]

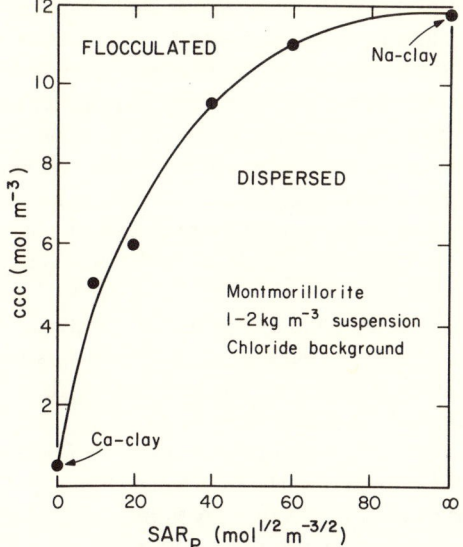

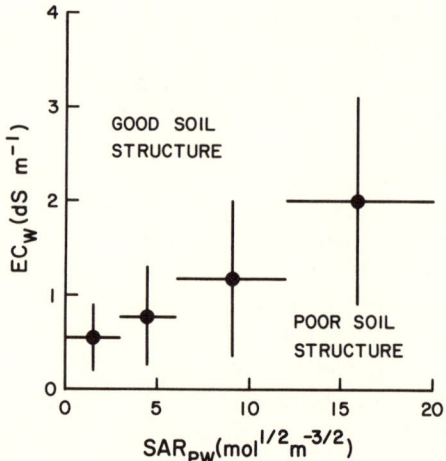

FIG. 12.3 A Quirk–Schofield diagram for California soils based on the properties of applied water for irrigation. [Based on data from R. S. Ayers and K. K. Tanji, reported by I. Shainberg and J. Letey, Response of soils to sodic and saline conditions, *Hilgardia* **52**(2):1–55 (1984).]

Table 10.2 also shows that Ca-saturated clay particles ($ESP = 0\%$) flocculate at electrolyte concentrations about 60 times smaller than those required to flocculate Na-saturated clay particles. It follows that the critical coagulation concentration (ccc) for soil clay particles should depend on the ESP and, according to Eq. 12.9, on the SAR as well. This kind of relationship between the ccc and ESP or SAR has been found in a number of experiments with soil and specimen clay minerals. An example is shown in Fig. 12.2. It is conventionally expressed as a *Quirk–Schofield diagram*, which is a graph of the electrolyte concentration or electrolytic conductivity, below which significant deterioration in soil structure should occur (as measured, for example, by a loss of permeability or an increase in solution turbidity), plotted against the ESP or SAR above which the same deterioration in soil structure should take place. Figure 12.2 is a Quirk–Schofield diagram for montmorillonite suspended in NaCl–CaCl$_2$ solutions. Figure 12.3 shows a Quirk–Schofield diagram based on field experiments and laboratory data relating to California soils. In this case, the electrolytic conductivity (EC_w) and SAR_p of irrigation water (SAR_{pw}) are the plotting variables. A "window," allowing for variability among soils, separates regions of expected good and poor soil structure. Note that any soil for which the EC_w–SAR_{pw} combination falls into the poor-soil structure region could be termed "sodic" insofar as soil permeability is concerned. A conventional SAR_p value of 13 would lead to poor soil structure only if the electrolytic conductivity dropped below 1.0 dS m^{-1}. On the other hand, an apparently low SAR_p value of 3.0 would lead to poor structure if the conductivity dropped below 0.2 dS m^{-1}. A saline soil as defined conventionally should not have poor structure unless the SAR_p value rises above 20.

12.3 Mineral Weathering

Soils in arid regions are often at the early stage of the Jackson–Sherman weathering sequence (Table 1.7), and, therefore, they contain silicate, carbonate, and sulfate minerals that are relatively susceptible to dissolution reactions in percolating water. The composition and structure of these minerals are described in Sections 2.2 and 2.5. Their dissolution reactions are discussed in Sections 5.1 and 5.2. Laboratory studies have shown that these reactions may add 3–5 mol_c m^{-3} in electrolyte charge concentration to percolating waters, with most of the addition coming from Ca, Mg, and HCO_3 under alkaline soil conditions (see, for example, Table 2.2). Possible dissolution reactions of the silicate minerals olivine and anorthite that can produce this effect are illustrated in Eq. 2.2 and Eq. 2.8, respectively.

Soil mineral weathering that increases the salinity of the soil solution and enriches it in Ca and Mg has important implications for the colloidal phenomena discussed in Section 12.2. If the water entering a soil has a very low electrolyte concentration (e.g., rainwater or irrigation water diverted from pristine surface waters), only a very small SAR_{pw} in the water would be needed to cause problems with soil structure and permeability (Fig. 12.3). For example, SAR_{pw} values as low as 3.0 can be deleterious to soil structure if the applied water EC_w is around 0.5 dS m^{-1}. If, on the other hand, infiltrating water causes soil minerals to dissolve and, say, raise the electrolytic conductivity of the soil solution to near 1.0 dS m^{-1}, then only SAR_p values > 5.0 would be of concern. Moreover, if most of the increase in electrolyte concentration comes from Ca^{2+} and Mg^{2+}, then the SAR_p value would *drop* in the equilibrated soil solution (Eq. 12.8d), further lessening the chance of adverse soil structure effects. The general conclusion to be drawn from this example is that *soils that contain abundant easily weatherable minerals will be less sensitive to percolating fresh waters in respect to colloidal phenomena than those that are depleted of easily weatherable minerals.*

Increasing salinity in the soil solution tends also to enhance the solubility of weatherable minerals. This effect can be predicted on the basis of the ionic strength dependence of single-ion activity coefficients (Eq. 4.23). Consider, for example, the mineral gypsum ($CaSO_4 \cdot 2H_2O$), whose solubility product constant is defined in Eq. 5.6:

$$K_{so} = (Ca^{2+})(SO_4^{2-}) = 2.4 \times 10^{-5} \tag{12.11}$$

The solubility of gypsum is related to the *concentration* of Ca^{2+}(aq) (see Problem 6 in Chapter 5) and, therefore, to the *conditional* solubility product, $^cK_{so}$:

$$^cK_{so} = [Ca^{2+}][SO_4^{2-}] = K_{so}/\gamma_{Ca}\gamma_{SO_4} \tag{12.12}$$

where the γ are single-ion activity coefficients. It follows from Eqs. 4.23 and

12.12 that:

$$\log {}^c K_{so} = \log K_{so} + 4.096 \left[\frac{\sqrt{I}}{1 + \sqrt{I}} - 0.3I \right] \qquad (12.13)$$

where I is ionic strength in moles per cubic decimeter. This equation shows that the logarithm of the conditional solubility product will *increase* with ionic strength, so long as the term in the square root of I exceeds $0.3I$ ($I < 2$ mol dm^{-3}). Experimental confirmation of the enhancement of gypsum solubility in NaCl solutions is shown in Table 12.2. As the concentration of NaCl increases from 0 to 0.548 mol dm^{-3}, the total concentration of Ca in solution more than doubles.

The weathering reactions of calcite ($CaCO_3$) are of great importance in arid-zone soils. As discussed in Section 5.1, the dissolution of this mineral is surface-controlled and, therefore, follows a zero-order kinetics expression (like Eq. 5.1), with Ca^{2+} replacing A in that equation. The net rate of precipitation/dissolution near equilibrium can be expressed analogously to Eq. 4.2 (see also Problem 4 in Chapter 5):

$$\frac{d[Ca^{2+}]}{dt} = k[K_{so} - (Ca^{2+})(CO_3^{2-})] \qquad (12.14a)$$

where k is a rate constant, $K_{so} = 3.3 \times 10^{-9}$ is the solubility product constant for the dissolution reaction:

$$CaCO_3(s) = Ca^{2+}(aq) + CO_3^{2-}(aq) \qquad (12.15a)$$

and $(Ca^{2+})(CO_3^{2-})$ is the *actual* ion activity product (IAP) in aqueous solution (Eq. 5.10). The rate constant k is a function of pH and specific surface area,

TABLE 12.2 Effect of electrolyte concentration on the solubility of gypsum in NaCl solutions[a]

$NaCl_T$	Ca_T	$NaCl_T$	Ca_T	$NaCl_T$	Ca_T
		mol m^{-3}			
0	15.10	51	19.40	200	27.05
12	16.20	75	20.90	300	30.70
25	17.90[b]	100	22.15	400	33.32
26	17.50[b]	115	23.10	500	35.44
50	19.41	192	26.60	548	37.20

[a]Data compiled from W. L. Marshall and R. Slusher, Thermodynamics of calcium sulfate dihydrate in aqueous sodium chloride solutions, 0–110°, *J. Phys. Chem.* **70**:4015–4027 (1966), and F. S. Nakayama, Calcium complexing and the enhanced solubility of gypsum in concentrated sodium-salt solutions, *Soil Sci. Soc. J.* **35**:881–883 (1971).
[b]Discrepancies between solubility values at essentially the same NaCl concentration reflect variability between two independent sets of measurements.

among other variables. At $pH > 8$, laboratory experiments suggest that $k = k_f S c_s$, where $k_f \approx 117$ mol m^{-2} s^{-1}, S is the specific surface area in m^2 kg^{-1}, and c_s is the calcite concentration (in suspension) in kg dm^{-3}. A representative value of k is 25 mol dm^{-3} s^{-1}. With the definition of the relative saturation (Eq. 5.11) introduced, Eq. 12.14a takes the form:

$$\frac{d[Ca^{2+}]}{dt} = kK_{so}(1 - \Omega) \tag{12.14b}$$

where $\Omega = (Ca^{2+})(CO_3^{2-})/K_{so}$. Equation 12.14b describes the net rate of precipitation of calcite if $\Omega > 1$ (supersaturation). For that case, if $\Omega < 10$ and the initial Ca^{2+} concentration is 0.001 mol dm^{-3}, Table 4.2 indicates a half-life of the order of thousands of seconds ($[Ca^{2+}]_0/2kK_{so}\Omega$).

A useful rate expression analogous to that in Eq. 12.14b is obtained by transforming the *IAP* according to the calcite dissolution reaction in Eq. 2.9b:

$$CaCO_3(s) + H^+(aq) = Ca^{2+}(aq) + HCO_3^-(aq) \tag{12.15b}$$

Given the equilibrium constant for the reaction:

$$H^+(aq) + CO_3^{2-}(aq) = HCO_3^-(aq) \qquad K_2 = 10^{10.3} \tag{12.16}$$

where $K_2 = (HCO_3^-)/(H^+)(CO_3^{2-})$, one can rewrite Eq. 12.14a in the form:

$$\frac{d[Ca^{2+}]}{dt} = k[K_{so} - (Ca^{2+})(HCO_3^-)/K_2(H^+)] \tag{12.14c}$$

Now the relationship, $(H^+) \equiv 10^{-pH}$, and the *Langlier Index*, $pH-pH_s$, where pH_s is defined by the equation:

$$(Ca^{2+})(HCO_3^-)/K_2 K_{so} \equiv 10^{-pH_s} \tag{12.17}$$

may be introduced to transform Eq. 12.14c into the expression:

$$\frac{d[Ca^{2+}]}{dt} = kK_{so}[1 - 10^{pH - pH_s}] \tag{12.14d}$$

Equation 12.14d permits estimates of the rate of calcite precipitation/dissolution based on the Langlier Index. To illustrate this relationship, consider the Ca and HCO_3 activities based on the speciation of the soil solution described in Table 12.1: $(Ca^{2+}) \approx 2.65 \times 10^{-3}$, $(HCO_3^-) \approx 2.35 \times 10^{-3}$, and $pH_s = 7.02$. It follows from Eq. 12.14d that, at pH 7.6, the Langlier Index equals 0.58 and the soil solution is supersaturated with respect to calcite. The same conclusion is reached by calculating Ω directly for the soil solution:

$$\Omega = \frac{(Ca^{2+})(CO_3^{2-})}{K_{so}} = \frac{2.65 \times 10^{-3} \times 4.69 \times 10^{-6}}{3.3 \times 10^{-9}} = 3.8$$

Since $\Omega > 1$, the soil solution is supersaturated.

The value of Ω just calculated leads to an IAP of 10^{-8} for calcite in the soil solution described in Table 12.1. This value has been observed consistently in a very large number of investigations of calcite solubility in arid-zone soils throughout the world. The cause of this ubiquitous supersaturation condition is the subject at present of intensive research. Careful experimentation has ruled out analytical error, microcrystalline disorder, and Mg substitution for Ca (Section 2.5) as principal factors leading to a persistently high IAP. Most recent studies have focused on kinetics-based mechanisms relating to Eq. 12.14 as the likely origin of the discrepancy between IAP and K_{so} in calcareous soils (see Problem 12 in Chapter 5). One possibility is a reduction in k produced by the adsorption of soluble organic matter on the surfaces of calcite particles. Laboratory research has shown that calcite precipitation is inhibited greatly by adsorbed fulvic acid. For example, when fulvic acid is present to make a soluble-carbon concentration of about 20 mmol m^{-3}, it is adsorbed and reduces the rate constant k_f from 117 mol m^{-2} s^{-1} to <1 mol m^{-2} s^{-1}. This kind of inhibition could operate in arid-zone soils. Another reasonable possibility is a sustained production of bicarbonate alkalinity through the oxidation of soil organic matter (Table 6.2). Arid-zone soils incubated with plant materials at ambient P_{CO_2} readily produce HCO_3^-(aq) that increases in concentration as the plant materials decompose. (In the same soils without added plant materials, the bicarbonate concentration achieves a constant value very rapidly that lies well below that in the amended soils.) The high bicarbonate concentration may be sustained under steady-state conditions, leading thereby to a persistently high IAP for calcite.

12.4 Boron Chemistry

Boron is a trace element in soils (Table 1.2) that occurs typically coprecipitated in secondary metal oxides and clay minerals (Table 1.5) and in mica (Table 1.4). Its occurrence as a separate solid phase (other than tourmaline) in arid-zone soils has not been established, but a number of Na, K, Ca, and Mg borates have been identified in saline geological environments. Among them are borax [$Na_2B_4O_5(OH)_4 \cdot 8H_2O$], nobleite [$CaB_6O_9(OH)_2 \cdot 3H_2O$], inyoite [$CaB_3O_3(OH)_5 \cdot 4H_2O$], McAllisterite [$MgB_6O_7(OH)_6 \cdot \frac{9}{2}H_2O$], and inderite [$MgB_3O_3(OH)_5 \cdot 5H_2O$]. Dissolution reactions and the corresponding equations that permit the calculation of the activity of $H_3BO_3^0$(aq) in equilibrium with these minerals are listed in Table 12.3. Given a representative value of 10^{-3} for the activity of Na^+(aq), Ca^{2+}(aq), or Mg^{2+}(aq) in arid-zone soil solutions (Table 12.1), the equations for ($H_3BO_3^0$) in Table 12.3 lead to the conclusion that the borate minerals would support very high B concentrations in the soil solution. For example, taking (Ca^{2+}) $= 2.7 \times 10^{-3}$ and pH $= 7.6$, one calculates for the mineral nobleite: $\log(H_3BO_3^0) = 1.28 + 0.43 + 2.53 = 4.24!$ This result means that nobleite ultimately would be dissolved

TABLE 12.3 Dissolution reactions of boron-bearing minerals[a]

Reaction	$\log K_{298}$
$\frac{1}{6}CaB_6O_9(OH)_2 \cdot 3H_2O(s) + \frac{1}{3}H^+(aq) + \frac{2}{3}H_2O(\ell) = H_3BO_3^0(aq) + \frac{1}{6}Ca^{2+}(aq)$	1.28
$\log(H_3BO_3^0) = 1.28 - \frac{1}{6}\log(Ca^{2+}) + \frac{1}{3}pH$	
$\frac{1}{3}CaB_3O_3(OH)_5 \cdot 4H_2O(s) + \frac{2}{3}H^+(aq) = H_3BO_3^0(aq) + \frac{1}{3}Ca^{2+}(aq) + H_2O(\ell)$	3.39
$\log(H_3BO_3^0) = 3.39 - \frac{1}{3}\log(Ca^{2+}) + \frac{2}{3}pH$	
$\frac{1}{12}Mg_2B_{12}O_{14}(OH)_{12} \cdot 9H_2O(s) + \frac{1}{12}H_2O(\ell) = H_3BO_3^0(aq) + \frac{1}{6}Mg^{2+}(aq) + \frac{1}{2}H^+(aq)$	1.59
$\log(H_3BO_3^0) = 1.59 - \frac{1}{6}\log(Mg^{2+}) - \frac{1}{2}pH$	
$\frac{1}{3}MgB_3O_3(OH)_5 \cdot 5H_2O(s) + \frac{2}{3}H^+(aq) = H_3BO_3^0(aq) + \frac{1}{3}Mg^{2+}(aq) + \frac{4}{3}H_2O(\ell)$	3.62
$\log(H_3BO_3^0) = 3.62 - \frac{1}{3}\log(Mg^{2+}) + \frac{2}{3}pH$	
$\frac{1}{4}Na_2B_4O_5(OH)_4 \cdot 8H_2O(s) + \frac{1}{2}H^+(aq) = H_3BO_3^0(aq) + \frac{1}{2}Na^+(aq) + \frac{5}{4}H_2O(\ell)$	2.87
$\log(H_3BO_3^0) = 2.87 - \frac{1}{2}\log(Na^+) + \frac{1}{2}pH$	

[a]Data for $\log K_{298}$ from S. V. Mattigod, A method for estimating the standard free energy of formation of borate minerals, *Soil Sci. Soc. Am. J.* **47**:654–655 (1983).

completely as water percolated into the soil. To achieve the concentration of $H_3BO_3^0(aq)$ indicated in Table 12.1 (36.5 mmol m^{-3}), only a very small amount of nobleite would have to be present in the soil, and this amount should be easily leached. The same conclusion would be drawn if the borate ion, $B(OH)_4^-(aq)$, or metal–borate complexes were considered in addition to $H_3BO_3^0(aq)$. The dissociation reaction:

$$H_3BO_3^0(aq) + H_2O(\ell) = B(OH)_4^-(aq) + H^+(aq) \qquad (12.18)$$

has an equilibrium constant equal to 5.8×10^{-10} ($\log K_{298} = 9.23$). Therefore, the $B(OH)_4^-(aq)$ species will not be significant in soil solutions (i.e., equal to the concentration of $H_3BO_3^0$) until the pH value approaches 9. This is true also for complexes like $CaB(OH)_4^+$ and $MgB(OH)_4^+$. For example, they account for only 0.3 and 0.1%, respectively, of the boron species in the soil solution described in Table 12.1 (pH 7.6).

Boron concentrations in arid-zone soil solutions can range up to 2 mol m^{-3}, depending on the mineralogy of the soil parent material or the presence of enriched groundwater. Sensitive crop plants are affected by concentrations >0.046 mol B m^{-3}, and all crops will be affected at concentrations >2.5 mol m^{-3} (Section 12.5). Leaching experiments with soil columns indicate that these higher concentrations cannot be reduced easily by percolating fresh water: their rate of removal is much less than that for chloride, and a kind of "regeneration" of B concentrations can occur a few weeks after they have been reduced by extensive leaching. This behavior suggests that soil boron is released slowly from the minerals in which it is a trace component (Tables 1.4 and 1.5) and that it adsorbs strongly onto soil particle surfaces.

An adsorption envelope for boron is shown in Fig. 8.4. It features a gradual rise as the pH value increases above 7, then a broad maximum near pH 9, and finally a sharp decline. These characteristics, as explained in Section 8.4, reflect the competition for H^+ between the soil adsorbent and the borate ion, $B(OH)_4^-$, which, according to Eq. 12.18, protonates significantly at pH < 9.2. The similarity in adsorption behavior between F^- and $B(OH)_4^-$ in soils, as well as studies of B adsorption by specimen metal oxides, have suggested that the principal adsorption mechanism is ligand exchange with surface hydroxyls (Eq. 8.11):

$$SOH(s) + H_3BO_3^0(aq) = SH_2BO_3^0(s) + H_2O(\ell) \qquad (12.19)$$

Indirect support for this mechanism has come from modeling studies wherein the constant capacitance model (Section 8.5) utilizing Eq. 12.19 was found to describe adsorption envelopes like that in Fig. 8.4 very well. These studies and the experimental investigations with specimen minerals indicate that surface aluminol groups in soils are the main reactive sites for B adsorption. The low leachability of adsorbed boron may derive from the inner-sphere surface complex formed in conjunction with the reaction in Eq. 12.19.

12.5 Irrigation Water Quality

The continued use of a water resource for the irrigation of agricultural land requires that no adverse effects from the applied water occur in the soil environment. From the perspective of soil chemistry, all irrigation waters are electrolyte solutions. Their chemical composition, which reflects their source and postwithdrawal treatment, may not be compatible with the suite of compounds and weathering processes that exist in the soils to which they are applied. Adding to this the salt-concentrating effects of crop extraction of water and fertilizer amendments, one sees the possibility that irrigated soils can become saline or sodic without careful management.

The chemical *properties* of irrigation water that must be identified and controlled in order to maintain the water suitable for agricultural use form part of the *irrigation water quality criteria*. The *numerical interpretation* of the water quality criteria to achieve goals in irrigation water quality management leads to *water quality standards*. These two distinct aspects of irrigation water quality are discussed in comprehensive detail in *Water Quality for Agriculture* (see the "For Further Reading" section at the end of this chapter). Suffice it to say here that water quality criteria are determined largely by the results of basic field and laboratory research, whereas water quality standards are the result of research data combined with the collective experience of extension scientists, farm advisers, and growers.

The three principal water quality–related problems in irrigated agriculture are *salinity hazard*, *sodicity hazard*, and *toxicity hazard*. Irrigation water quality standards to control salinity hazard are listed in Table 12.4. They are designated preferentially by three classes of electrolytic conductivity (EC_w)

TABLE 12.4 Irrigation water quality standards to control salinity and sodicity hazards[a]

	Restriction on water use		
	None	Slight to moderate	Severe
Salinity hazard			
EC_w (dS m^{-1})	<0.7	0.7–3.0	>3.0
Sodicity hazard			
SAR_{pw} range[b]		EC_w (dS m^{-1})	
(mol$^{\frac{1}{2}}$ m$^{-\frac{3}{2}}$)			
0–3	>0.7	0.7–0.2	<0.2
3–6	>1.2	1.2–0.3	<0.3
6–12	>1.9	1.9–0.5	<0.5
12–20	>2.9	2.9–1.3	<1.3

[a]Adapted from R. S. Ayers and D. W. Wescot, *Water Quality for Agriculture*, FAO Irrigation and Drainage Paper No. 29, Rev. 1. FAO, Rome, 1985.
[b]SAR_{pw} defined in Eq. 12.8d for total concentrations in *irrigation water*.

measured in decisiemens per meter. These classes correspond approximately to groupings of agricultural crops into sensitive, relatively sensitive, and relatively tolerant categories, respectively. Thus, for example, sensitive crops require $EC_w < 0.7$ dS m^{-1}, and only relatively tolerant crops can withstand $EC_w > 3$ dS m^{-1} without significant yield reduction. According to Eq. 4.22, the three EC_w ranges in Table 12.4 are equivalent to the ionic strength ranges: $I < 10$ mol m^{-3}, $10 < I < 44$ mol m^{-3}, and $I > 44$ mol m^{-3}.

The definition of a saline soil refers to the electrolytic conductivity of the soil saturation extract (EC_e), not that of applied water (Section 12.1). Even though EC_w is recommended to be <3 dS m^{-1}, the validity of this restriction depends on knowing precisely the relationship between EC_w and EC_e in the root zone. This relationship continues to be the subject of much research in the chemistry of soil salinity, since many complicated factors enter into it, even in the absence of effects from rainwater and shallow groundwater. As a rule of thumb, the steady-state value of EC_e that results from irrigation with water of conductivity EC_w is estimated from a knowledge of the *leaching fraction* (LF) of the applied water. The leaching fraction is defined by the equation:

$$LF = \frac{\text{volume of water leached below root zone}}{\text{volume of water applied}} \tag{12.20}$$

Typically, LF is in the range 0.15–0.20, meaning that 15–20% of the water applied leaches below the root zone while 80–85% is used in evapotranspiration processes. Once the value of LF is known, the average value of EC_e in the root zone is estimated with the formula:

$$EC_e = X(LF) \cdot EC_w \tag{12.21}$$

where $X(LF)$ is a factor whose dependence on LF has been worked out empirically on the basis of experience with typical irrigated, cropped soils. The function $X(LF)$ is given in numerical form in Table 12.5. As an example of its use, if water with $EC_w = 1.2$ dS m^{-1} is applied and $LF = 0.25$, then EC_e is predicted to be 1.44 dS m^{-1}, on the average, over all the root zone. Note that $LF > 0.3$ results in $EC_e < EC_w$, and that $LF < 0.1$ would produce a saline soil if water with $EC_w > 2$ dS m^{-1} is applied.

Irrigation water quality standards to control sodicity hazard are also listed in Table 12.4. They reflect the interplay between electrolyte concentration and exchangeable-cation composition discussed in Section 12.2. Thus, for example, if SAR_{pw} is in the range of 3–6 mol$^{1/2}$ m$^{-3/2}$ and EC_w is > 1.2 dS m^{-1}, the development of poor soil structure from exchangeable sodium is unlikely because the electrolyte concentration in the applied water is deemed large enough to maintain flocculated soil colloids. It is instructive in this sense to compare Table 12.4 with Fig. 12.3.

The cation exchange relationship on which the use of SAR_{pw} is based refers to SAR in the soil solution, not in the applied water. Like EC_w and EC_e, the relationship between SAR_{pw} and the soil solution SAR is the subject of much current research. The conversion of SAR_{pw} to an SAR_w value involving *free-cation* concentrations can be made with the help of speciation calculations like that in Table 12.1, or with the "12% rule of thumb" mentioned in Section 12.2. More serious usually is the need to account for calcite precipitation or dissolution as the irrigation water percolates into soil under the influence of ambient $CO_2(g)$ pressures. For this purpose, the *Suarez adjusted sodium adsorption ratio* may be calculated and used to estimate ESP with Eq. 12.9. This parameter, denoted "adj RNa," is defined by the equation:

$$\text{adj } R\text{Na} \equiv c_{\text{Naw}}/[c_{\text{Mgw}} + c_{\text{Ca}}^{\text{eq}}]^{1/2} \tag{12.22}$$

TABLE 12.5 The factor $X(LF)$ in Eq. 12.21[a]

LF	X(LF)	LF	X(LF)
0.05	3.2	0.30	1.0
0.10	2.1	0.40	0.9
0.15	1.6	0.50	0.8
0.20	1.3	0.60	0.7
0.25	1.2	0.70	0.6

[a]Adapted from R. S. Ayers and D. W. Wescot, *Water Quality for Agriculture*, FAO Irrigation and Drainage Paper No. 29, Rev. 1. FAO, Rome, 1985.

where c is a free-cation concentration in moles per cubic meter, and c_{Ca}^{eq} is the concentration of $Ca^{2+}(aq)$ in a soil solution having the same ratio $(HCO_3^-)/(Ca^{2+})$ as the irrigation water when it is in equilibrium with calcite at the soil value of P_{CO_2}.

The relationship between c_{Ca}^{eq} and the $[HCO_3^-]/[Ca^{2+}]$ ratio in irrigation water, which is necessary in order to apply Eq. 12.22, can be derived from the expression for the relative saturation in terms of the Langlier Index (see Eqs. 12.14c and 12.14d):

$$\Omega_w = 10^{pH_w - pH_s} \tag{12.23}$$

where pH_w is the pH value of the irrigation water. Equation 12.2 can be combined with Eq. 12.17 to derive from Eq. 12.23 the alternative expression:

$$\Omega_w = (Ca^{2+})_w (HCO_3^-)_w^2 / 10^{2.5} K_{so} P_{CO_2} \tag{12.24}$$

where the numerical factor is $K_2/10^{7.8}$. The denominator in Eq. 12.24 is equal to $(Ca^{2+})_{eq}(HCO_3^-)_{eq}^2$, as can be seen by setting $\Omega_w = 1.0$ in the equation and changing "w" to "eq" for that case. It follows that

$$(Ca^{2+})_{eq} = \frac{(Ca^{2+})_w (HCO_3^-)_w^2}{(HCO_3^-)_{eq}^2 \Omega_w}$$

which is transformed to the equation:

$$(Ca^{2+})_{eq}^3 = (Ca^{2+})_w^3 / \Omega_w \tag{12.25}$$

on multiplying by $(Ca^{2+})_{eq}^2$ on both sides, then multiplying by $[(Ca^{2+})_w/(Ca^{2+})_w]^2$ on the right side only, and using the condition, $(HCO_3^-)_{eq}/(Ca^{2+})_{eq} = (HCO_3^-)_w/(Ca^{2+})_w$, assumed by hypothesis. The substitution of Eq. 12.24 into Eq. 12.25 yields the expression desired:

$$(Ca^{2+})_{eq} = \left\{ \frac{10^{2.5} K_{so}}{[(HCO_3^-)_w/(Ca^{2+})_w]^2} \right\}^{\frac{1}{3}} P_{CO_2}^{1/3} \tag{12.26}$$

Equation 12.26 can be used to calculate c_{Ca}^{eq} (which differs by a factor of 10^3 from $[Ca^{2+}]_{eq}$) once the values of P_{CO_2}, $[HCO_3^-]/[Ca^{2+}]$, the activity coefficients of $Ca^{2+}(aq)$ and $HCO_3^-(aq)$, and K_{so} have been chosen. A table of values of $2c_{Ca}^{eq}$ (in $mol_c\ m^{-3}$) for given values of $[HCO_3^-]/[Ca^{2+}]$ and EC_w (in dS m^{-1}) is presented in *Water Quality for Agriculture*, based on the choices $P_{CO_2} = 10^{-3.15}$ atm and $K_{so} = 10^{-8}$. For these choices, Eq. 12.26 can be reduced to the form:

$$c_{Ca}^{eq} = \{0.447 \gamma_{HCO_3} \gamma_{Ca}([HCO_3^-]/[Ca^{2+}])^2\}^{-\frac{1}{3}} \tag{12.27}$$

where the single-ion activity coefficients may be calculated with Eqs. 4.22 and 4.23, given EC_w. For example, if $EC_w = 1$ dS m^{-1}, then $\gamma_{HCO_3} \gamma_{Ca} = 0.532$ (by

TABLE 12.6 Irrigation water quality standards to control toxicity hazards[a]

	Restriction on water use		
Water quality criterion	None	Slight to moderate	Severe
SAR_{pw} (mol$^{\frac{1}{2}}$ m$^{-\frac{3}{2}}$)[b]	<3	3–9	>9
Cl_T (mol m^{-3})	<4	4–10	>10
B_T (mmol m^{-3})	<65	65–277	>277

[a]Adapted from R. S. Ayers and D. W. Wescot, *Water Quality for Agriculture,* FAO Irrigation and Drainage Paper No. 29, Rev. 1. FAO, Rome, 1985.
[b]To control sodium toxicity, primarily for perennial plants.

Eqs. 4.22 and 4.23) and, for $[HCO_3^-]/[Ca^{2+}] = 2$, Eq. 12.27 predicts $c_{Ca}^{eq} = 1.02$ mol m^{-3}. This value can be used in Eqs. 12.22 and 12.9 to predict the soil *ESP* value.

Irrigation water quality standards to control toxicity hazards are listed in Table 12.6. They involve the three elements of most concern in arid-zone soils for sensitive plant species. The toxicity hazard of trace elements other than boron is discussed in Section 13.1.

FOR FURTHER READING

R. Keren and F. T. Bingham, Boron in water, soils, and plants, *Adv. Soil Sci.* **1**:229–276 (1985). A comprehensive review of the soil chemistry and plant uptake of boron.

J. D. Oster, Gypsum usage in irrigated agriculture: A review, *Fertilizer Res.* **3**:73–89 (1982). An advanced discussion of the soil chemistry aspects of gypsum amendment to improve sodic soils.

J. P. Quirk, Soil permeability in relation to sodicity and salinity, *Phil. Trans. R. Soc. (London)* **A316**:297–317 (1986). A fine essay on the basic chemistry involved in the reclamation of sodic soils.

I. Shainberg and J. Letey, Response of soils to sodic and saline conditions, *Hilgardia* **52**(2):1–57 (1984). A comprehensive monograph on the physical chemistry and physics of soil permeability.

I. Shainberg and J. Shalhevet (eds), *Soil Salinity under Irrigation*, Springer-Verlag, New York, 1984. This excellent collection of position papers provides up-to-date discussions of all aspects of soil salinity—from physics to management strategies.

PROBLEMS

The more difficult problems are indicated by an asterisk.

1. In a study of sodic soil solutions, the statistical relationship:

$$pH = 4.9 - \log P_{CO_2}$$

was observed, where P_{CO_2} is in atmospheres. Estimate the average bicarbonate alkalinity for these soils in moles per cubic meter.

2. In the accompanying table are data pertaining to the saturation extract of an Aquic Natrusalf. Use these data to calculate the corresponding equilibrium $CO_2(aq)$ pressures, in atmospheres.

pH	EC_e (dS m^{-1})	Alkalinity (mol m^{-3})	pH	EC_e (dS m^{-1})	Alkalinity (mol m^{-3})
8.05	0.709	1.13	8.25	0.930	1.75
8.10	0.849	1.50	8.30	1.279	1.63
8.20	0.954	1.25	8.35	1.012	1.88

3. In the accompanying table are *ESP* values measured in the upper 0.3 m of an Alfisol irrigated for 8 years with waters of varying *SAR* values. Use these data to calculate an average value of K_{ex}^V for the soil. Take $\Gamma \approx 1.3$.

	Irrigation water			
	"Gage Canal"	"Colorado R."	"Sulfate"	"Chloride"
ESP value:	2.1	3.4	4.4	2.7
SAR (mol$^{1/2}$ m$^{-3/2}$):	1.30	2.92	4.85	3.31

4. Explain conceptually, using Fig. 9.2, and quantitatively, using $K_{ex}^V = 2.2$ for Na→Ca exchange and $K_{ex}^V = 2.0$ for Na→Mg exchange, why soil structure may become adversely affected as the $Mg^{2+}(aq)$ concentration increases in a soil solution at the expense of $Ca^{2+}(aq)$.

5. In a study of soil permeability, it was found that the relationship between *SAR* and the ionic strength above which good soil structure existed was:

$$I = 0.6 + 0.56 \, SAR$$

where I is in mol m^{-3}, and *SAR* is in mol$^{1/2}$ m$^{-3/2}$. Prepare a Quirk–Schofield plot like that in Fig. 12.3 based on this empirical relationship and Eq. 4.22. (You may equate the factor before log κ in Eq. 4.22 with 1.00 instead of 1.009.)

*6. Derive a relationship between $[SO_4^{2-}]$ and the *ESP* of a soil containing gypsum. Calculate the *ESP* resulting from $c_{Na} = 5$ mol m^{-3} and $[SO_4^{2-}] = 0.0053$ mol dm^{-3} using $K_{ex}^V = 2.2$ for Na → Ca exchange. Ignore $Mg^{2+}(aq)$ in these calculations, but consider an ionic strength of 20 mol m^{-3}.

(*Hint*: Combine Eqs. 12.9 and 12.12.)

*7. Gypsum is applied to a soil irrigated with water in which $EC_w = 1.3\,dS$ m^{-1}. Given that $c_{Na} = 12\,mol\;m^{-3}$, $c_{Mg} = 5.2\,mol\;m^{-3}$, and $[SO_4^{2-}] = 0.014\,mol\;dm^{-3}$ in the soil solution at steady state, calculate the steady-state SAR of the soil solution if the leaching fraction was 0.20. (*Hint*: Use Eq. 12.13 in the calculation.)

8. Estimate the net rate of decrease of the $Ca^{2+}(aq)$ concentration in the soil solution described in Table 12.1. Use Eq. 12.14b to make the estimate, and express the result in $mol\;m^{-3}\;s^{-1}$.

9. Derive Eq. 12.14d from Eq. 12.14b and the definition of Ω given below it. Calculate the half-life for calcite precipitation from a solution whose initial $Ca^{2+}(aq)$ concentration is $1\;mol\;m^{-3}$ and whose Langlier Index is 0.7. Take the specific surface area of calcite equal to $120\;m^2\;kg^{-1}$ and its suspension concentration equal to $1.59\;g\;dm^{-3}$.

10. The equilibrium constant for the boron adsorption reaction in Eq. 12.19 is $K = 10^{5.5}$ according to the constant capacitance model as applied to a variety of arid-zone soils. Given $[SOH] = 5 \times 10^{-6}\;mol\;dm^{-3}$, calculate the concentration of $SH_2BO_3^0(s)$ that is in equilibrium with a $H_3BO_3^0(aq)$ concentration at the maximum permitted for unrestricted use of irrigation water with respect to boron toxicity hazard (Table 12.6). Convert your result to an amount adsorbed in $mmol\;g^{-1}$ given $c_s = 200\;kg\;m^{-3}$ as the density of a soil suspension.

11. Given the accompanying table of EC_w values, indicate which irrigation waters are likely to result in a saline root zone if a leaching fraction of 0.2 is used. What maximum SAR_{pw} values would be acceptable for these waters?

	Water		
	Salt R.	Colorado R.	Sevier R.
EC_w (dS m^{-1})	1.56	1.27	2.03

12. Give a rationale why the equation

$$SAR_{dw} = (c_{Naw}/LF)/[(c_{Mgw}/LF) + c_{Ca}^{eq}]^{\frac{1}{2}}$$

should provide a reasonably accurate estimate of the SAR value for water draining from the root zone. Explain carefully why LF appears in the equation and why c_{Ca}^{eq} is used.

13. Derive Eq. 12.25 from Eq. 12.24. Indicate precisely where the assumption $(HCO_3^-)_{eq}/(Ca^{2+})_{eq} = (HCO_3^-)_w/(Ca^{2+})_w$ is involved in the derivation.

*14. Evaluate the factor before $P_{CO_2}^{\frac{1}{3}}$ in Eq. 12.26 for $K_{so} = 10^{-8}$, $I = 20\;mol$ m^{-3}, and $[HCO_3^-]_w/[Ca^{2+}]_w = 1.0$, after converting the equation to an expression for c_{Ca}^{eq} using γ_{HCO_3} and γ_{Ca}. Compare your result with the appropriate entry in Table 1 of D. Suarez, Relation between pH_c and sodium adsorption ratio (SAR) and an alternative method of estimating SAR of soil or drainage waters, *Soil Sci. Soc. Am. J.* **45**:469–475 (1981).

*15. The "Colorado River" irrigation water referred to in Problem 11 has $EC_w = 1.3\,dS\ m^{-1}$, $[HCO_3^-]/[Ca^{2+}] = 1.12$, $c_{Naw} = 5\,mol\ m^{-3}$, and $c_{Mgw} = 1.3\,mol\ m^{-3}$. Calculate the value of adj RNa for this water using Eq. 12.27. Compare your result with the value of SAR based on $c_{Caw} = 2.6\,mol\ m^{-3}$.
(*Answer*: $c_{Ca}^{eq} = 1.95\ mol\ m^{-3}$ and adj RNa = 2.77 mol$^{\frac{1}{2}}$ m$^{-\frac{3}{2}}$)

13

Soil Fertility

13.1 Essential and Toxic Elements

The chemical elements essential to plant growth in soil are H, B, C, N, O, Mg, P, S, Cl, K, Ca, Mn, Fe, Cu, Zn, and Mo. Of these 16 elements, B, Cl, Mn, Fe, Cu, Zn, and Mo are *micronutrients* (absorbed in very small amounts) and Mg, S, and Ca are *secondary nutrients*. The remaining six elements are *macronutrients*. The essential elements—with one exception, molybdenum—have the distinguishing property that their atomic number in the Periodic Table is 30 or less. Thus they are among the "light elements" that have relatively small ionic radii (see also Problem 4 in Chapter 1). The biological significance of this property of essential elements can be seen readily by reference to a *Banin–Navrot plot* (Fig. 13.1). A Banin–Navrot plot is a log-log graph of the biological enrichment factor (see also Table 1.2):

$$EF_B = \frac{\text{element concentration in organism}}{\text{element concentration in crustal rock}} \tag{13.1}$$

plotted against the ionic potential (Section 7.5):

$$IP = \frac{\text{valence of element free cation}}{\text{radius of element free cation}} \tag{13.2}$$

For example, according to the data in Table 1.2 and in Problem 5 of Chapter 1, EF_B for Mg is 0.183 ($= 4.2 \text{ g kg}^{-1}/23 \text{ g kg}^{-1}$). According to Table 2.1, IP for Mg is 27.8 nm^{-1} ($= 2/0.072$ nm). These data, when converted to common logarithms, correspond to the point (-0.74, 1.44) in a Banin–Navrot plot, as indicated in Fig. 13.1.

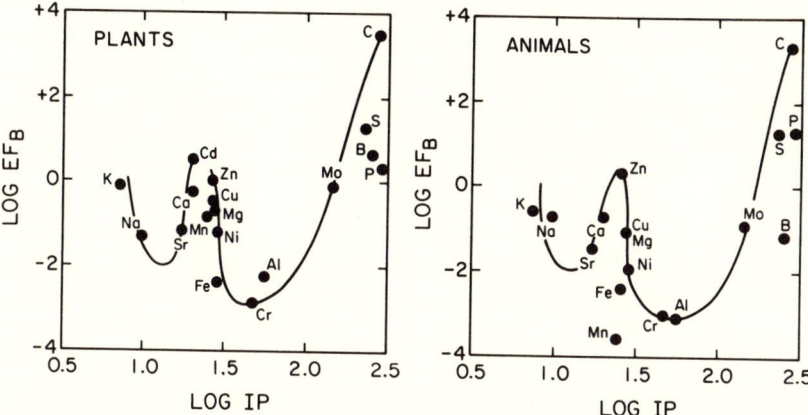

FIG. 13.1 Banin–Navrot plots for terrestrial plants and animals. [Based on enrichment factor data compiled by A. Banin and J. Navrot, Origin of life: Clues from relations between chemical compositions of living organisms and natural environments, *Science* **189**:550–551 (1975).]

The two Banin–Navrot plots in Fig. 13.1 refer to terrestrial higher plants and animals. Their shapes are remarkably similar and, in turn, are very much like those of Banin–Navrot plots for soil microflora (bacteria and fungi). In general, the biological enrichment factor displays a shallow minimum for *IP* in the range 11–15 nm^{-1} (Li$^+$, Ba^{2+}), a maximum near *IP* $\approx$ 22 nm^{-1}, then a precipitous drop to a deep minimum for *IP* in the range 45–60 nm^{-1}, and, finally, a gradual rise toward very large values for *IP* > 100 nm^{-1}. The sharp drop in *EF*$_B$—by almost three orders of magnitude—occurs at *IP* $\approx$ 30 nm^{-1}, which is the ionic potential value used commonly in geochemistry to separate cations that merely solvate in aqueous solution from those that hydrolyze extensively at pH 7. A marked rise in *EF*$_B$ after this drop occurs at *IP* $\approx$ 95 nm^{-1}, which is the ionic potential value that separates ions that only hydrolyze from those that repel the protons in solvating water molecules strongly enough to form oxyanion species in aqueous solution. For example, Ca^{2+} (*IP* = 20 nm^{-1}) is a solvating cation, Al^{3+} (*IP* = 56 nm^{-1}) is a hydrolyz-ing cation, and Mo^{6+} (*IP* = 143 nm^{-1}) is an oxyanion-forming cation. Therefore, one infers from Fig. 13.1 that *EF*$_B$ increases in the order: hydrolyzed cations ≪ solvated cations < oxyanion-forming ions. This ordering reflects the concept that essential elements are those that tend *not* to form hydrolyzed species that could adsorb or precipitate from neutral aqueous solutions. The macronutrients are almost all oxyanion formers (H and K are exceptions), whereas the micronutrient metals are perched along the "hydrolysis cliff" in a Banin–Navrot plot. [The shallow minimum that separates K from Ca may be associated with the relatively low crustal abundance of metals like Li, Ba, and Sr (Table 1.2).] These trends support the notion that *solubility*—and therefore mobility—are determining factors in essentiality.

"Optimal" soil solution concentrations of the essential elements for the nutrition of higher plants have not been found because of interdependencies among the biochemical functions of these elements. At a sufficiently high concentration of any element, however, deleterious effects on plant growth can be expected. This conclusion applies particularly to the micronutrients, since the need for them is small to begin with. Phytotoxic concentrations of B and Cl are described in Sections 12.4 and 12.5. Similar considerations apply to the micronutrient metals: at some soil solution concentration, which is plant species–dependent, each of these elements can produce phytotoxic effects. The general soil chemical conditions that lead to toxicity (or deficiency) are discussed in Section 13.2. In the remainder of the present section, attention will be focused on the electronic properties of the chemical elements themselves that promote phytotoxicity.

One of these properties is the ionic potential (Sections 7.5 and 8.3). Values of IP in the range 30–$100 \, nm^{-1}$ define a set of hydrolyzing elements whose natural abundance in plants is very low (Fig. 13.1). Increases in the soil solution concentrations of these elements [e.g., Cu ($IP = 27 \, nm^{-1}$) or Al ($IP = 56 \, nm^{-1}$)] above background levels ultimately must be hazardous to a biosphere that evolved under conditions in which the elements were relatively depleted. This perspective is just the "flip side" of the essentiality–solubility concept: if plants have evolved under conditions wherein an element is at low soil solution concentrations, they will not have developed many biochemical pathways by which to detoxify the elements when it is at high soil solution concentrations.

The mechanisms underlying phytotoxicity are not understood completely, but a consensus does exist as to the importance of the following possibilities for potentially toxic *metals*: (1) displacement, by a nonessential metal (e.g., Cd), of an essential metal (e.g., Ca) bound to a bioligand, (2) complexation of a metal by a functional group in a biomolecule (Section 3.1) that effectively blocks the group from reacting further, and (3) modification, by interaction with a metal, of the conformation of a biomolecule that is critical to its biochemical function. All of these mechanisms are related to complex formation between a potentially toxic metal ion and the functional groups on a biomolecule. Indeed, they suggest that strong complex formers are the more likely metals to induce phytotoxicity.

Metals that tend to form bonds with an important covalent character are candidates for strong complex formation (Section 2.1). This property of a metal ion can be quantified with the *Misono softness* parameter (Y), which is defined by the equation:

$$Y = 10 \frac{I_z R}{\sqrt{Z} I_{z+1}} \tag{13.3}$$

where R is the ionic radius of a metal cation whose valence is Z and whose ionization potential (*not* ionic potential) is I_z. For example, with the data in

Table 2.1 and ionization potentials taken from the *Handbook of Chemistry and Physics*, one calculates for Cu^{2+} (with coordination number 6): $Y = 10(1958)(0.073)/\sqrt{2}(3554) = 0.284$ nm, where $I_2 = 1958$ kJ mol^{-1} and $I_3 = 3554$ kJ mol^{-1}. Values of Y smaller than about 0.25 nm correspond to metal ions that have high electronegativity and low polarizability; they tend to form ionic instead of covalent chemical bonds (Sections 2.1 and 8.3). Values of $Y > 0.32$ nm correspond to metal ions that have low electronegativity and high polarizability; they tend to form covalent chemical bonds. Values of Y between 0.25 and 0.32 nm correspond to "borderline" metal ions whose tendency to covalency depends on whether specific solvent, stereochemical, and electronic configurational factors are present.

Figure 13.2 shows the classification of a number of metal ions according to their *IP* and *Y* values. Those with small *Y* values are sometimes termed "hard" Lewis acids, whereas those with large *Y* values are "soft" Lewis acids (Section 7.1). Hardness and softness are picturesque equivalents of low and high polarizability, respectively. The "hard" Lewis acids include the macronutrient and secondary-nutrient metal cations. The "borderline" Lewis acids include the micronutrient elements, Mn(II), Fe(II), Cu, and Zn. (Note that metal cations whose *IP* values fall into the "insolubility chasm" in Fig. 13.1 are found in the upper left portion of Fig. 13.2.)

The role of covalency as epitomized quantitatively in the Misono softness parameter can be appreciated after examination of the toxicity sequences in Table 13.1. For each class of plants, the ordering of metals from left to right reflects an increasing concentration of the metal (in moles per cubic meter) required to produce a substantial toxic effect, with the *smallest* concentration

FIG. 13.2 Classification of metals according to ionic potential and Misono softness.

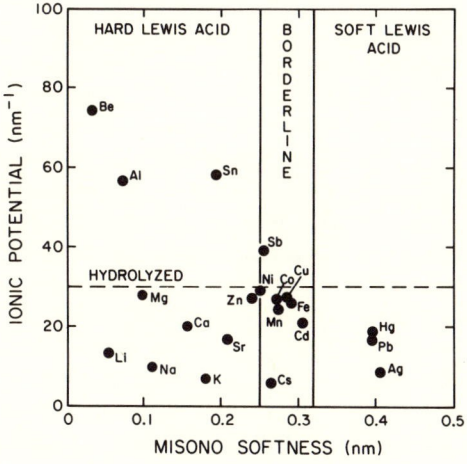

TABLE 13.1 Representative metal toxicity sequences[a]

Organisms	Toxicity sequence[b]
Algae	$Hg > Cu > Cd > Fe > Cr > Zn > Co > Mn$
Flowering plants	$Hg > Pb > Cu > Cd > Cr > Ni > Zn$
Fungi	$Ag > Hg > Cu > Cd > Cr > Ni > Pb > Co > Zn > Fe$
Phytoplankton (freshwater)	$Hg > Cu > Cd > Zn > Pb$

[a]Based on data compiled by E. Nieboer and D. H. S. Richardson, The replacement of the nondescript term 'heavy metals' by a biologically and chemically significant classification of metal ions, *Environ. Pollution* **B1**:3–26 (1980), and by E. Eichenberger, The interrelation between essentiality and toxicity of metals in the aquatic ecosystem, *Metal Ions Biol. Systems* **20**:67–100 (1986).
[b]$Hg = Hg(II)$, $Fe = Fe(II)$, $Cr = Cr(III)$, $Co = Co(II)$, $Mn = Mn(II)$, $Pb = Pb(II)$.

associated with the *most* toxic metal. The toxicity sequences in Table 13.1 are remarkably similar. In the context of Fig. 13.2, they show that *"soft" metal cations are more toxic to plants than "borderline" metal cations, which are more toxic than "hard" metal cations.* With the notable exception of Pb(II), this conclusion indicates that "soft" metal cations—the strong complex formers—are the most likely to participate in the three toxicity mechanisms enumerated previously.

From the point of view of electronic characteristics, then, essential elements can be characterized by ionic potentials that fall outside the "hydrolysis valley" ($30 < IP < 95$ nm^{-1}) and, if they are metals, by Misono parameters less than the "softness threshold", $Y < 0.32$ nm. Elements for which $IP > 100$ nm^{-1} will exist as oxyanions in the soil solution (B, C, N, P, S, and Mo), whereas elements for which $IP < 30$ nm^{-1} will exist as solvated cations (Mg, K, Ca, Mn, Fe, Cu, and Zn). Elements for which either $30 < IP < 95$ nm^{-1} *or* $Y > 0.25$ nm are potentially phytotoxic in dissolved form [e.g., Al(III), Ni(II), Hg(II), Cd(II)]. It is likely that the low solubility of these elements during the evolution of plant species has caused the latter to develop without selective means of detoxification when soil solution concentrations of the former are increased above the range of background levels.

13.2 Bioavailability

The absorption of essential and toxic elements by plants grown in soil continues to be the subject of intensive research in chemistry and plant biochemistry. Fundamental to these studies is a gradual refinement of the definition of *bioavailability* in mechanistic terms. Broadly speaking, *an essential or a toxic element is bioavailable if it is in a chemical form that plants can absorb readily and if, once absorbed, it affects the life cycle of the plant.* The adverb "readily" means "on a time scale relevant to the continued growth of a plant." The phrase "affects the life cycle of the plant" means "produces growth and

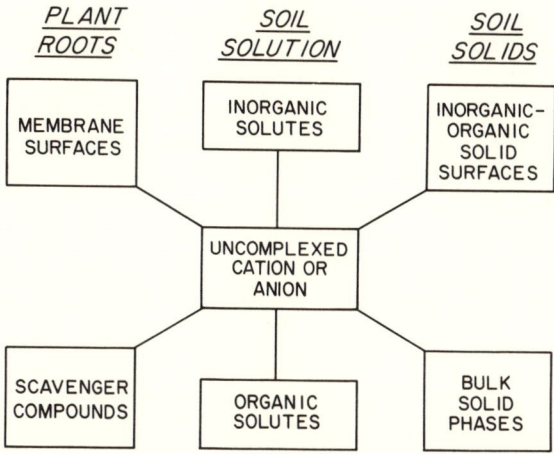

FIG. 13.3 Schematic illustration of the competition among functional groups associated with plant roots, the soil solution, and soil solid phases for free aqueous ions.

development" in the case of an essential element, whereas for a toxic element it means "produces phytoxicity."

From the perspective of soil chemistry, the bioavailability of an element is determined by a competition among the plant root system, the soil solution, and soil solid phases. This point of view is illustrated in Fig. 13.3 for an element that can exist as an ionic solute in the soil solution (e.g., Ca or Al, S or Se). With the solvated-ion species taken as a reference, competition for the element in the soil solution itself involves complexation reactions with both inorganic and organic ligands (cation) or with metals (anion). The typical effects of this kind of competition are illustrated in Table 4.1, which shows the speciation of metals in acid and alkaline soils. The general trend seen in this table is that complexation of the free ion becomes more pronounced as either its ionic potential or its Misono softness increases (e.g., Na^+ vs. Ca^{2+}; Ca^{2+} vs. Cu^{2+}). The existence of these complexes tends, at the least, to complicate the mechanisms by which plant roots can absorb the elements. With the very important exception of "scavenger compounds," like the amino acid siderophores that complex Fe and Cu, the ligands that complex metal ions must be shed before the metal can be absorbed by plant roots.

Bulk solid phases compete effectively for free ions if they form rapidly and weather slowly. Inorganic solids that have small solubility product constants, like clay minerals, phosphates, and metal oxides (Chapters 2 and 5), fall into this category. Soil organic matter, the chief repository of C, N, P, and S in the upper soil profile, can retain essential elements strongly against dissolution into ionic form if humification has been significant (Chapter 3). The functional groups protruding from either inorganic or organic solid surfaces in soil constitute a set of obvious competitors with aqueous solutes and plant-root surface functional groups (or scavenger compounds) for free-ion forms of chemical elements. The mechanistic underpinnings of this competition are based on inner-sphere surface complexation ("specific adsorption"), as discussed in Chapters 7 and 8. For metal cations, inner-sphere surface complexation should increase with ionic potential or Misono softness (e.g., K deficiency induced by adsorption onto 2:1 layer-type clay minerals, Cu deficiency induced by Cu adsorption onto soil organic matter, Al toxicity reduced by the specific adsorption of Al-hydroxy species). For anions, the importance of

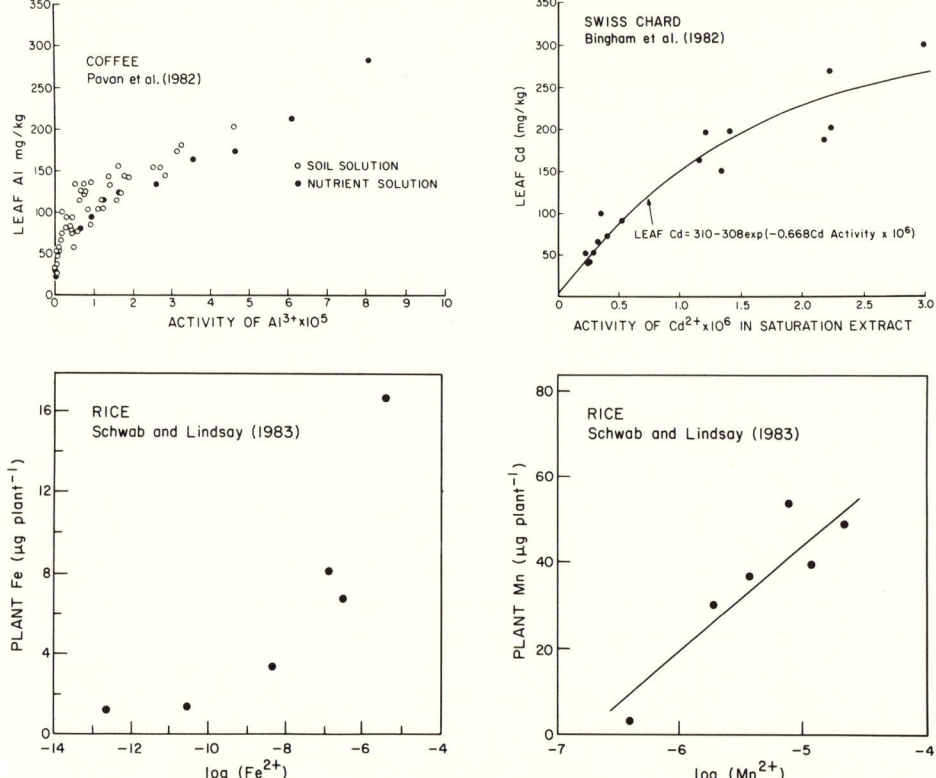

FIG. 13.4 The relationship between metal leaf concentration or uptake and the aqueous free metal ion activity. *Coffee* [M. A. Pavan and F. T. Bingham, Toxicity of aluminum to coffee seedlings grown in nutrient solution, *Soil Sci. Soc. Am. J.* **46**:993–997 (1982); M. A. Pavan, F. T. Bingham, and P. F. Pratt, Toxicity of aluminum to coffee in Ultisols and Oxisols amended with $CaCO_3$, $MgCO_3$, and $CaSO_4 \cdot 2H_2O$, *Soil Sci. Soc. Am. J.* **46**:1201–1207 (1982)]. *Swiss chard* [F. T. Bingham, J. E. Strong, and G. Sposito, Influence of chloride salinity on cadmium uptake by Swiss chard, *Soil Sci.* **136**:160–165 (1983)]. *Rice* [A. P. Schwab and W. L. Lindsay, Effect of redox on the solubility and availability of iron, *Soil Sci. Am. J.* **47**:201–205 (1983); The effect of redox on the solubility and availability of manganese in a calcareous soil, *Soil Sci. Soc. Am. J.* **47**:217–220 (1983)].

ligand exchange (Section 8.4) as an adsorption mechanism determines the extent of inner-sphere surface complexation (e.g., B, Mo, and P, as opposed to anionic N and Cl).

Central to this picture of a three-way competition for a dissolved chemical element is the hypothesis that the free-ion species influences plant absorption, regardless of whether the absorption process is "active," that is, mediated by respiration processes. Experimental data that support this hypothesis are presented in Fig. 13.4, which shows positive correlations between either metal uptake or concentration in plants and the thermodynamic activity (Section 4.5) of the free-ion species of the metal in aqueous solution. Similar correlations have been observed for the micronutrient cations, Cu^{2+} and Zn^{2+}. This evidence and the scheme in Fig. 13.3 can be used to refine somewhat the definition of bioavailability given at the beginning of this section: *A chemical element is bioavailable if it is present as, or can be transformed readily to, the free-ion species, if it can move to plant roots on a time scale that is relevant to plant growth and development, and if, once absorbed by the root, it affects the life cycle of the plant.*

13.3 Nutrient Uptake Kinetics

The definition of bioavailability given in Section 13.2 requires that a chemical element move to a plant-root surface on a time scale relevant to plant growth and development. Within the rhizosphere (Section 3.1), the movement of nutrient elements (and nonessential elements as well) involves diffusion and convective flow in water over distances in the range 0.1–15 mm, depending on the root population density in the soil. Convective flow of a nutrient occurs only when a plant transpires and its roots absorb water from the rhizosphere. The rate of water absorption (volume per unit root surface area per unit time) is typically on the order of micrometers per second (approximately the same as decimeters per day). Given a soil solution concentration of the order of moles per cubic meter, a nutrient uptake flux on the order of micromoles per square meter per second (tens of micromoles per square centimeter per day) is expected. (The *flux* is the amount absorbed per unit cross-sectional area of root surface per unit time and is equal numerically to the product of nutrient concentration and the water absorption rate.) For relatively mobile nutrients, like NO_3^-, $H_3BO_3^0$, and Ca^{2+}, uptake by convective flow can be a dominant mechanism in fertile soils. Taking 3 mm as a representative distance over which convective flow to the root occurs, one estimates a time scale of hours $[= 3 \, mm/10^{-3} \, mm \, s^{-1}] \cdot (1 \, h/3600 \, s)]$ for convective nutrient uptake. This time scale is consistent with the fact that water absorption takes place during daylight only.

The diffusive uptake of nutrients is usually a much slower process than convective uptake, but it can be very important for relatively immobile elements, like P and Cu, especially when their soil solution concentrations are low. The mathematical description of nutrient diffusion in the rhizosphere can be understood in a basic way by reference to the discussion in Special Topic 2 (Chapter 3). Equation s2.1 is a useful point of departure, with the following interpretation: c_s is the bulk *soil* concentration (mol m^{-3} soil) of the nutrient, assumed uniform initially over the rhizosphere distance δ ($\approx$ 3 mm); c_s' is the time-dependent, soil nutrient concentration at the root–soil solution interface; and D is replaced by the *Nye diffusivity parameter, D_e*:

$$D_e = \theta f D_\ell / \beta_e \qquad (13.4)$$

where D_ℓ is the diffusion coefficient of the nutrient in the soil solution, θ is the volumetric water content of the soil, f is an "impedance factor," and

$$\beta_e = dc_s / dc_\ell \qquad (13.5)$$

is the *nutrient buffer power*. The product, θf, in Eq. 13.4 corrects the liquid-phase diffusion coefficient for the reduced volume of diffusion space (θ) in a soil and for the obstacle to diffusion (f is a "tortuosity") presented by its solid phases. Usually θ is assumed to correspond to the water content at field capacity and f is about 1.5 times the value of θ ($0.1 < f < 0.6$). The parameter β_e represents the influence of soil solid phases on the diffusion of a nutrient to a root (Fig. 13.3), *under the assumption that the solid phases equilibrate rapidly with the soil solution.*

The nutrient buffer power can be measured in experiments on the solubility (c_ℓ, the *soil solution* concentration, mol m^{-3} solution) of an element in soil. If adsorption controls the *rapid* equilibrium of a nutrient element between soil aqueous and solid phases, then

$$\beta_e = \theta \left(1 + \rho_b \frac{dq}{dc_\ell} \right) \qquad (13.6)$$

where q is the surface excess of the nutrient (Eq. 8.1), and ρ_b is the dry bulk density of the soil. In this case, β_e is related directly to the slope of the nutrient adsorption isotherm (Section 8.2). For nutrient species that essentially do not react with soil solid phases and are readily soluble in the soil solution (e.g., NO_3^- and Cl^-), $\beta_e = \theta$ according to Eq. 13.6 and the concept of c_s. In general, as can be inferred, for example, from Fig. 8.1, β_e will be a function of c_ℓ and, therefore, D_e will vary with the soil solution concentration of a nutrient, even if D is a constant.

Table 13.2 lists representative values of c_ℓ, β_e, and D_e for several plant nutrients. The range of c_ℓ values applies to agricultural soils in temperate climatic zones. Lower values would be found in highly leached, tropical soils (Chapter 11), and higher values could be observed in saline soil environments (Chapter 12). This point notwithstanding, the typical range of c_ℓ is still over at least one order of magnitude. The same applies to β_e, even when normalized by

TABLE 13.2 Representative values of the soil parameters c_ℓ, β_e, D_e, and τ_D for plant nutrient ions[a]

Nutrient species	c_ℓ (mmol m^{-3})	β_e/θ	D_e (10^{-10} m^2 s^{-1})	τ_D^b (days)
NO_3^-	3,000–11,000	1.0	2.5	1
$H_2PO_4^-$, HPO_4^{2-}	0.6–5.0	10–10^3	10^{-4}–10^{-2}	2×10^4
K^+	100–2,500	1–50	10^{-2}–10^{-1}	200
Ca^{2+}	500–5,000	10–100	10^{-3}–10^{-2}	2×10^3
Mg^{2+}	500–5,000	1–60	10^{-2}	200
SO_4^{2-}	700–1,500	2–10	1	1
$H_3BO_3^0$	10–3,500	1–3	2–4	1
Fe^{2+}	0.1–1.0	10^3	10^{-4}	2×10^4
Mn^{2+}	0.1–100	1–100	10^{-3}–10^{-1}	2×10^3
MoO_4^{2-}	0.01–0.2	10–2,000	10^{-2}–1.0	200

[a]Compiled from S. A. Barber, *Soil Nutrient Bioavailability*, Wiley, New York, 1984.
[b]$\tau_D = 2\delta^2/D_e$, $\delta = 3$ mm, $D_e =$ lower value in column 4.

division with the volumetric water content. Low values of β_e/θ imply a relatively soluble, mobile nutrient (NO_3^- and $H_3BO_3^0$), whereas high values imply a large affinity of the nutrient for soil solid phases ($H_2PO_4^-$ and Fe^{2+}). Besides depending on c_ℓ, β_e may also be sensitive to soil pH. For example, β_ℓ for $H_3BO_3^0$ is low because the adsorption of B by soils is maximum at a very high soil pH, while β_e for $H_2PO_4^-$ and MoO_4^{2-} is high because these anions adsorb strongly at normal soil pH values (Fig. 8.4). Similarly, β_e for Fe^{2+} is larger than for Mn^{2+} because Fe hydrolyzes and precipitates as an hydroxy solid more readily than Mn at typical soil pH values (see Fig. 6.3 and Problem 9 in Chapter 6). Moreover, lower pE values in soil are required to solubilize the hydroxy solids of Fe as compared with those of Mn (Fig. 6.2). Thus β_e is expected to vary considerably among soils, as indicated in Table 13.2. Since D_e depends on β_e, it, too, will vary considerably. At present, many of the data on either β_e or D_e are sufficient only to provide approximate estimates of a representative value or range of values. As a general rule, D_e varies inversely with β_e, with strongly adsorbing nutrient species showing the smallest values ($H_2PO_4^- < SO_4^{2-} < NO_3^-$).

The time scale for nutrient diffusion can be estimated on the basis of the expression:

$$x_t^2 = \tfrac{1}{2}D_e t \tag{13.7}$$

where x_t^2 is the mean-square distance diffused during the elapsed time t. Equation 13.7, a fundamental relationship in the theory of diffusion, is often used to define a (constant) diffusion coefficient in terms of observations of ion travel distances and times. If $x_t = \delta$ (≈ 3 mm), then Eq. 13.7 can be rearranged to define the *diffusion time constant*, τ_D:

$$\tau_D = 2\delta^2/D_e \qquad (13.8)$$

The values of τ_D in Table 13.2 range over four orders of magnitude, from being comparable with the convective-uptake time scale (1 day) to the order of hundreds or even thousands of days. For nutrient species whose diffusive motions in the rhizosphere are on a time scale of nearly 1 year or more, deficiencies are likely in the absence of significant convective uptake, or uptake mediation by the plant root itself (scavenger compounds and direct-contact uptake).

13.4 Soil pE and pH Effects

The role of pE and pH as "master variables" controlling the chemistry of soils is emphasized in Chapters 6 and 11. Figure 6.1, which shows the ranges of aqueous electron and proton activity common to soils (and accessible to microorganisms), underscores the interrelatedness of these two variables. This connection can also be inferred from the general reduction half-reaction in Eq. 6.9 and its equilibrium constant in Eq. 6.10. The *product* of proton and electron activities appears in the denominator of K_R in Eq. 6.10. Therefore, if pE increases, pH must decrease, in order to maintain constant the ratio of reduced to oxidized species activities, and vice versa. This inverse relationship between pE and pH is reflected in Fig. 6.1 by the fact that the lines separating "oxic" from "suboxic" and "suboxic" from "anoxic" soils are not horizontal (see also Problem 4 in Chapter 6). If a fixed ratio of redox species activities is to be maintained, a higher pE is needed for oxic conditions in acid soils than in alkaline soils. Conversely, anoxic soils that are alkaline support lower pE values than those that are acid, with the *same* relative redox speciation in both.

The effects of pH alone on soil fertility are described in Sections 11.1 and 11.3, and the use of lime to mitigate soil acidity that is deleterious to plant growth is discussed in a soil chemistry context in Section 11.5. Section 11.4 outlines the important redox factors in soil acidity, while Sections 6.2–6.4 discuss the influence of pH on soil redox speciation. Throughout these sections, the mediating role of organic matter as the chief repository of protons and electrons in fertile soils is a recurring theme. In the present section, these topics are brought together again with a focus on plant nutrient bioavailability as defined in Section 13.3.

Aside from direct toxicity effects on plant roots (loss of root membrane integrity, cation exchange capacity, etc.) and on soil microorganisms, the principal influence of soil pH on bioavailability is indirect, through proton competition with metal cations for dissolved ligands or surface functional groups, and through proton attack on soil minerals. Both of these latter mechanisms tend to produce free metal cations in the soil solution or on the soil exchange complex and, therefore, to enhance the metal bioavailability, insofar as its mobility aspect is concerned. (The aspect dealing with metal utilization by the plant, however, may be affected adversely by *direct* proton toxicity; for example, the Ca requirement can be altered if root membrane integrity is threatened.) The Al, Fe, or Mn toxicity effects observed in acid soils come about in this way (Sections 11.2 and 11.3). For nonhydrolyzing nutrient metals (Fig. 13.2), the increased mobility also means increased leachability that ultimately can produce deficiency, if replacement through mineral weathering or soil amendments does not occur. For the anionic nutrients, the protonation of mineral surface functional groups will increase both specific and nonspecific adsorption (Section 8.4), thereby diminishing the free-ion concentration in the soil solution, and the same will result from mineral dissolution if the released metal cations can form stable precipitates with the anions (e.g., *o*-phosphate and sulfate). Thus decreasing soil pH can increase the bioavailability of metals while reducing that of ligands, if their uptake by plant roots is not diminished significantly by direct proton intervention.

The immediate effects of soil pE on the bioavailability of plant nutrient elements are summarized in Table 13.3. Except for O, which participates in microbial respiration and the formation of oxidized species of the other five principal redox elements (Section 6.1), the two most important direct results of increasing soil pE are the mineralization of soil organic matter and the precipitation of hydroxy solids. The former process results in an increase in bicarbonate alkalinity and in the concentration of the free-ion species of N, P, and S in the soil solution, thus enhancing their bioavailability. The latter process immobilizes Fe and Mn, thus decreasing their bioavailability (see also Problem 11 in Chapter 11). In the case of C and N, changing soil pE leads to a complicated series of redox species (Table 6.2) that play varying roles in plant nutrition. For example, organic acids, $CO_2(g)$, and $NH_4^+(aq)$ affect rhizosphere pH, cation–anion uptake balance (Section 11.4), and metal nutrition. The microbially catalyzed processes of nitrification (Problem 7 in Chapter 6), nitrogen fixation, denitrification (Section 6.1), and ammonia volatilization can alter N bioavailability dramatically. These processes are most active for pE

TABLE 13.3 Direct effects of soil pE on nutrient elements

Element	Effects of pE on bioavailability
C	Conversion of organic forms to HCO_3 and $CO_2(g)$; $CO_2(aq)$ to $CH_4(g)$
N	Conversion of organic forms to NH_4; NH_4 to NO_2 and NO_3; NO_3 to $N_2(g)$ and $N_2O(g)$; NH_4 and $NH_3(g)$; $N_2(g)$ to organic forms
O	Conversion to $H_2O(\ell)$ and other oxide forms
P	Conversion of organic forms to H_2PO_4 and HPO_4
S	Conversion of organic forms to SO_4
Mn	Precipitation of Mn(IV) hydroxy solids
Fe	Precipitation of Fe(III) hydroxy solids

values in the range 3–9 (Table 6.1), the transition interval for oxygen–nitrogen respiration by soil microorganisms.

Table 13.4 summarizes the important indirect influences soil pE can have on nutrient bioavailability. Aside from H, which plays a very general role in the formation of reduced species for the six principal redox elements (Eq. 6.9 and Table 6.2), the most significant indirect effects of decreasing soil pE have to do with adsorption phenomena. Low pE values (i.e., pE < 3, suboxic to anoxic conditions) destabilize Fe and Mn hydrous oxide adsorbents while stabilizing soil humus adsorbents (see Fig. 6.2 and Section 3.2). The disappearance of the hydrous oxides tends to increase the free-ion concentrations of anionic nutrients (especially B, inorganic C, P, and Mo) and micronutrient metal cations that were in adsorbed forms. The preservation of organic adsorbents tends to decrease the free-ion concentrations of the micronutrient metals via surface complexation (Section 8.3). Thus low pE indirectly enhances the bioavailability of anionic nutrients, but may have little net effect on micronutrient metals, if organic matter stabilization offsets metal hydrous oxide dissolution.

Low pE values produce increased concentrations of the nutrient cations, NH_4^+, Mn^{2+}, and Fe^{2+} (Table 13.3). These cations, in turn, will compete as adsorptives with the nutrient cations K^+, Mg^{2+}, and Ca^{2+}, which may be readily exchangeable on soil particle surfaces (Section 9.1). Since the affinity of soil adsorbents for NH_4^+, Mn^{2+}, and Fe^{2+} is about the same as for the macro- and secondary-nutrient cations (Section 8.3), the possibility of increased soil solution concentrations of the latter resulting from exchange for the former will depend sensitively on the relative activities of these aqueous cations (see Section 9.4). As a general rule, it follows from Eq. 9.25 that the likelihood of significant displacement of the nutrient cation N^{Z_N} (N = K, Mg, or Ca) by the low-pE cation P^{Z_P} (P = NH_4, Mn, or Fe) will be proportional to $^cK_{ex}(P^{Z_P})/(N^{Z_M})$ for homovalent exchange. Thus, for example, if the dissolution of Mn hydrous oxides results in $(Mn^{2+}) = 10^{-4}$ and the ambient activity of $Ca^{2+}(aq)$ is 10^{-3}, then $^cK_{ex} > 10$ is required to produce substantial amounts of adsorbed Mn relative to Ca.

TABLE 13.4   Indirect effects of soil pE on nutrient elements

Element	Effect of pE on bioavailability
B, C, P, S, Mo, Cu, Zn	Dissolution of Fe(III) and Mn(IV) hydroxy solid adsorbents (specific adsorption)
N, S, Cl	Dissolution of Fe(III) and Mn(IV) hydroxy solid adsorbents (nonspecific adsorption)
Mn, Fe, Cu, Zn	Stabilization of polymeric, metal-complexing organic matter; production of inorganic sulfides
K, Mg, Ca	Competition with NH_4^+, Mn^{2+}, and Fe^{2+} for cation exchange sites
H	Conversion to $H_2O(\ell)$ via the reduction of oxide forms of N, S, Mn, and Fe

FOR FURTHER READING

F. Adams (ed.), *Soil Acidity and Limiting*, 2nd Ed., American Society of Agronomy, Madison, WI, 1984. Chapters 2–9 in this standard reference provide in-depth discussions of the effects of acidity and pH on soil fertility.

S. A. Barber, *Soil Nutrient Bioavailability*, Wiley, New York, 1984. This book gives a comprehensive chemical and mathematical discussion of diffusion-controlled nutrient uptake kinetics, with 12 chapters devoted to individual plant nutrient elements.

M. Bernhard, F. E. Brinckman, and P. J. Sadler (eds.), *The Importance of Chemical "Speciation" in Environmental Processes*, Springer-Verlag, Berlin, 1986. This advanced collection of reviews covers all aspects of the relation between chemical form and bioavailability in natural aqueous systems.

P. H. Nye, Diffusion of ions and uncharged solutes in soils and soil clays, *Adv. Agron.* **31**:225–272 (1979). An outstanding, comprehensive exposition on ion diffusion from the point of view of soil chemistry.

K. R. Reddy and W. H. Patrick, Jr., Effects of aeration on reactivity and mobility of soil constituents, in D. W. Nelson (ed.), *Chemical Mobility and Reactivity in Soil Systems*, pp. 11–33. American Society of Agronomy, Madison, WI, 1983. A fine survey of the effects of pE on soil fertility.

H. Sigel (ed.), *Concepts on Metal Ion Toxicity* Vol. 20, *Metal Ions in Biological Systems*, Marcel Dekker, New York, 1986. This collection of advanced reviews provides a useful survey of metal toxicity characteristics and mechanisms, with the first five chapters devoted particularly to soil and water systems.

PROBLEMS

The more difficult problems are indicated by an asterisk.

1. Use the data in Table 2.1 to calculate the ionic potentials (for $CN = 6$) of the metal cations in the three adsorption selectivity sequences grouped together in Section 8.3. Show that ionic potential is inversely related to cation adsorption selectivity.

2. Combine the ionic potentials calculated in Problem 1 with the crustal abundance data in Table 1.2 to make a log-log diagram with log(crustal abundance) as the y-axis variable. Denote essential elements with a different symbol from nonessential elements. Can you detect any pattern in your graph that can help to explain the evolution of essentiality?

*3. Use the data in Table 2.1 and in the accompanying table to calculate ionic potentials and Misono parameters for the metal cations in the Irving–Williams selectivity sequence (Section 8.3). Plot the results as in Fig. 13.2, and use the graph to interpret the Irving–Williams ordering.
(*Hint*: Consider ionic potential as a measure of the tendency to ionic bonding, the Misono parameter as a measure of the tendency to covalent

bonding, and selectivity as a reflection of strong bonding.)

	Metal cation				
	Cu^{2+}	Ni^{2+}	Co^{2+}	Fe^{2+}	Mn^{2+}
I_2/I_3:	0.551	0.517	0.509	0.528	0.465

4. Comment on the following statement, using Fig. 13.3 as a point of departure: "The importance of metal complexes in the soil solution is not that they are absorbed by plant roots, but that they increase the total metal concentration and, therefore, the possible flux of the metal toward the root surface, where the free-ion concentration is maintained by complex dissociation."

*5. Estimates of bioavailable concentrations of nutrient elements in soils are often made on the basis of extractions with special reagents or ion exchange resins. For example, micronutrient metals are extracted with organic ligands like DTPA (1,4,7-triazaheptane-1,1,7,7-pentaacetic acid), and nutrient anions are extracted with anion exchange resins. Discuss the conceptual basis of this approach to measuring bioavailability in the context of the discussion in Section 13.2.
(*Hint*: What critical assumptions must be made in order to equate extractability with bioavailability?)

6. The table accompanying contains data on the concentration of Zn in nutrient solutions and in corn plants grown in the solutions at pH 7.5. Plot plant Zn against the concentration of Zn^{2+}, and predict the plant concentration expected from a soil solution in which $[Zn^{2+}] = 10$ nmol m^{-3}.

Zn_T (mmol m^{-3})	α_{10}^{a}	Zn (plant) (mg kg^{-1})
1.54	7.63×10^{-7}	8.6
7.65	8.25×10^{-7}	14.1
15.40	1.63×10^{-6}	19.5
22.9	2.76×10^{-6}	45.7

[a]Distribution coefficient for Zn^{2+}(aq) (see Section 4.3).

*7. Precipitates of $CaC_2O_4 \cdot 2H_2O(s)$ have been observed in the rhizosphere of forest soils. Given $K_{so} = 2.5 \times 10^{-9}$ for this solid phase and the information about oxalic acid in Section 3.1, determine the relative saturation of the soil solution described in Table 11.3 with respect to solid calcium oxalate. Take $I = 0.227$ mol m^{-3}.
(*Hint*: Review Section 5.1.)

8. The cell walls of roots contain surface functional groups that can complex cations in the rhizosphere solution. Go to the library and read the article by J. E. Dufey and R. Braun, Cation exchange capacity of roots: Titration, sum of exchangeable cations, copper adsorption, *J. Plant Nutr.* 9:1147–1155 (1986). Compare the authors' methodologies with the concepts developed in Sections 3.3, 7.2, 8.3, 9.1, and 11.2; then interpret the results in Table 2 of their article according to these concepts.

9. Use the data in Problem 2 of Chapter 8 to estimate β_e/θ for Cu in a soil whose dry bulk density is 1.3×10^3 kg m^{-3}. Plot the values of β_e/θ against c_{Cu}, and compare them with the entries for Fe and Mn in Table 13.2.

*10. The adsorption of *o*-phosphate by a loam soil was found to conform to the Langmuir expression (Eq. 8.6):

$$q_P = \frac{bKc_P}{1 + Kc_P}$$

where $b = 16$ mmol kg^{-1} and $K = 65$ m^3 mol^{-1}. Calculate β_e/θ as a function of c_P for $0 < c_P < 5$ mmol m^{-3} and $\rho_b = 1.4 \times 10^3$ kg m^{-3}; then compare the results with the entry for *o*-phosphate in Table 13.2. (*Hint*: Show that $dq/dc_P = bK/(1 + Kc_P)^2$, and express this derivative in m^3 kg^{-1}.)

11. Given the typical value for D_ℓ of 10^{-9} m^2 s^{-1} and the typical f–θ relationship, calculate a range of D_e values for Cu based on the results of Problem 9 and $\theta = 0.2$ m^3 m^{-3}. What is the corresponding τ_D value for the smaller D_e value? How does it compare with those for Fe^{2+} and Mn^{2+}?

12. Diffusion theory provides an equation for the amount of substance taken through a plane surface from a reservoir at fixed concentration. If the root surface is idealized as this plane surface and the concepts summarized in Table 13.2 are applied, the amount of nutrient taken up is given by the equation

$$M(t) = 2c_\ell (D_e t/\pi)^{\frac{1}{2}}$$

where $M(t)$ is the number of moles per unit root surface area taken up by the time t has elapsed. The half-life for the uptake process can be defined as the time at which $M(t) = \frac{1}{2}c_\ell \delta$. Use this definition and the lower estimate of D_e for each nutrient in Table 13.2 to compute the uptake half-life for each, in days. Compare your results with the values of τ_D in the table.

*13. Suppose that the solubility of Mn in a neutral soil is controlled by MnO_2(s) according to the seventh half-reaction in Table 6.2. This soil also adsorbs $Mn^{2+}(aq)$ according to the Langmuir expression:

$$q_{Mn} = \frac{bKc_{Mn}}{1 + Kc_{Mn}}$$

where $b = 0.1 \text{ mol kg}^{-1}$ and $K = 1.0 \text{ m}^3 \text{ mol}^{-1}$, with c_{Mn} expressed in mol m^{-3}. Use this information to calculate β_e/θ for Mn^{2+} as a function of pE in the range 8.0–9.0, for a soil whose dry bulk density is $1.3 \times 10^3 \text{ kg m}^{-3}$. [*Hint*: Review Section 6.3 and the hint in Problem 10 here. Ignore the activity coefficient of $Mn^{2+}(aq)$.]

*14. The sixth half-reaction in Table 6.2 relates the two nitrogen species, $NO_3^-(aq)$ and $NH_4^+(aq)$. The value of β_e/θ is 1.0 for NO_3^-, whereas for NH_4^+ it is about the same as for K^+, 1–50. This difference results in an order-of-magnitude increase in τ_D. Use the data in Table 6.2 to establish the pE/pH range over which the transition in β_e/θ from 1.0 to up to 50 for N species is expected. Assume that a species is "depleted" if its aqueous concentration is 10^{-6} that of the dominant species, and consider an *overall* pE/pH region consistent with the shaded portion of Fig. 6.1. Present your results graphically as a "belt" of pE–pH values within this soil-accessible region.

15. Suppose that the solubility of P is controlled by adsorption onto $Fe(OH)_3(s)$ in a soil at pH 6.0. Use the ninth half-reaction in Table 6.2 to estimate the pE value at which β_e/θ would drop from the range of values listed in Table 13.2 for *o*-phosphate to near 1.0, the value representing a nonadsorbing ion. [Assume that $(Fe^{2+}) = 10^{-6}$ and that *o*-phosphate would behave like nitrate once $Fe(OH)_3(s)$ disappears.] Discuss the implications of your result for *o*-phosphate uptake kinetics.

APPENDIX

Units and Physical Constants in Soil Chemistry

The chemical properties of soils are measured in units related to *le Système Interna-tional d'Unités*, abbreviated SI. This system of units is organized around seven *base physical quantities*, six of which are listed in Table A.1. (The seventh base physical quantity, luminous intensity, is seldom used in soil chemistry.) The definitions of the SI units of the base physical quantities have been established by international agreements. They are as follows.

One *meter* is a length equal to 1,650,763.73 wavelengths in vacuum of the radiation corresponding to the transition between the levels $2p_{10}$ and $3d_5$ in ^{86}Kr atoms.

One *kilogram* is the mass of the international metal prototype mass reference.

One *second* is the duration of 9,192,631,770 periods of the radiation corresponding to the transition between two hyperfine levels of the ground state in ^{133}Cs atoms.

One *ampere* is the electric current that, if maintained constant in two straight, parallel conductors, of infinite length and negligible cross section and placed 1 mm apart in vacuum, would produce between them a force of $0.2 \, \mu N$ per meter of length.

One *kelvin* is 1/263.16 of the absolute temperature at which water vapor, liquid water, and ice coexist at equilibrium ("the triple point").

One *mole* is the amount of any substance that contains as many elementary particles as there are atoms in 0.012 kg of ^{12}C.

Fractions and multiples of the SI base units are assigned conventional prefixes, as indicated in Table A.2. Thus, for example, $0.1 \, m = 1 \, dm$, $0.01 \, m = 1 \, cm$, $10^{-3} \, m = 1 \, mm$, $10^{-6} \, m = 1 \, \mu m$ (*not* $1 \, \mu$!), and $10^{-9} \, m = 1 \, nm$. An exception to this procedure is made for the unit of mass, since it already contains the prefix *kilo*. Fractions and multiples of the kilogram are denoted by adding the appropriate prefix to the mass in units of grams. For example, $10^{-6} \, kg = 1 \, mg$, *not* $1 \, \mu kg$, and $10^3 \, kg = 1 \, Mg$, *not* $1 \, kkg$.

Several important units of measure in laboratory work or fieldwork are defined directly in terms of the SI base units. The time units, minute (1 min = 60 s), hour (1 h = 3600 s), and day (1 d = 86,400 s), are examples, as are the liter (1 L = 1 dm^3), the

TABLE A.1 Base units in the *Système International*

Property	SI Unit	Symbol
Length	Meter	m
Mass	Kilogram	kg
Time	Second	s
Electric current	Ampere	A
Temperature	Kelvin	K
Amount of substance	Mole	mol

TABLE A.2 Prefixes for units in the *Système International*

Fraction	Prefix	Symbol	Multiple	Prefix	Symbol
10^{-1}	deci	d	10	deca	da
10^{-2}	centi	c	10^2	hecto	h
10^{-3}	milli	m	10^3	kilo	k
10^{-6}	micro	μ	10^6	mega	M
10^{-9}	nano	n	10^9	giga	G
10^{-12}	pico	p	10^{12}	tera	T

coulomb (the quantity of electric charge transferred by a current of 1 A during 1 s), and the degree Celsius ($°C$), which is equal to the temperature in kelvins minus 273.15. The pressure units, atmosphere (1 atm = 101.325 kPa) and bar (1 bar = 10^5 Pa), are common alternatives in the laboratory to the small SI unit, pascal (Pa). Other important units related to the SI base units are listed in Table A.3.

The unit of mole is closely related to the concept of *relative molecular mass*, M_r. The relative molecular mass of a substance of definite composition is the ratio of the mass of 1 mol of the substance to the mass of 1/12 mol of ^{12}C (i.e., 0.001 kg). For example, the relative molecular mass of $H_2O(\ell)$ is 18.015, which means that the absolute mass of 1 mol of water is 0.018015 kg. The relative molecular mass of the smectite, montmorillonite, with the chemical formula $Na_{0.9}[Si_{7.6}Al_{0.4}]Al_{3.5}Mg_{0.5}O_{20}(OH)_4$, is the weighted sum of M_r for each element in the solid: 0.9×22.991 (Na) + 7.6×28.086 (Si) + 3.9×26.982 (Al) + 0.5×24.312 (Mg) + 24×15.999 (O) + 4×1.008 (H) = 739.54. The same method of calculation applies to any other substance of known composition.

The preferred SI unit of concentration is moles per cubic meter, which is equal numerically to millimoles per liter, a convenient measure in laboratory work. The unit *molality* is preferred for measurements made at several temperatures, since it is a ratio of the amount of substance to the mass of solvent, neither of which is affected by changes in temperature. The "concentration" of adsorbed charge in a soil is measured in moles of charge per kilogram of soil (Table A.3). For example, if a soil contains 49 mmol adsorbed Ca per kilogram, then it also contains $0.098 \; mol_c \; kg^{-1}$ contributed by adsorbed Ca. In general, the moles of adsorbed charge equal the absolute value of the

TABLE A.3 Units related to the SI base units

Property	Unit	Symbol	SI Relation
Area	hectare	ha	10^4 m^2
Charge concentration	moles of charge per cubic meter	mol$_c$ m^{-3}	
Concentration	moles per cubic meter	mol m^{-3}	
Electric capacitance	farad	F	m^{-2} kg^{-1} s^4 A^2
Electric charge	coulomb	C	A s
Electrical potential difference	volt	V	m^2 kg s^{-3} A^{-1}
Electrolytic conductivity	siemens per meter	S m^{-1}	m^{-3} kg^{-1} s^3 A^2
Energy	joule	J	m^2 kg s^{-2}
Force	newton	N	m kg s^{-2}
Mass density	kilogram per cubic meter	kg m^{-3}	
Molality	moles per kilogram of solvent	mol kg^{-1}	
Pressure	pascal	Pa	m^{-1} kg s^{-2}
Specific adsorbed charge[a]	moles of charge per kilogram of adsorbent	mol$_c$ kg^{-1}	
Specific surface area[a]	hectare per kilogram	ha kg^{-1}	10^4 m^2 kg^{-1}
Viscosity	newton-second per square meter	N s m^{-2}	
Volume	liter	L	10^{-3} m^3

[a]"Specific" means "divided by mass."

TABLE A.4 Values of selected physical constants

Name	Symbol	Value
Atmospheric pressure	P_0	101.325 kPa (exactly)
Atomic mass unit[a]	u	1.6606×10^{-27} kg
Avogadro constant	N_A	6.022×10^{23} mol^{-1}
Boltzmann constant	k_B	1.3807×10^{-23} J K^{-1}
Diffuse double-layer parameter[b]	β	1.084×10^{16} m mol^{-1}
Faraday constant	F	9.6485×10^4 C mol^{-1}
Molar gas constant	R	8.3144 J K^{-1} mol^{-1}
Permittivity of vacuum	ε_0	8.8542×10^{-12} C^2 J^{-1} m^{-1}
Zero of the Celsius temperature scale	T_0	273.15 K (exactly)

[a]1 u $\equiv 10^{-3}/N_A$ is 1/12 the mass of a ^{12}C atom.
[b]Value at 298.15 K (Problem 14 in Chapter 8).

valence of the adsorbed ion times the number of moles of adsorbed ion per kilogram of soil. The cation exchange capacity of a soil is expressed in the units of specific adsorbed charge. In a similar manner, the concentration of ion charge in a soil solution is measured in moles of charge per cubic meter ($mol_c \ m^{-3}$) and is equal numerically to the absolute value of the ion valence times the ion concentration in moles per cubic meter.

The values of the most important physical constants used in soil chemistry are listed in Table A.4. These fundamental constants appear in theories of molecular behavior in soils. Note that $R = N_A k_B$ and that $\beta = 2F^2/\varepsilon_0 DRT$, where D is the relative permittivity of liquid water ($D = 78.3$ at 298.15 K).

FOR FURTHER READING

International Union of Pure and Applied Chemistry, Manual of symbols and terminology for physicochemical quantities and units, *Pure and Applied Chem.* **51**:1–41 (1979).

Soil Science Society of America, *Glossary of Soil Science Terms*, Soil Science Society of America, Madison, WI, 1987.

S. J. Thien and J. D. Oster, The international system of units and its particular application to soil chemistry, *J. Agron. Educ.* **10**:62–70 (1981).

PROBLEMS

1. Show that a pascal is the same as a force of 1 N acting on 1 m^2.
   (*Hint*: Use Table A.3 to express pascals in terms of newtons.)

2. Using the information in Table A.3, show that a volt is the same as a joule per coulomb. Calculate the "diffuse double-layer potential," RT/F (see Eq. 8.15), in volts at 298.15 K.
   (*Answer*: 0.02569 V)

3. Use the data in Table A.4 to calculate the mass of N_A atoms of ^{12}C.
   (*Answer*: 0.012 kg)

4. Calculate the relative molecular mass of Ca-vermiculite with the chemical formula $Ca_{0.7}[Si_{6.6}Al_{1.4}]Al_4O_{20}(OH)_4$.
   [*Answer*:
   $M_r = 0.7(40.08) + 6.6(28.086) + 5.4(26.982) + 24(15.999) + 4(1.008) = 747$]

5. Calculate the relative molecular mass of a "fulvic acid molecule" with the chemical formula $C_{135}H_{182}O_{95}N_5S$.
   [*Answer*: $M_r = 135(12.011) + 182(1.008) + 95(15.999) + 5(14.007) + 32.064 = 3427$]

6. Calculate the mass of 1 mol of humic acid with the chemical formula $C_{187}H_{186}O_{89}N_9S$.
   (*Answer*: 4.016 kg)

7. Use the result of Problem 6 to calculate the concentration of humic acid in a solution containing 0.5 g humic acid per liter.
   (*Answer*: $0.1245 \ mol \ m^{-3}$)

8. Ten milliliters of soil solution contain 5.5 mg $CaCl_2$. Calculate the concentration of $CaCl_2$ and the charge concentration of Cl^- in the solution.
   (*Answer:* The concentration of $CaCl_2$ is 4.96 mol m^{-3}, and the charge concentration of Cl^- is 9.92 mol$_c$ m^{-3}. Note that 49.6 μmol of $CaCl_2$ are dissolved in the 10 mL of water.)

9. Given that the mass density of liquid water is 10^3 kg m^{-3}, calculate the molality of $CaCl_2$ in the solution described in Problem 8.
   (*Hint:* Derive the relation: concentration = molality $\times$ mass density of solvent.)

10. Calculate the specific adsorbed charge of Ca on the vermiculite whose chemical formula was given in Problem 4.
    (*Answer:* 1.87 mol$_c$ kg^{-1}. *Hint:* What is the mass of 1 mol of Ca-vermiculite? How many moles of Ca charge does 1 mol of the clay mineral contain?)

INDEX